Instabilities and Nonequilibrium Structures II

Mathematics and Its Applications

Managing Editor:

M. HAZEWINKEL

Centre for Mathematics and Computer Science, Amsterdam, The Netherlands

Editorial Board:

F. CALOGERO, *Università degli Studi di Roma, Italy*
Yu. I. MANIN, *Steklov Institute of Mathematics, Moscow, U.S.S.R.*
A. H. G. RINNOOY KAN, *Erasmus University, Rotterdam, The Netherlands*
G.-C. ROTA, *M.I.T., Cambridge, Mass., U.S.A.*

Volume 50

Instabilities and Nonequilibrium Structures II

Dynamical Systems and Instabilities

edited by

Enrique Tirapegui
Facultad de Ciencias Físicas y Matemáticas,
Universidad de Chile, Santiago, Chile

and

Danilo Villarroel
Facultad de Ciencias, Universidad Técnica Federico Santa María,
Valparaíso, Chile

KLUWER ACADEMIC PUBLISHERS

DORDRECHT / BOSTON / LONDON

Library of Congress Cataloging in Publication Data

Instabilities and nonequilibrium structures II : dynamical systems and
   instabilities / edited by Enrique Tirapegui and Danilo Villarroel.
      p.   cm. -- (Mathematics and its applications ; 50)
      "A selection of the seminars given at the Second International
Workshop on Instabilities and Nonequilibrium Structures in
Valparaíso, Chile, in December 1987 ... organized by Facultad de
Ciencias Físicas y Matemáticas of Universidad de Chile and by
Universidad Técnica Federico Santa María"--Foreword.
      Includes index.

      1. Fluid dynamics--Congresses.   2. Stochastic processes-
-Congresses. 3. Stability--Congresses.   I. Tirapegui. Enrique.
II. Villarroel, D.   III. International Workshop on Instabilities and
Nonequilibrium Structures (2nd : 1987 : Valparaíso, Chile)
IV. Universidad de Chile.   Facultad de Ciencias Físicas y
Matemáticas.   V. Universidad Técnica Federico Santa María.
VI. Title: Instabilities and nonequilibrium structures 2.
VII. Series: Mathematics and its applications (Kluwer Academic
Publishers) ; 50.
QA911.I522   1989
532'.05--dc20   89-15408

ISBN-13: 978-94-010-7535-0   e-ISBN-13: 978-94-009-2305-8
DOI: 10.1007/978-94-009-2305-8

Published by Kluwer Academic Publishers,
P.O. Box 17, 3300 AA Dordrecht, The Netherlands.

Kluwer Academic Publishers incorporates
the publishing programmes of
D. Reidel, Martinus Nijhoff, Dr W. Junk and MTP Press.

Sold and distributed in the U.S.A. and Canada
by Kluwer Academic Publishers,
101 Philip Drive, Norwell, MA 02061, U.S.A.

In all other countries, sold and distributed
by Kluwer Academic Publishers Group,
P.O. Box 322, 3300 AH Dordrecht, The Netherlands.

printed on acid free paper

TABLE OF CONTENTS

FOREWORD

We present here a selection of the seminars given at the Second International Workshop on Instabilities and Nonequilibrium Structures in Valparaíso, Chile, in December 1987. The Workshop was organized by Facultad de Ciencias Físicas y Matemáticas of Universidad de Chile and by Universidad Técnica Federico Santa María where it took place.

This periodic meeting takes place every two years in Chile and aims to contribute to the efforts of Latin America towards the development of scientific research. This development is certainly a necessary condition for progress in our countries and we thank our lecturers for their warm collaboration to fulfill this need. We are also very much indebted to the Chilean Academy of Sciences for sponsoring officially this Workshop.

We thank also our sponsors and supporters for their valuable help, and most especially the Scientific Cooperation Program of France, UNESCO, Ministerio de Educación of Chile and Fundación Andes. We are grateful to Professor Michiel Hazewinkel for including this book in his series and to Dr. David Larner of Kluwer for his continuous interest and support to this project.

E. Tirapegui
D. Villarroel

LIST OF SPONSORS OF THE WORKSHOP

- Academia Chilena de Ciencias

- Academia de Ciencias de América Latina

- Academia de Ciencias Exactas, Físicas y Naturales de Argentina

- Facultad de Ciencias Físicas y Matemáticas de la Universidad de Chile

- Universidad Técnica Federico Santa María

- Ministere Francais des Affaires Etrangeres

- UNESCO

- Fundación Andes (Chile)

- Ministerio de Educación (Chile)

- Departamento Técnico de Investigación y de Relaciones Internacionales de la Universidad de Chile

- Universidad de Santiago de Chile

- Sociedad Chilena de Física

LIST OF SUPPORTERS OF THE WORKSHOP

- Nestlé Chile S.A.

- Embotelladora Williamson, Coca-Cola-Chile.

PREFACE

The articles we present here are devoted to the study of macroscopic systems both from a mathematical and a phenomenological point of view.

In Part I we have included works which emphasize the mathematical aspects of the subject, namely the theory of dynamical systems and statistical mechanics. The last articles study the notion of quantum chaos and the problem of the reduction of quantum noise in atomic physics.

In Part II the works are of much more phenomenological nature and explore new developments in some fundamental open questions. The first articles treat the old problem of the transition to turbulence. Next we include works on the role of topological defects in phase transitions and articles on instabilities and pattern formation. Finally the last works study the role of fluctuations in nonequilibrium systems.

PART I

DYNAMICAL SYSTEMS
AND
STATISTICAL
MECHANICS

ERGODIC THEORY AND SOBOLEV SPACES

André AVEZ
Département de Mécanique
et U.A. 213 du C.N.R.S.
Université Pierre et Marie Curie
4 Place Jussieu
75252 PARIS CEDEX 05, FRANCE

ABSTRACT.   A volume-preserving diffeomorphism φ of a compact Riemannian manifold M induces a Koopman operator $\hat{\varphi}$ on the Sobolev spaces : $\hat{\varphi}f = f \circ \varphi$ for $f \in W_{1,p}(M)$.

 Relationships between φ and $\hat{\varphi}$ are studied. It turns out that spectral invariants of $\hat{\varphi}$ are sharper than the invariants of the classical Koopman operator acting on L^2. In particular they are related to the entropy of Kolmogorov-Sinaï and they provide new criteria of ergodicity.

1. INTRODUCTION

1.1. The Koopman operator

A probability space is a set M together with a specified sigma-algebra of subsets of M, called the measurable sets, and a measure μ defined on that algebra. The measure is assumed to be normalized : $\mu(M) = 1$.

   An automorphism of (M, μ) is a bijection $\varphi : M \to M$ such that the image of every measurable set is measurable and has the same measure as the original set. The collection (M, μ, φ) is called a dynamical system.

   A generalization is in order. Instead of φ we consider a one-parameter group of automorphisms, i.e. a collection of automorphisms φ_t, where t is a real number and where $\varphi_s \varphi_t = \varphi_{t+s}$. The collection (M, μ, φ_t) is called a flow.

   Let $L^2 = L^2(M, \mu)$ be the Hilbert space of the complex-valued functions defined on M, with μ-integrable square. The inner product

5

E. Tirapegui and D. Villarroel (eds.), Instabilities and Nonequilibrium Structures II, 5–25.
© *1989 by Kluwer Academic Publishers.*

of two functions f and g of L^2 is $<f,g> = \int_M f.\bar{g}. d\mu$, or
simply $\int f.\bar{g}$. The norm is denoted by $\|f\|$. If $f \in L^2$, another func-
tion $\hat{\varphi} f$ is defined by writing $\hat{\varphi} f(x) = f(\varphi x)$. It turns out that
$\hat{\varphi} f \in L^2$ and $\|\hat{\varphi} f\| = \|f\|$. Therefore $\hat{\varphi}$ is an invertible isometry of L^2 :
this is the so-called unitary operator of Koopman.

Many properties of φ are reflected by $\hat{\varphi}$ (see Halmos [8]), as we
are going to see.

1.2. Ergodicity

We say that φ is ergodic if no measurable set A is invariant, except
M and the empty set : $\varphi(A) = A$ implies $\mu(A) = 0$ or $\mu(M-A) = 0$.

A reformulation of ergodicity is this : φ is ergodic if and
only if the number 1 is a simple eigenvalue of the unitary operator $\hat{\varphi}$.

1.3. Mixing

We say that φ is mixing if

$$\lim_{n = \infty} \mu(\varphi^n A \cap B) = \mu(A).\mu(B)$$

holds for every pair A,B of measurable sets.

This may be translated in terms of Koopman operator : φ is mixing
if and only if $\lim_{n = \infty} <\hat{\varphi}^n f , g> = <f,1> . <1,g>$ for every $f,g \in L^2$.

1.4. Weak mixing

Mixing implies obviously ergodicity (take $\varphi A = A = B$; then
$\mu(A) = 0$ or 1). Between these two concepts there is room for another
one. We say that φ is weakly mixing if

$$\lim_{n = \infty} n^{-1} . \sum_{r=o}^{n-1} |\mu(\varphi^r A \cap B) - \mu(A). \mu(B)| = 0$$

for every pair A,B of measurable sets.

Weak mixing has the following functional form : φ is weakly
mixing if and only if

$$\lim_{n = \infty} n^{-1} . \sum_{r=o}^{n-1} |<\varphi^r f , g> - <f,1> <1,g>| = 0$$

for every $f,g \in L^2$.

The "mixing theorem" gives two alternative definitions : φ is weakly mixing if and only if its Cartesian square is ergodic, or, alternatively, if and only if the only eigenvalue of $\hat{\varphi}$ is the number 1 and if that eigenvalue is simple (see Halmos [8] p. 39).

1.5. Discrete spectrum

We say that φ has discrete spectrum if there is a complete orthonormal basis $\{f_r\}$ of L^2 of eigenfunctions of $\hat{\varphi}$.

The mixing theorem shows that in some sense weak-mixing runs the opposite way of discrete spectrum.

1.6. The classification

When two automorphisms φ and φ' may be considered as equivalent ? The proper answer is the following : φ and φ' are isomorphic if there exists an automorphism $F : M \to M$ such that $\varphi' = F \varphi F^{-1}$.

If we look to φ through the Koopman operator $\hat{\varphi}$, we may say that φ and φ' are spectrally equivalent if there exists a unitary operator $U : L^2 \to L^2$ such that $\hat{\varphi}' = U \hat{\varphi} U^{-1}$. It is easy to see that isomorphism implies spectral equivalence. The converse is false.

In the hope of solving the isomorphism problem, several invariants have been proposed, e.g. the entropy of Kolmogorov. What I want to do is to introduce some new "Koopman operators" giving rise to new invariants. The following examples give the motivation.

1.7. Automorphisms of the torus

The 2×2 matrix

$$A = \begin{pmatrix} 1 & 1 \\ 1 & 2 \end{pmatrix}$$

can be thought of as a linear mapping of the plane $\mathbb{R}^2$ which preserves the lattice Z^2 of points with integer coordinates. This matrix induces an automorphism φ of the quotient group $M = \mathbb{R}^2/Z^2$. Since $\det(A) = 1$, φ is an area-preserving diffeomorphism of the torus M : (M, φ) is a dynamical system.

Let us select two constant vector fields $u \neq 0$ and $v \neq 0$

such that the first is parallel to the eigendirection of A corresponding to the eigenvalue $\lambda > 1$, the second being parallel to the
eigendirection corresponding to the eigenvalue λ^{-1} . Let us select
the Riemannian metric $u \otimes u + v \otimes v$ on M and let us denote by ∇f
the gradient of $f \in C^1(M, \mathbb{R})$. We have $\int |\nabla(f \circ \varphi^n)|^2 = \int (A^n u , df)^2 +$

$$\int (A^n v, df)^2 = \lambda^{2n} . \int (u, df)^2 + \lambda^{-2n} . \int (v, df)^2 \quad .$$

Since each orbit of u (or v) is everywhere dense ([1] , p. 5),
(u, df) and (v, df) cannot vanish identically if f is non constant.
Therefore, if $f \neq$ constant,

$$\lim_{|n| \to \infty} \int |\nabla(f \circ \varphi^n)|^2 = + \infty \quad .$$

1.8. Geodesic flow

The unit tangent bundle M of a Riemannian manifold S is the
union of the unit tangent vectors. We shall assume, to simplify, that
S is compact and two dimensional and we prove that M has a natural
structure of 3-dimensional compact manifold. Indeed, given isothermal
coordinates (x^1, x^2) on S , the Riemannian metric reads

$$g = e^{2A} . [dx^1 \otimes dx^1 + dx^2 \otimes dx^2].$$

The components of a unit vector $u \in M$ satisfy

$$e^{2A} [(u^1)^2 + (u^2)^2] = 1$$

and we may write $u^1 = e^{-A} . \cos \theta$, $u^2 = e^{-A} . \sin \theta$. This defines
local coordinates (x^1, x^2, θ) on M .

Given a unit vector u tangent to S at x , there is a unique
geodesic γ , parametrized by arc-length s , such that $\gamma(0) = x$,
$\dot{\gamma}(0) = u$ [$\dot{\gamma}$ means $\frac{d}{ds} \gamma$] . If s is the arc-length counted from x ,
the map $\mathbb{R} \times M \to M$ which sends (s,u) to $\dot{\gamma}(s)$ is a one-parameter
group φ_s of diffeomorphisms of M , the so-called geodesic flow.
Its infinitesimal generator X is a vector field on M , the geodesic
field.

What are the components of X in the coordinates (x^1, x^2, θ) ?
In isothermal coordinates the equations of the geodesics are

$$\frac{d}{ds}(e^{2A} \cdot \dot{x}^r) - \partial_r A = 0 \quad , \quad r = 1,2 \ .$$

Since $|\dot{\gamma}(s)| = 1$, $\dot{x}^1 = e^{-A} \cos \theta$, $\dot{x}^2 = e^{-A} \cdot \sin \theta$ and these
equations give $\dot{\theta} = e^{-A} \cdot (\cos \theta \cdot \partial_2 A - \sin \theta \cdot \partial_1 A)$. Then
$X = (e^{-A} \cdot \cos \theta, \ e^{-A} \cdot \sin \theta, \ e^{-A}(\cos \theta \cdot \partial_2 A - \sin \theta \cdot \partial_1 A))$.

If we change θ into $\theta + \frac{\pi}{2}$ we get the second geodesic field
Y :

$$Y = (-e^{-A} \cdot \sin \theta, \ e^{-A} \cdot \cos \theta, \ - e^{-A}(\sin \theta \cdot \partial_2 A + \cos \theta \cdot \partial_1 A)) \ .$$

Finally let us introduce the tangent field $T = (0, 0, 1)$. Since the
Gaussian curvature K of S is $- \text{div} \nabla A$ in isothermal coordinates,
direct computations yields the Lie Brackets $[Y,T] = X$, $[T,X] = Y$,
$[X,Y] = K.T$. They are the structure equations of the Riemannian
connection written in vector notations (see [13]).

Now select on M the Riemannian metric which makes (X, Y, T)
an orthonormed frame. We find easily that X, Y, T are divergence-free
(see [4]). In particular the geodesic flow φ_s is volume-preserving
(Liouville's theorem) and (M, φ_s) is a dynamical system.

Take $f \in C^2(M, \mathbb{R})$. The square of its gradient is

$$|\nabla f|^2 = (Xf)^2 + (Yf)^2 + (Tf)^2$$

Our purpose is to study

$$F(s) = \int |\nabla(f \circ \varphi_s)|^2 \cdot d\mu \ ,$$

where $d\mu$ is the Riemannian measure of M . We write $f_s = f \circ \varphi_s$ and

$$G(s) = F(s) - \int (X f_s)^2 \ .$$

Using the structural equations and differentiation under integration
we find easily

$$G'(s) = 2 \int (1-K) \cdot Tf_s \cdot Yf_s \ ,$$

$$G''(s) = 2 \int (1-K) \cdot [-K(Tf_s)^2 + (Yf_s)^2] \ .$$

Now let us assume that $K = -1$.

We obtain $G'' = 4G$, which can be solved : $G(s) = A e^{2s} + B e^{-2s}$ with

$$A = 2\,G(0) + G'(0) = 2\,\int (Yf + Tf)^2\,,$$
$$B = 2\,G(0) - G'(0) = 2\,\int (Yf - Tf)^2\,.$$

I claim that both of these expressions are positive if $f \neq$ constant.
Indeed, suppose for instance that $B = 0$, then $Yf = Tf$. But, if
$Z = Y-T$, the structural equations imply $ZTf = Xf$, $ZXf = 0$ and
$TXf - XTf = Tf$. Since Z is divergence-free

$$0 = \int Z\,Xf . Tf = -\int Xf . Z\,Tf = -\int (Xf)^2\,,$$

and $Xf = 0$. Therefore $-XTf = Tf$ and, since X is divergence-free,

$$0 = \int X\,Tf . Tf = -\int (Tf)^2\,.$$

Finally $\nabla f = 0$, a contradiction. Thus we proved :

__Proposition__ : If S is a compact connected surface of curvature
$K = -1$ and if φ_s is the geodesic flow on the unit tangent Bundle M,
then

$$\int \left| \nabla (f \circ \varphi_s) \right|^2 = \int \left| X(f_s) \right|^2 + G(s) \to +\infty$$

as $|s| \to \infty$ for any non-constant $f \in C^2(M, \mathbb{R})$.

    With a little more work we could have proved that the same result
holds if $K < 0$ (non constant). A more involved result is the following :

__Proposition__ : Let S be a compact connected Riemannian manifold with
negative Ricci curvature. If φ_s is the geodesic flow on the unit
tangent Bundle M, then $\displaystyle \lim_{|s| \to \infty} \int \left| \nabla (f \circ \varphi_s) \right|^2 = +\infty$ for any
non-constant function $f \in C^2(M, \mathbb{R})$. (to be published).

1.9. The setting

    Examples 1.7 and 1.8 suggest to study the action of the Koopman
operator on spaces of differentiable functions.

    In the future M will be a compact connected Riemannian mani-
fold. The measure μ will be the Riemannian measure : if $A \subset M$ is

measurable, $\mu(A)$ is the Riemannian volume of A . Up to a constant
we may suppose that $\mu(M) = 1$.

The gradient operator will be denoted by ∇ . The Laplacian is
$\Delta = -$ div ∇ (the sign "minus" makes Δ positive definite).

The Sobolev space $W_{1,p}$ is the Banach space of the functions
$f : M \to \mathbb{C}$ such that f and its weak gradient ∇f belong to
$L^p(M, \mu)$. The norm of f is $\|f\|_{1,p}$. If $p = 2$, this is a Hilbert
space with inner product $< f , g >_{1,2} = < f , g > + < \nabla f , \nabla g >$.

If $\varphi : M \to M$ is a volume-preserving diffeomorphism, it is clear
that $f \in W_{1,p}$ implies $\hat{\varphi} f = f \circ \varphi \in W_{1,p}$. However $\hat{\varphi}$ does not need
to be a unitary operator of $W_{1,2}$.

My purpose is to study the relationships between the ergodic
properties of φ and the spectral properties of $\hat{\varphi}$ acting on the
Sobolev spaces $W_{1,p}$. Some of the results were announced in [3] .

2. Mixing

2.1 Mixing and $W_{1,2}$.

Assume φ is mixing. Then, for any non-constant $f \in W_{1,2}$,
$$\lim_{|r| \to \infty} \|\hat{\varphi}^r f\|_{1,2} = + \infty .$$

Proof - Let us begin with an easy observation. We know that L^2 possesses a complete orthonormal basis $\{e_r\}$, the elements of which are
eigenfunctions of Δ :

$$\Delta e_r = \lambda_r \cdot e_r , \quad 0 = \lambda_o < \lambda_1 \leqslant \ldots \leqslant \lambda_n \to \infty$$

as $n \to \infty$. Adding a constant to f we may suppose $< f , 1 > = 0$.
Therefore

$$\|\nabla f\|_2^2 = \sum_{i=1}^{\infty} \lambda_i \cdot |< f , e_i >|^2$$

and we have

$$\sum_{n+1}^{\infty} |< f , e_i >|^2 \leqslant \lambda_n^{-1} \cdot \|\nabla f\|^2 .$$

Now the Parseval-Bessel inequality gives

$$\| f \|^2 \leq \sum_1^n |< f, e_i >|^2 + \lambda_n^{-1} \cdot \| \nabla f \|^2 \quad .$$

Replace f by $\hat{\varphi}^r f$ and observe that $\hat{\varphi}$ is unitary on L^2 :

(a) $\| f \|^2 - \sum_{i=1}^n |< \hat{\varphi}^k f, e_i >|^2 \leq \lambda_n^{-1} \cdot \| \nabla (\hat{\varphi}^k f) \|^2 \quad .$

Since φ is mixing and $< f, 1 > = 0$, $\lim\limits_{|k| = \infty} < \hat{\varphi}^k f, e_i > = 0$
and we obtain

$$\lambda_n \cdot \| f \|^2 \leq \lim_{|k| = \infty} \inf \| \nabla (\hat{\varphi}^k f) \|^2 \quad .$$

Since n is arbitrary and $\lim\limits_{n = \infty} \lambda_n = + \infty$ the result follows.

2.2 Weak-mixing and $W_{1,2}$

Assume φ is weakly mixing. Then, for any non-constant $f \in W_{1,2}$, we have

$$\lim_{m = \infty} m^{-1} \cdot \sum_{k=1}^m \| \nabla (\hat{\varphi}^k f) \|^2 = + \infty \quad .$$

In particular $\lim\limits_{|k| = \infty} \sup \| \hat{\varphi}^k f \|_{1,2} = + \infty \quad .$

Proof - The inequality (a) holds. Averaging these inequalities for $k = 1, \ldots, m$ gives

$$\| f \|^2 - \sum_{i=1}^n [m^{-1} \cdot \sum_{k=1}^m |< \hat{\varphi}^k f, e_i >|^2] \leq \lambda_n^{-1} \cdot m^{-1} \sum_{k=1}^m \| \nabla (\hat{\varphi}^k f) \|^2 \quad .$$

Let us fix n . Since φ is weakly mixing and $< f, 1 > = 0$ we have

$$\lim_{m = \infty} m^{-1} \cdot \sum_{k=1}^m |< \hat{\varphi}^k f, e_i > = 0 \quad .$$

But since the sequence $|< \hat{\varphi}^k f, e_i >|$ is bounded by $\| f \|$, strong Cesaro convergence implies quadratic strong Cesaro convergence, i.e.

$$\lim_{m = \infty} m^{-1} \cdot \sum_{k=1}^m |< \hat{\varphi}^k f, e_i >|^2 = 0 \quad .$$

Thus

$$\lambda_n \cdot \|f\|^2 \leqslant \liminf_{m = \infty} \ m^{-1} \cdot \sum_{k=1}^{m} \|\nabla(\hat{\varphi}^k f)\|^2 \quad .$$

Since n is arbitrary and $\lim_{n = \infty} \lambda_n = + \infty$, the theorem is proved.

Comment. I was unable to prove (or disprove) that

$$\lim_{m = \infty} \|\nabla(\hat{\varphi}^k f)\| = \infty \quad .$$

Let us show that the conclusion of 2.1 holds under an (apparently) weaker condition than mixing.

2.3. The property P

We say that φ satisfies the property P if

$$\limsup_{n = \infty} |< \hat{\varphi}^n f , f >| < \|f\|^2$$

for any non-constant $f \in L^2$.

Mixing implies obviously the property P . On the other hand the property P implies weak-mixing. Indeed, let f be an eigenfunction of $\hat{\varphi} : \hat{\varphi} f = \omega . f$, $|\omega| = 1$. Then $|< \hat{\varphi}^n f , f >| = \|f\|^2$ and the property P implies $f = $ constant. Weak-mixing follows from the "mixing theorem".

Theorem - Assume that φ has the property P . Then, for any non-constant $f \in W_{1,2}$, we have $\lim_{n = \infty} \|\hat{\varphi}^n f\|_{1,2} = + \infty$.

Proof - Once more it is enough to prove it for real-valued functions f such that $< f, 1 > = 0$.

Assume the result is false. There exists a subsequence of $\hat{\varphi}^n f$, still denoted by $\hat{\varphi}^n f$, and a constant $M > 0$ with $\|\hat{\varphi}^n f\|_{1,2} \leqslant M$. But the closure of the ball of center 0 and radius M is weakly compact. We can extract from $\hat{\varphi}^n f$ a subsquence, again denoted by $\hat{\varphi}^n f$, which converges weakly in $W_{1,2}$. According to the Kondrachov's theorem for compact Riemannian manifolds (see [2] , p. 55) this subsequence converges strongly in L^2 . Therefore, since $\hat{\varphi}$ is a unitary operator of L^2 ,

$$< \hat{\varphi}^p f , \hat{\varphi}^q f > = \| f \|^2 - 2^{-1} . \| \hat{\varphi}^p f - \hat{\varphi}^q f \| \to \| f \|^2 \quad \text{as} \quad p,q \to \infty \text{ , and}$$

$$\lim_{p,q = \infty} < \hat{\varphi}^{p-q} f , f > = \| f \|^2 \quad .$$

This contradicts the property P .

Comments $-$ If U is an arbitrary unitary operator of L^2 we saw that : mixing $\Rightarrow$ P $\Rightarrow$ weak-mixing.

It is not difficult, as pointed out G. Choquet, to prove that weak-mixing does not imply the property P . But $\hat{\varphi}$ is not just an arbitrary unitary operator, it is induced by a measure-preserving diffeomorphism. In such a case I do not know if one of the following implications is true :

weak-mixing $\Rightarrow$ P   ,   P $\Rightarrow$ mixing.

2.4   Mixing and $W_{1,1}$

Assume φ is mixing. Then, for any non-constant $f \in W_{1,1}$,
$$\lim_{n = \infty} \| \hat{\varphi}^n f \|_{1,1} = + \infty .$$

Proof $-$ We may suppose that f is real-valued and, adding a constant, that $f > 0$. Let g be the positive square root of f .

Suppose the theorem is false. There exists a subsequence of $\hat{\varphi}^n f$, still denoted by $\hat{\varphi}^n f$, and a constant M such that $\| \nabla (\hat{\varphi}^n f) \|_1 \leqslant M$. But, according to the Rellich's lemma (see [2] , p. 55) the inclusion $W_{1,1} \subset L^1$ is compact. We may therefore extract a subsequence, once again denoted by $\hat{\varphi}^n f$, which converges in L^1 :

$$\lim_{i,j = \infty} \| \hat{\varphi}^i g^2 - \hat{\varphi}^j g^2 \|_1 = 0 \quad .$$

But $\hat{\varphi}^n g^2 = (\hat{\varphi}^n g)^2$ and $| a^2 - b^2 | \geqslant (a-b)^2$ for any positive numbers a and b . Therefore $\| \hat{\varphi}^i g^2 - \hat{\varphi}^j g^2 \|_1 \geqslant \| \hat{\varphi}^i g - \hat{\varphi}^j g \|_2^2$ and

$$\lim_{i,j=\infty} \|\hat{\varphi}^i g - \hat{\varphi}^j g\|_2 = 0 .$$ Since $\hat{\varphi}$ is a unitary operator on L^2 this implies $$\lim_{i,j=\infty} < \hat{\varphi}^{i-j} g, g > = \|g\|_2^2 .$$

Since φ is mixing and g is non-constant we deduce the contradiction

$$\int g^2 = \lim_{i,j=\infty} < \hat{\varphi}^{i-j} g, g > = (\int g)^2 .$$

2.5. Area-dilating diffeomorphisms

Let $f : M \to \mathbb{R}$ be a C^1-function. The level surfaces of f are $f^{-1}(t), t \in \mathbb{R}$.

Note that by Sard's theorem the set of singular values for which $\nabla f = 0$ somewhere on $f^{-1}(t)$ has measure zero. For all other values of t, $f^{-1}(t)$ is a regular manifold bounding the domain $\{x : f(x) > t\}$.

Using orthogonal coordinates in the neighbourhood of $f^{-1}(t)$, we see that the volume element of M is $|\nabla f|^{-1} . d\sigma . dt$, where $d\sigma$ is the area element of $f^{-1}(t)$. It follows that for an arbitrary function h,

$$\int_M h . |\nabla f| = \int (\int_{f^{-1}(t)} h) \, dt .$$

This is the so-called coarea formula ([6]).

Now let $\varphi : M \to M$ be a volume-preserving diffeomorphism such that $\|\hat{\varphi}^n f\|_{1,1} \to \infty$ as $|n| \to \infty$ for any non-constant $f \in W_{1,1}$. Take a smooth function $F : \mathbb{R} \to \mathbb{R}$, take $h = 1$ and replace f by $F(f \circ \varphi^{-n})$ in the coarea formula :

$$\int |F'(f \circ \varphi^{-n})| . \text{Area}(\varphi^n f^{-1}(t)).dt =$$

$$\int_M |F'(f \circ \varphi^{-n})| . |\nabla(f \circ \varphi^{-n})| =$$

$$\int_M |\nabla(F(f \circ \varphi^{-n})| \to \infty \text{ as } |n| \to \infty .$$

Since F is arbitrary this implies readily that

$$\lim_{|n|=\infty} \text{Area} [\varphi^n(f^{-1}(t))] = + \infty .$$

In particular mixing diffeomorphisms dilate to infinity the area of one-codimensional submanifolds of M, as theorem 2.4. proves.

3. Discrete spectrum

If φ is an isometry of the Riemannian manifold M it is well-known that $\Delta \hat{\varphi} = \hat{\varphi} \Delta$. Since $\hat{\varphi}$ is a unitary operator of L^2, we have

$$\| \nabla (\hat{\varphi}^n f) \|^2 = < \hat{\varphi}^n f, \Delta(\hat{\varphi}^n f) > =$$

$$< \hat{\varphi}^n f, \hat{\varphi}^n \Delta f > = < f, \Delta f > = \| \nabla f \|^2,$$

and $\hat{\varphi}$ is a unitary operator of $W_{1,2}$.

What happens if, more generally, $\{\hat{\varphi}^n f : n \in \mathbb{Z}\}$ is uniformely bounded for any $f \in W_{1,2}$? Via Banach-Steinhaus this is equivalent to the existence of a constant C such that $\| \hat{\varphi}^n \|_{1,2} \leqslant C$ for any n.

3.1   Theorem - If the norms of the operators $\hat{\varphi}^n$, $n \in \mathbb{Z}$, acting on $W_{1,2}$ are uniformly bounded, then φ has discrete spectrum.

Proof - Assume the result is false. According to the "mixing theorem" there exists $g \in L^2$, $\| g \|_2 = 1$, $< g, 1 > = 0$, such that

$$\lim_{n = \infty} n^{-1} . \sum_1^n | < \hat{\varphi}^k g, g > | = 0 .$$

Since the sequence $| < \hat{\varphi}^k g, g > |$ is bounded by 1, $< \hat{\varphi}^k g, g > \to 0$ as $n \to \infty$ and does not take values from a subset $D \subset \mathbb{Z}$ of density zero.

Take $\varepsilon > 0$ small enough. Since $W_{1,2}$ is dense in L^2 there exists $f \in W_{1,2}$ such that $\| g - f \|_2 < \varepsilon$. Therefore, since $\hat{\varphi}$ is a unitary operator of L^2, $| < \hat{\varphi}^n f, f > | \leqslant | < \hat{\varphi}^n g, g > | + 3\varepsilon$. and $\lim \sup | < \hat{\varphi}^n f, f > | \leqslant 3.\varepsilon$ as $n \to \infty$ and $n \notin D$.

Thus we can find n_1 such that $| < \hat{\varphi}^{n_1} f, f > | < 4\varepsilon$.

Using the fact that the union of a finite number of sets of density zero has density zero, we can find n_2 such that

$|< \hat{\varphi}^{n_2} f , f >|< 4\varepsilon$ and $|< \hat{\varphi}^{n_1} f , \hat{\varphi}^{n_2} f > | < 4\varepsilon$. Similarly we find n_3 , n_4 ... such that

$$|< \hat{\varphi}^{n_i} f , \hat{\varphi}^{n_j} f > | < 4\varepsilon \quad \text{for} \quad i \neq j .$$

In particular

(b) $(1 - \varepsilon)^2 \leqslant \| f \|^2 < 2^{-1} . \| \hat{\varphi}^{n_i} f - \hat{\varphi}^{n_j} f \|_2^2 + 4\varepsilon$.

But the sequence $\hat{\varphi}^{n_i} f$ is bounded in $W_{1,2}$ and the inclusion $W_{1,2} \subset L^2$ is compact (see [2]). Therefore we can extract a subsequence of $\hat{\varphi}^{n_i} f$ which converges strongly in L^2. This gives the contradiction $(1-\varepsilon)^2 < 4\varepsilon$.

I shall give now a second and more instructive proof. It rests on the following folkloric result (see [7] , p. 536) :

3.2. Lemma – The Koopman operator $\hat{\varphi}$ on L^2 has discrete spectrum if and only if the closure in L^2 of the orbit $\{\hat{\varphi}^n f : n \in \mathbb{Z}\}$ is compact for every $f \in L^2$.

Proof – If φ has discrete spectrum there exists a complete orthonormal basis u_r of eigenfunctions of $\hat{\varphi} : \hat{\varphi} u_r = \omega_r . u_r$, $|\omega_r| = 1$. Take $f \in L^2$; then $\hat{\varphi}^n f = \Sigma < f , u_r >. \omega_r^n . u_r$ and the closure of $\{\hat{\varphi}^n f : n \in \mathbb{Z}\}$ is contained in the set

$$\{ \Sigma_r a_r . u_r : |a_r| \leqslant | < f , u_r > | \}$$

which is compact by Tychonov's theorem.

Let us prove the converse per absurdo. If φ has no discrete spectrum there exists $f \in L^2$, $f \neq 0$, orthogonal to all eigenfunctions of $\hat{\varphi}$. The "mixing theorem" implies $\lim < \hat{\varphi}^n f , f > = 0$ as $n \to \infty$ and does not take values from a subset $D \subset \mathbb{Z}$ of density zero. Now, using the argument of the first proof, we can find a sequence $n_1 < n_2 < ...$ such that $| < \hat{\varphi}^{n_i} f , \hat{\varphi}^{n_j} f > | < \varepsilon$ for $i \neq j$.

This implies

$$\| \hat{\varphi}^{n_i} f - \hat{\varphi}^{n_j} f \|^2 \geqslant 2 \| f \|^2 - 2 \mid < \hat{\varphi}^{n_i} f , \hat{\varphi}^{n_j} f > \mid \geqslant 2 \| f \|^2 - 2 \epsilon$$

for $i \neq j$. Therefore $\{ \hat{\varphi}^n f : n \in \mathbb{Z} \}$ is not precompact : a contradiction.

3.3. Second proof

Let us recall a useful criterium of precompactness (Riesz-Tamarkin) : If $f \in L^2 (\mathbb{R}^d)$ and $h \in \mathbb{R}^d$, we define $T_h f(x) = f(x+h)$. Then, a subset $A \subset L^2$ is precompact if and only if

a) A is bounded ;

b) for every $\epsilon > 0$ there exists $\delta > 0$ such that $\| T_h f - f \| < \epsilon$ for every $f \in A$ when $\| h \| < \delta$.

Since T_h is meaningless on a Riemannian manifold, we reformulate this criterium.

Let $T_1 M$ be the unit tangent bundle of M , $\Pi : T_1 M \to M$ the canonical projection and g_s the geodesic flow on $T_1 M$. Then $A \in L^2(M)$ is precompact if and only if

a) A is bounded ;

b) for every $\epsilon > 0$ there exists $\delta > 0$ such that, $\| f \circ \Pi \circ g_s - f \circ \Pi \|'_2 < \epsilon$ for every $f \in A$ when $|s| < \delta$ (The notation $\| \quad \|'_2$ means the norm in $L^2(T_1 M)$).

The proof is easy. We just have to replace the translation T_h in $\mathbb{R}^d$ by the parallel transport in M .

Let us come to the second proof.

1st step. If $f \in W_{1,2}$ we assumed that $\| \hat{\varphi}^n f \|_{1,2} < C$ for some constant C and for any integer n . Since the inclusion $W_{1,2} \subset L^2$ is compact this shows that $\{ \hat{\varphi}^n f : n \in \mathbb{Z} \}$ is precompact in L^2 . According to the Riesz-Tamarkin's criterium, given $\epsilon > 0$ there exists $\delta > 0$ such that

$$\| \hat{\varphi}^n f \circ \Pi \circ g_s - \hat{\varphi}^n f \circ \Pi \|' < \epsilon$$

for any integer n when $|s| < \delta$.

2^d **step.** Let h be in $L^2(M)$. Since $W_{1,2}$ is dense in L^2 there exists $f \in W_{1,2}$ such that $\|h-f\| < \varepsilon$, or $\|h \circ \Pi - f \circ \Pi\|' < D.\varepsilon$, where D is the volume of the unit ball of $\mathbb{R}^d$, $d = \dim M$.

Obviously $\{\hat{\varphi}^n h : n \in \mathbb{Z}\}$ is bounded in $L^2(M)$. On the other hand since the geodesic flow preserves the canonical volume element of $T_1 M$ we have

$$\|\hat{\varphi}^n \, h \circ \Pi \circ g_s - \hat{\varphi}^n \, h \circ \Pi\|' \leqslant$$

$$\|\hat{\varphi}^n \, h \circ \Pi \circ g_s - \hat{\varphi}^n \, f \circ \Pi \circ g_s\|' +$$

$$\|\hat{\varphi}^n \, f \circ \Pi \circ g_s - \hat{\varphi}^n \, f \circ \Pi\|' +$$

$$\|\hat{\varphi}^n \, f \circ \Pi - \hat{\varphi}^n \, h \circ \Pi\|' \leqslant$$

$$2D \cdot \|h-f\| + \varepsilon \leqslant (2D + 1).\varepsilon$$

when $|s| < \delta$.

This proves that $\{\hat{\varphi}^n h : n \in \mathbb{Z}\}$ is precompact and, according to the lemma, that φ has discrete spectrum.

I do not know how far a reciprocal of this theorem is true.

4. Entropy

4.1. Brief review of entropy theory.

Let (M, μ, φ) be a dynamical system. We introduce the function $z : [0,1] \to \mathbb{R}$ defined by $z(t) = - t . \text{Log } t$ if $0 < t \leqslant 1$ and prolonged by $z(0) = 0$.

If $a = \{A_i : i \in I\}$ is a finite collection of disjoint measurable sets of M whose union is M, the entropy of the measurable partition a is defined by $h(a) = \sum_i z[\mu(A_i)]$.

We denote by $\varphi^n a$ the measurable partition $\{\varphi^n(A_i) : i \in I\}$ and by $a \vee \ldots \vee \varphi^n a$ the measurable partition

$$\{A_{i_0} \cap \varphi A_{i_1} \cap \ldots \cap \varphi^n A_{i_n} : i_0, \ldots, i_n \in I\}.$$

Its turns out that $\lim_{n = \infty} n^{-1} . h(a \vee \ldots \vee \varphi^n a) = h(\varphi, a)$ exists (see [1] or [5]). It is called the entropy of φ relative to a.

Finally the entropy of φ is $H(\varphi) = \sup h(\varphi , a)$, where the supremum is taken over all the partitions a of M .

It is easy to see that $H(\varphi)$ is an invariant : two isomorphic systems have the same entropy. The importance of this concept is due, among others, to the fact that two systems may be spectrally equivalent but with different entropies.

Since we are interested in smooth manifold M and smooth mapping φ , more can be proved. Let us say that a is a classical partition if the boundary of each element A_i of a is the finite union of smooth submanifolds of codimension one of M . Then Kouchnirenko [10] proved that $H(\varphi) = \sup h(\varphi , a)$, where the supremum is taken over all the classical partitions, and he deduced that $H(\varphi)$ is finite.

4.2. An inequality of Kouchnirenko

Let $a = \{A_i : i \in I\}$ be a finite classical partition. Consider $I_n = - n^{-1} . \Sigma \mu( ) . \mathrm{Log} \ \mu( )$, where the sum is taken over all the sets

$$( ) = A_{i_0} \cap ... \cap \varphi^n A_{i_n} \ ; \ i_0 ,..., i_n \in I \ .$$

Since the function Log is concave Jensen's inequality yields $I_n \leqslant d . n^{-1} . \mathrm{Log} \ \Sigma [\mu( )]^{1-1/d}$, where $d = \dim M$. Now there exists a constant C , which depends only on the Riemannian manifold M (see [9]) such that, if the $\mu( )$ are small enough, $[\mu( )]^{1-1/d} \leqslant C . \mathrm{Area} \ \mathrm{bd}( )$, where $\mathrm{bd}( )$ is the boundary of $( )$. Therefore $I_n \leqslant d . n^{-1} . \mathrm{Log} [C . \Sigma \ \mathrm{Area} \ \mathrm{bd}( )]$. But, obviously,

$$\Sigma \ \mathrm{Area} \ \mathrm{bd}( ) = \sum_{k=1}^{n} \ \Sigma_i \ \mathrm{Area} \ \mathrm{bd}(\varphi^k A_i) \ .$$

We deduce

$$h(\varphi , a) \leqslant d . \liminf_{n = \infty} \mathrm{Log} [\sum_{k=1}^{n} \ \Sigma_i \ \mathrm{Area} \ \mathrm{bd}(\varphi^k A_i)]^{1/n}$$

This is in substance an estimate of Kouchnirenko ([10]).

4.3. Diffeomorphisms of Kolmogorov

Let φ be a K-diffeomorphism (see [1] for this notion). Then, for any non-constant function $f \in W_{1,1}$ or $W_{1,2}$, we have

$$\lim_{n = \infty} \sup [\| \hat{\varphi}^n f \|_{1,p}]^{1/n} = + \infty .$$

Proof. Let us take a non-constant smooth function $f : M \to \mathbb{R}$. For each $t \in \mathbb{R}$ we define a partition a_t whith two elements

$$A_1 = \{x \in M : f(x) \leqslant t\} \quad \text{and} \quad A_2 = \{x \in M : f(x) > t\} .$$

We want to use the estimate of 4.2.

According to the theorem of Shannon, Mc. Millan and Breiman (see [5]) the functions $I_n(a_t) = - n^{-1} . \Sigma 1_{(\ )} . \text{Log } \mu(\ )$, where the sum extends to all the sets

$$(\ ) = A_{i_0} \cap \ldots \cap \varphi^n A_{i_n}$$

and $1_{(\ )}$ is the characteristic function of $(\ )$, converge to a φ-invariant, μ-integrable function f_a almost everywhere, and $\int f_a = h(\varphi, a)$.

Since φ is a K-diffeomorphism it is ergodic and $f_a = h(\varphi, a)$. Therefore

$$I_n(a_t)(x) = \text{Log } [\mu(\ )]^{-1/n} \to .h(\varphi, a)$$

almost everywhere as $n \to \infty$. Now, according to a theorem of Pinsker, $h(\varphi, a_t) > 0$ for a_t is a non trivial partition if $\inf f < t < \sup f$. This proves that $\mu(\ )$ is small enough if n is large enough, and we may apply the estimate of 4.2.

For this purpose we observe that

$$\Sigma_i \text{ Area bd}(\varphi^k A_i) = 2 . \text{Area } (\varphi^k S_t) ,$$

where $S_t = f^{-1}(t)$. We obtain

$$0 < h(\varphi, a_t) \leqslant d . \liminf_{n = \infty} \text{Log } [\Sigma_1^n \text{ Area } \varphi^k S_t]^{1/n} .$$

The left hand side of this inequality is integrable, for a_t depends continuously on t and $h(\varphi, a_t)$ depends continuously of a_t. We apply the Fatou-Lebesgue lemma to the right hand side and we get

$$0 < \int h(\varphi, a_t) \cdot dt \leqslant d \cdot \liminf_{n = \infty} \int \text{Log} [\quad]^{1/n} \cdot dt \quad .$$

Since Log is concave we deduce

$$0 < \liminf_{n = \infty} \text{Log} [\int (\sum_{k=1}^{n} \text{Area } \varphi^k \, S_t) \, dt]^{1/n} \quad .$$

Finally we use the coarea formula

$$\int \text{Area } (\varphi^k \, f^{-1}(t)) \, dt = \int_M | \nabla(\hat{\varphi}^{-k} f)| \quad .$$

This gives

$$\liminf_{n = \infty} \text{Log} [\sum_{k=1}^{n} \int_M | \nabla(\hat{\varphi}^{-k} f)|]^{1/n} > 0 \quad .$$

In particular

$$\limsup_{n = \infty} | \int_M | \nabla(\hat{\varphi}^n f) |]^{1/n} > 1 \quad .$$

Since the C^1-functions are dense in $W_{1,1}$ this inequality still holds
in $W_{1,1}$. Schwarz' inequality shows that it is also true in $W_{1,2}$.

Ergodic automorphisms of the torus and the geodesic flow on
manifolds of negative curvature are K-systems. This theorem explains
the results of 1.7 and 1.8.

4.4. An estimate from above of the entropy

Let r be the spectral radius of $\hat{\varphi}$ acting on $W_{1,2}$ and
$d = \dim M$. Then $2 \cdot H(\varphi) \leqslant d \cdot \log r$.

Proof. From the very definition of r we have

$$(c) \qquad \limsup_{n = \infty} [\int_M | T \varphi^n \nabla f |^2]^{1/2n} \leqslant r \quad ,$$

where $T\varphi$ is the tangent mapping of φ and $f \in W_{1,2}$. We want to
replace ∇f by an arbitrary vector field X in this inequality.

Since M is compact we can find $f_1, \ldots, f_N \in C^1(M, \mathbb{R})$ such
that their gradients span the tangent space $T_m M$ for every $m \in M$.

Therefore we may write $X = \sum_1^N g_k \cdot \nabla f_k$ where $g_k \in C^1(M, \mathbb{R})$.

Given $\varepsilon > 0$ and n large enough the relation (c) and the Schwarz' inequality imply

$$\int_M |T \varphi^n X|^2 \leq \sup_M \sum_1^N g_k^2 \cdot \sum_1^N \int_M |T \varphi^n \nabla f_k|^2 \leq N \cdot (\sup_M \sum_1^N g_k^2) \cdot (r + \varepsilon)^{2n} .$$

Since Log is concave the Jensen's inequality gives

$$\int_M \text{Log } |T \varphi^n X|^{1/n} \leq \text{Log } [\int_M |T \varphi^n X|^2]^{1/2n}$$

Combining the previous relations we obtain

(d) $\lim_{n = \infty} \sup \int_M \text{Log } |T \varphi^n X|^{1/n} \leq \text{Log } r$.

Now Oseledec ([11]) proved that for almost every $m \in M$ the tangent space $T_m M$ splits into $Z^o(m) \oplus Z^+(m) \oplus Z^-(m)$ and that $\lim_{n = \infty} n^{-1} \text{Log } |T \varphi^n X|$ exists and is o if $X \in Z^o$, positive if $X \in Z^+$ and negative if $X \in Z^-$. Furthermore

$$s(m) = \lim_{n = \infty} n^{-1} \cdot \text{Log } |\det(T \varphi^n|_{Z^+})|$$

exists and D. Ruelle ([12]) proved that $H(\varphi) \leq \int_M s$.

We may suppose $\dim Z^+(m) \leq d/2$; otherwise we should change φ into φ^{-1}. Now take an orthonormal basis $u_1, \ldots, u_k$ of $Z^+(m)$. Hadamard's inequality gives $|\det(T \varphi^n|_{Z^+})| \leq |T \varphi^n u_1| \ldots |T \varphi^n u_k|$. Therefore $s(m) \leq \lim_{n = \infty} \sup \sum_1^k \text{Log } |T \varphi^n u_i|$ and we conclude with the formula (d) and Ruelle's inequality since $k \leq d/2$.

This theorem relies the entropy $H(\varphi)$ to a spectral property of $\hat{\varphi}$ acting on $W_{1,2}$, although $H(\varphi)$ is not related to the usual Koopman operator of L^2 .

5. Final comments

Let φ be a volume-preserving diffeomorphism of a compact connected Riemannian manifold. We shall say that φ is a G_p-diffeomorphism if $\lim_{|n| = \infty} \|\hat{\varphi}^n f\|_{1,p} = \infty$ for every non-constant function $f \in W_{1,p}$.

If φ is a G_p-diffeomorphism the Koopman operator $\hat{\varphi}$ on L^2 cannot have non-trivial eigenfunction $f \in W_{1,p}$ since $|\hat{\varphi}^n f| = |f|$. In fact we can prove the following result :

Theorem. If $p = 1$ or 2 a G_p-diffeomorphism is weakly mixing.

The proof is rather involved and will be given in a subsequent article. A corollary which does not seem to be treatable by the Anosov-Sinai method is :

Corollary. The geodesic flow on the unit tangent bundle of a compact Riemannian manifold with strictly negative curvature is weakly-mixing.

Finally I cannot prove or disprove the following conjecture.

Let φ and φ' be two volume-preserving diffeomorphisms of a compact Riemannian manifold. Assume that $\hat{\varphi}$ and $\hat{\varphi}'$ acting on $W_{1,2}$ are unitarily equivalent. Is it true that φ and φ' are isomorphic ?

References

[1]   V. Arnold and A. Avez, Ergodic problems of classical Mechanics, Benjamin (1967).

[2]   T. Aubin, Non linear analysis on manifolds. Monge-Ampère equations, Springer (1980).

[3]   A. Avez, C.R. Acad. Sci. Paris. t. 306. Serie I, 239-241 (1988).

[4]   A. Avez, Simple closed geodesics on surfaces, Dynamical systems and microphysics. Edited by A. Avez, A. Blaquière, A. Marzollo, Academic Press (1982).

[5]   P. Billingsley, Ergodic theory and information, Wiley (1965).

[6]   H. Federer, Geometric measure theory, Springer-Verlag (1969).

[7]   H. Furstenberg, Y. Katznelson and D. Orsnstein, Bull. Amer. Mat. Soc., Vol. 7, n° 3 (1982).

[8]   P. Halmos, Lectures on Ergodic theory, Math. Soc. of Japan (1956).

[9]   D. Hoffman and J. Spruck, Proc. of Symposia in pure Math., Amer. Math. Soc. Vol. 27, 139-141 (1975).

[10] A.G. Kouchnirenko, Dokl. Akad. Nauk. SSSR. 6, 37-38 (1965).
        English transl : Sov. Math. Dokl. Vol. 6, N 2, 360-362 (1965).

[11] V.I. Oseledec, Trudy Moskov. Mat. Obsc., 19, 179-210 (1968).
        English transl : Trans. Moscow Mat. Soc., 19, 197-231 (1969).

[12] D. Ruelle, Bol. Soc. Bras. Mat., 9, 83-87 (1978).

[13] I.M. Singer and J.A. Thorpe, <u>Lectures notes on elementary
        topology and geometry</u>, Scott-Foresman (1967).

ABSOLUTELY CONTINUOUS INVARIANT MEASURE FOR EXPANDING PICK MAPS OF THE INTERVAL EXCEPT AT A MARGINAL FIXED POINT*.

P. COLLET
Centre de Physique Théorique
Ecole Polytechnique
F91128 Palaiseau CEDEX (France)

P.FERRERO
C.P.T.-C.N.R.S.
Centre de Luminy
case 907
F13288 Marseille CEDEX 2 (France)

I Introduction.

It is an important problem in the theory of concrete dissipative dynamical systems to construct the relevant invariant measure. A rather complete theory has been developed for the case of uniformly hyperbolic dynamical systems following the initial ideas of Sinaï Bowen and Ruelle. They proved an equivalence between questions about invariant measures and problems in statistical mechanics. In particular, the interesting invariant measures are Gibbs states of a one dimensional spin system.

Up to now, these ideas have been successfully developed mostly in the context of uniformly hyperbolic dynamical systems. There are however interesting situations where the dynamical system is not uniformly hyperbolic. We shall consider below such a situation which is related to the intermittency transition. More precisely we shall consider a dynamical system which is defined by a mapping of the interval $[0,1]$ which is piecewise C^3, and has slope everywhere larger than 1 except at 0 which is a fixed point of the map. We shall also assume that the mapping has a Markoff property, namely the interval $[0,1]$ can be written as a finite union

---

* Talk given by P.Collet

27

E. Tirapegui and D. Villarroel (eds.), Instabilities and Nonequilibrium Structures II, 27–36.
© 1989 by Kluwer Academic Publishers.

of closed intervals which only intersect at their boundary and whose image by the map is the whole interval $[0, 1]$.

For these dynamical systems it is known that the Bowen-Ruelle measure is the Dirac measure at the fixed point 0 [Me.]. In other words, the large time statistics of Lebesgue almost every initial condition is trivial. However the transient behavior is more interesting. This was first analysed by P.Manneville [Ma.] who gave strong arguments for the existence of a non integrable absolutely continuous invariant measure. As a consequence one can observe in these systems non trivial transient behaviors for the laws of occupation time and also slow decay of correlations.

The problem of existence of a non integrable invariant measure was first solved by R.Bowen [B.] using return maps. In [C.F.] we proved existence and uniqueness for these measures using a direct Perron-Frobenius method under a C^3 condition on the map. We shall present below a much simpler proof which is applicable to situations where the mapping has some strong regularity.

II Invariant measure for Pick functions.

We shall consider maps f of the interval $[0, 1]$ of a more restricted class. We shall indeed see that under these special hypothesis the proof of the existence of an absolutely continuous invariant measure becomes much simpler.

Our first hypothesis concerns the dynamics of the map:

H1) f is defined and C^3 on $[0, 1/2]$ and $[1/2, 1]$. These two branches are diffeomorphisms with images $[0, 1]$. We also impose $f(0) = 0$, $f(1/2^-) = 1$ and $f(1/2^+) = 0$, and moreover $f'(0) = 1$, $f'(x) > 1$ if $x \neq 0$ and $f''(0) > 0$.

Our second hypothesis is of a more technical nature. As already mentioned the result holds under the above condition only but this second condition will make the proof much simpler. We shall denote by f_- and f_+ the two inverse branches of f corresponding to the two intervals $[0, 1/2]$ and $[1/2, 1]$ respectively.

H2) There is a finite number $b > 1$ such that f_+ and f_- belong to the class $P(0, b)$ of Pick functions [D].

All results in this section are valid under the weaker assumption $b \geq 1$. We shall only need the stronger assumption $b > 1$ in the next section. We recall that a function belongs to the Pick class $P(a, b)$, where $a, b \in \mathbf{R}$ if it is analytic in the cut plane $\mathbf{C} \backslash [a, b]$ and maps the upper (respectively lower)half plane into itself.

We shall denote by P the class of functions which are derivatives of a function in $P(0, b)$. We observe that any Moëbius transformation with real coefficients defines a function belonging to some Pick class. The same is true for its inverse. This leads to the following simple example of a map that satisfies our hypothesis:

$$f_{[0,1/2]}(x) = x/(1 - x) ,$$
$$f_{[1/2,1]}(x) = 2x - 1 .$$

We now recall that if ϕ is the density of an absolutely continuous invariant measure it must satisfy

$$\int_0^1 g \circ f(x)\, \phi(x)\, dx = \int_0^1 g(x)\, \phi(x)\, dx$$

for any measurable function g for which the above integrals are finite. It follows easily from this formula that ϕ must be an eigenvector with eigenvalue one of the so called Perron-Frobenius operator P which is defined by

$$Ph(x) = \sum_{f(y)=x} \frac{h(f(y))}{|f'(y)|} .$$

The relation between the above operator and Pick functions is obvious from the following trivial lemma.

Lemma 2.1. $PP \subset P$.

The proof follows at once if we notice that the action of P on a function h is given by

$$Ph(x) = \sum_{\epsilon=\pm} f'_\epsilon(x) h(f_\epsilon(x)) .$$

This Lemma suggests to look for a fixed point of P in P. The most naive way to find a fixed point is of course to iterate P acting on a function in the space (1 for example). In the usual case one can show that the Cesaro average

$$n^{-1} \sum_0^{n-1} P^j 1$$

converges to a fixed point. We now show that this cannot be the case here. Let h be a continuous function on the interval $[0,1]$. Using the Lebesgue dominated convergence Theorem and the fact that the Bowen-Ruelle measure is the Dirac measure at 0, we have

$$\int_0^1 h(x)\, n^{-1} \sum_0^{n-1} P^j 1(x)\, dx = \int_0^1 n^{-1} \sum_0^{n-1} h(f^j(x))dx \xrightarrow[n\to\infty]{} h(0) .$$

If the above Cesaro average were to converge to a function F in L^1 this function would satisfy

$$\int F(x)\, h(x)\, dx = h(0)$$

for any continuous function h which is impossible. In other words, the sequence of measures

$$n^{-1} \sum_0^{n-1} P^j 1(x)\, dx$$

converges to the delta measure at zero. Similarly, if we assume that P has a fixed point F in L^1, we get a contradiction by replacing 1 by F in the above argument.

Intuitively, the normalization by n^{-1} of the Cesaro average is too large to obtain a non trivial limit, and we are going to use a different one. We shall fix the normalization at one point in the interior of the interval. For convenience we shall choose the point $1/2$ although, as we shall see later on, any other choice would give an equivalent result. More precisely we shall consider the sequence of functions $h_n(x)$ defined by

$$h_n = \frac{\sum_0^{n-1} P^j 1(x)}{\sum_0^{n-1} P^j 1(1/2)} \, . \tag{1}$$

We now have the following compactness result.

Lemma 2.2. *The sequence h_n is precompact for the topology of compact convergence on* $\mathbf{C} \backslash [0, 1]$.

Proof. $(h_n)_{n \in \mathbf{N}}$ is a sequence of functions in P normalized by $h_n(1/2) = 1$. The lemma follows from the precompactness theorem for normalized derivatives of Pick functions (see [D.]).

It follows at once from Lemma 2.2 that the sequence h_n has accumulation points in P. We now have to show that we can construct fixed points from these accumulation points (this would have been almost obvious for the Cesaro average, however we cannot argue like this here because of the different normalization).

Lemma 2.3. *P has non trivial fixed points in P.*

Proof. Let $(h_{n_i})_{i \in \mathbf{N}}$ be a subsequence of the sequence $(h_n)_{n \in \mathbf{N}}$ which converges to a function h. We have

$$P h_n = h_n + \frac{P^n 1 - 1}{\sum_0^{n-1} P^j 1(1/2)} \, ,$$

and we conclude that if $\sum_0^{n-1} P^j 1(1/2)$ diverges, then $Ph = h$. The proof of divergence of the normalization requires more involved estimates and we refer the reader to [C.F.] for a complete proof. However in the present situation we do not have to use these finer estimates to complete the argument. Assume that the sequence $\sum_0^{n-1} P^j 1(1/2)$ converges to a finite (necessarily positive) number ω. It follows from the compactness theorem for functions in P that the increasing sequence of functions $\sum_0^{n-1} P^j 1$ converges in $(0, 1)$ to a function ϕ in P. It is easy to verify by a limiting argument that ϕ (which is normalized by $\phi(1/2) = 1$) satisfies the identity

$$P\phi = \phi - 1/\omega \, .$$

Using recursively this relation, we obtain

$$\sum_0^{n-1} P^j \phi = n \left(\phi - \omega^{-1} \sum_0^{n-2} P^j 1 \right) + \omega^{-1} \sum_0^{n-2} (1 + j) P^j (1) \, . \tag{2}$$

The sequence $\omega^{-1} \sum_0^{n-2} P^j 1$ is increasing and converges to ϕ. Therefore the first term of the right hand side of (2) is positive. We shall now give a lower bound on the second term. From the definition of P it follows immediately that

$$P^n 1(x) \geq \mathcal{O}(1) f_-^{n\,'}(x) \, .$$

Using the chain rule one derives the following lower bound

$$P^n 1(1/2) \geq O(1)n^{-2} .$$

We now have using equation (2)

$$\sum_0^n P^j \phi(1/2) \geq O(1) \log n .$$

Since ϕ belongs to P the argument of Lemma 2.2 is still valid with 1 replaced by ϕ in the definition of h_n. We now have in this case

$$Ph_n = h_n + \frac{P^n \phi - 1}{\sum_0^{n-1} P^j \phi(1/2)} .$$

From $P\phi \leq \phi$ we have $P^n \phi \leq \phi$, and we can now complete the proof as before since the denominator diverges.

As explained above, none of the above fixed points can be in L^1. In other words, we have constructed non normalizable absolutely continuous invariant measures.

III Some properties of the absolutely continuous invariant measure.

In this section we shall derive some properties of the invariant measure constructed in section 2 and also some ergodic properties of the associated dynamical system. In all the following statements we shall assume that the hypothesis H1) and H2) of the previous section are satisfied. We shall first give a precise estimate for the singularity of the invariant measure (see [T.] for related results).

Proposition 3.1. *Uniformly in the index n the elements of the sequence $(h_n)_{n \in \mathbb{N}}$ defined in equation (1) satisfy*

$$h_n(x) \leq O(1)x^{-1} .$$

Moreover any accumulation point h of this sequence satisfies

$$|h(x) - f'_+(0)| \leq O(1)x^{-1/2} .$$

Proof. We first need a ratio estimate for functions in P. From the Herglotz representation theorem (see [D.]), for any function h in P there is a positive measure μ on the real line with support outside (a, b) such that

$$\int_{-\infty}^{+\infty} \frac{d\mu(t)}{1 + t^2} < \infty ,$$

and for any x in $[0, 1]$ (and in fact for any x in $\mathbb{C} \backslash [0, b]$)

$$h(x) = \int_{-\infty}^{+\infty} \frac{d\mu(t)}{(t - x)^2} .$$

From the positivity of the measure μ we obtain immediately for any x and y in $(0,1)$ the inequality

$$\frac{h(y)}{h(x)} \leq \max\left(\frac{x^2}{y^2}, \frac{(b-x)^2}{(b-y)^2}\right).$$

It is at this point that we use explicitly the hypothesis $b > 1$. Applying this result to the function $P^n 1$ we get in particular for x in $[1/2, 1]$

$$\mid P^n 1(x) - P^n 1(1/2) \mid \leq \mathcal{O}(1) P^n 1(1/2).$$

It is easy to show recursively that

$$P^n 1 = f_-^{n'} + \sum_{l=0}^{n-1} [f_+ \circ f_-^l]' [P^{n-l-1} 1 \circ f_+ \circ f_-^l].$$

We now use this decomposition in order to estimate $\sum_{q=0}^n P^q 1(x)$. We shall first replace the contributions of the sum in the above expression by the more manageable quantity (this corresponds to preimages whose orbit is not entirely contained in $[0, 1/2]$)

$$\sum_{q=0}^n \sum_{l=0}^{q-1} f_-^{l'}(x) f_+'(0) P^{q-l-1} 1(1/2).$$

We first estimate the correction coming from this replacement. It is given by

$$\sum_{q=0}^n \sum_{l=0}^{q-1} f_-^{l'}(x) \mid f_+' \circ f_-^l(x) P^{q-l-1} 1 \circ f_+ \circ f_-^l(x) - f_+'(1/2) P^{q-l-1} 1(1/2) \mid$$

$$\leq \mathcal{O}(1) \sum \sum f_-^{l'}(x) (\mid P^{q-l-1} 1 \circ f_+ \circ f_-^l(x) - P^{q-l-1} 1(1/2) \mid$$
$$+ \mid f_+'(0) - f_+' \circ f_-^l(x) \mid P^{q-l-1} 1(1/2))$$

$$\leq \mathcal{O}(1) \sum \sum f_-^{l'}(x) (\mid f_+ \circ f_-^l(x) - 1/2 \mid^{1/2}$$
$$+ \mid f_+ \circ f_-^l(x) - f_+(0) \mid) P^{q-l-1} 1(1/2)$$

$$\leq \mathcal{O}(1) \sum \sum f_-^{l'}(x) [f_-^l(x)]^{1/2} P^{q-l-1} 1(1/2)$$

$$\leq \mathcal{O}(1) \sum_{k=0}^{n-1} P^k 1(1/2) \sum_{l=0}^{\infty} [f_-^l(x)]^{1/2} f_-^{l'}(x).$$

Using the fact that the slope at 0 of f is 1 and the second derivative a is non zero it follows by an easy recursion that $f_-^l(x) \leq \mathcal{O}(1) x/(alx + 1)$, and $f_-^{l'}(x) \leq \mathcal{O}(1)/(1 + alx)^2$. We now have an estimate on the sequence of functions h_n

$$h_n = \frac{\sum_0^{n-1} \left(f_-^{l'}(x) + \sum_0^{l-1} f_-^{j'}(x) f_+'(0) P^{l-j-1} 1(1/2) \right)}{\sum_0^{n-1} P^l 1(1/2)} + \mathcal{O}(1) x^{-1/2}.$$

From the above estimate on $f''_-(x)$ it follows that

$$\frac{\sum_0^{n-1} f''_-(x)}{\sum_0^{n-1} P^l 1(1/2)} \le \frac{O(1)}{x \sum_0^{n-1} P^l 1(1/2)} .$$

Therefore the dominant contribution to h_n is given by the ratio

$$f'_+(0) \frac{\sum_0^{n-2} P^r 1(1/2) \sum_0^{n-2-r} f''^{j}_-(x)}{\sum_0^{n-1} P^r 1(1/2)} .$$

It is easy to see that this quantity is bounded by $O(1)/x$. We can however obtain a better bound. The above estimates imply that

$$|h_n(x) - f'_+(0)/x| \le \frac{O(1)}{x \sum_0^{n-1} P^l 1(1/2)} + O(1) x^{-1/2} ,$$

and the result follows from the divergence of $\sum_0^{n-1} P^l 1(1/2)$.

Now that we have identified the true singularity of the invariant measure, we can prove uniqueness because we can exactly cancel this singularity by subtraction.

Theorem 3.2. *There is a unique absolutely continuous invariant measure with density h satisfying $h(1/2) = 1$ and*

$$|h(x) - C/x| \le O(1) x^{-1/2} ,$$

where C is a positive constant.

Proof. Assume that h_1 and h_2 are two such densities corresponding to the constants C_1 and C_2 respectively. We shall show that these two functions are proportional. Let g be the function defined by $g = h_1 - C_1 h_2/C_2$. Then obviously $Pg = g$, and from Proposition 3.1 we have

$$|g(x)| \le O(1) x^{-1/2} .$$

This implies using elementary properties of the operator P

$$\int_0^1 x|g(x)| \, dx = \int_0^1 x|P^j g(x)| \, dx \le \int_0^1 x P^j |g(x)| \, dx = \int_0^1 f^j(x) |g(x)| \, dx .$$

Summing over j we obtain

$$\int_0^1 x|g(x)| \, dx \le n^{-1} \int_0^1 \sum_0^{n-1} f^j(x) |g(x)| \, dx \le O(1) \int_0^1 |x|^{-1/2} n^{-1} \sum_0^{n-1} f^j(x) \, dx .$$

The sequence of functions $n^{-1} \sum_0^{n-1} f^j(x)$ is uniformly bounded and converges to zero almost everywhere. Therefore it follows from the Lebesgue dominated convergence theorem that $g = 0$.

The following lemma is the first step in the proof of ergodicity of the mapping f with respect to the unique absolutely continuous invariant measure. From now on, we shall denote by λ the normalized Lebesgue measure of the interval $[0, 1]$.

Lemma 3.3.   *Any measurable f invariant set A satisfying*

$$\int_A x^{-1}\,dx < \infty$$

has Lebesgue measure zero.

Proof. From the invariance of the characteristic function χ_A of A, one easily deduces

$$\lambda(A) = n^{-1} \sum_{0}^{n-1} P^j 1(1/2) \int_0^1 \chi_A(x) h_j(x)\,dx$$

$$\leq O(1) n^{-1} \sum_{0}^{n-1} P^j 1(1/2) \int_0^1 \chi_A(x) x^{-1}\,dx \ .$$

If we let n tend to infinity we obtain $\lambda(A) = 0$.

We now come to the proof of the ergodicity of the dynamical system with respect to the absolutely continuous invariant measure.

Proposition 3.4.   *There is a probability measure σ such that for any function u in $L^1(dx)$*

$$\frac{\sum_0^{n-1} P^j u}{\sum_0^{n-1} P^j 1} \xrightarrow[n\to\infty]{} \sigma(u) \quad a.e.$$

Proof. We shall only give the proof fro the case where u belongs to P, and satisfies $C > u > C^{-1}$ for some positive constant C. The reader is referred to [C.F.] for a complete proof.

From the uniqueness of the limit point (Theorem 3.2) it follows that the ratio

$$\frac{\sum_0^{n-1} P^j 1}{\sum_0^{n-1} P^j 1(1/2)}$$

converges to a function e. Note also that form the Herglotz representation Theorem, this function e has no zero in $(0,1)$.

Form the Chacon-Ornstein ergodic theorem (see [K.]) applied to the operator P in the space $L^1([0,1], dx)$ we deduce that the ratio

$$\frac{\sum_0^{n-1} P^j u(x)}{\sum_0^{n-1} P^j 1(x)}$$

converges Lebesgue almost everywhere. Therefore the ratio

$$\frac{\sum_0^{n-1} P^j u}{\sum_0^{n-1} P^j 1(1/2)}$$

converges also almost everywhere. It is easy to show that the limit is again the density of an absolutely continuous invariant measure. It can be shown that this density satisfies the hypothesis of Theorem 3.2. (see [C.F.]). Therefore the limit must be proportional to the function e. The proportionality factor will be denoted by $\sigma(u)$. In order to show that σ is a measure, one has to prove the above results for a class of functions which is larger than P. We refer to [C.F.] for the details.

We observe that it follows from similar argument and the Chacon-Ornstein Theorem that the normalization factor which ensures the convergence to a non trivial function is essentially universal. More precisely, assume that u is a function in P satisfying as above $C > u > C^{-1}$ for some positive constant C. Then we obviously have

$$C^{-1} < \frac{\sum_0^{n-1} P^j u}{\sum_0^{n-1} P^j 1} < C .$$

Since this ratio converges almost everywhere by the Chacon-Ornstein ergodic Theorem, we can find a point $a \in (0,1)$ where the ratio converges. We now observe that the sequences

$$\frac{\sum_0^{n-1} P^j u}{\sum_0^{n-1} P^j u(a)} \quad \text{and} \quad \frac{\sum_0^{n-1} P^j 1}{\sum_0^{n-1} P^j 1(a)}$$

are two sequences of normalized (in a) derivatives of Pick functions. Therefore by the compactness argument we can extract convergent subsequences. In fact, using the uniqueness of the absolutely continuous invariant measure again, it follows that these two sequences converge. This now implies that the ratio

$$\frac{\sum_0^{n-1} P^j u(x)}{\sum_0^{n-1} P^j 1(x)}$$

converges for every x in $(0,1)$.

Theorem 3.5.   *The transformation f is ergodic for the absolutely continuous invariant measure constructed above.*

Proof. Let $\mathbf{E}(\ |\Sigma_i)$ denote the conditional expectation on the sigma-algebra of measurable invariant sets. By the Chacon-Ornstein identification Theorem [K.] we have

$$\lim_{n \to \infty} \frac{\sum_0^{n-1} P^j u}{\sum_0^{n-1} P^j 1} = \frac{\mathbf{E}(u|\Sigma_i)}{\mathbf{E}(1|\Sigma_i)}$$

almost everywhere for any function u in L^1. From Proposition 3.3 we have almost everywhere and for any function in L^1

$$\mathbf{E}(u|\Sigma_i) = \sigma(u)\mathbf{E}(1|\Sigma_i) .$$

Therefore the function $\mathbf{E}(1|\Sigma_i)$ is almost surely constant since all the characteristic functions of its level sets are proportional to itself. This implies immediately that the sigma-algebra Σ_i is trivial.

Remark. It follows from a result of Aaronson [A.] that for a measurable function u integrable with respect to the invariant measure, one cannot have convergence almost everywhere of the ergodic average

$$\frac{\sum_0^{n-1} u\left(f^j(x)\right)}{\sum_0^{n-1} P^j 1(1/2)} \, .$$

The sequence converges however in some weaker sense.

References.

[A.]   J.Aaronson. Journal d'Analyse Mathematique **39**, 203 (1981).

[B.]   R.Bowen. Commun. Math. Phys. **69**, 1 (1979).

[C.F.] P.Collet, P.Ferrero. To appear.

[D.]   W.F.Donoghue. *Monotone Matrix Functions and Analytic Continuation.* Springer Verlag, Berlin Heidelberg New York 1974.

[K.]   U.Krengel. *Ergodic Theorems.* De Gruyter Studies in Mathematics **6**, Walter de Gruyter, Berlin New York 1985.

[Ma.] P.Manneville. Journal de Physique **41**, 1235 (1980).

[Me.] C.Meunier. Journ. Stat. Phys. **36**, 3 (1984).

[T.]   M.Thaler. Isr. Journ. Math. **37**, 303 (1980).

TRANSPORT OF PASSIVE SCALARS: KAM SURFACES AND DIFFUSION IN THREE-DIMENSIONAL LIOUVILLIAN MAPS.

Mario Feingold, Leo P. Kadanoff and Oreste Piro[†]
The James Franck Institute
The University of Chicago
5640 S. Ellis Ave.
Chicago, IL 60637
U.S.A.

ABSTRACT. The global aspects of the motion of passive scalars in time-dependent incompressible fluid flows are well described by volume-preserving (Liouvillian) three-dimensional maps. In this paper the possible invariant structures Liouvillian maps and the two most interesting nearly integrable cases are investigated in detail.

1. Introduction.

Particles suspended in a fluid and moving with it according to the velocity field u of the flow are known in hydrodynamics as *passive scalars*.[1] On one hand, such particles should be small not to disturb the fluid flow. On the other hand, they should be big enough to avoid being disturbed by the microscopic molecular motion. Besides, their mass should be small enough to make inertia negligible. Particles obeying these requirements are the tracers used for flow visualization in fluid mechanics.[2] Moreover, scalar quantities such as temperature, the density of a second fluid and the magnitude of the magnetic field in a highly conducting fluid can be described as passive scalars. The understanding of their dynamics is important in problems like convection,[3] theory of mixing[4] and the fast dynamo effect.[5, 6] The same description applies also to the case of a charged particle in a highly viscous fluid which moves under the influence of an external electric field. Here, the force field becomes equivalent to a velocity field because in this medium the particle quickly reaches a saturation velocity which is proportional to the force. We confine our attention to passive scalars in incompressible fluid flows for which the velocity field $u(x,y,z,t)$ satisfies

$$\nabla \cdot u = 0 \tag{1}$$

Since the velocity field is given, the motion of such particles in the real space is governed by the following set of differential equations

$$\dot{r} = u(x,y,z,t) \tag{2}$$

Let us quickly review some simple formulations of this problem.

[†] Also Departamento de Física, Universidad Nacional de La Plata, C.C. 67, (1900)La Plata, Argentina.

E. Tirapegui and D. Villarroel (eds.), Instabilities and Nonequilibrium Structures II, 37–51.
© *1989 by Kluwer Academic Publishers.*

When the flow is two-dimensional and stationary, Eq. (1) implies that the velocity field $\mathbf{u}(x,y)$ can be derived, from a *stream function* $\Psi(x,y)$

$$u_x = \frac{\partial \Psi}{\partial y} \quad ; \quad u_y = -\frac{\partial \Psi}{\partial x} \tag{3}$$

In this case the equations of motion (2) have a Hamiltonian structure with Ψ playing the role of the Hamilton function. It is important to notice that the phase space of this Hamiltonian system is actually the real configuration space in which the motion of the passive scalars takes place. Hamiltonian systems with only one degree of freedom are always integrable. That means that all the trajectories in the phase space lie on closed (or periodic) curves. Accordingly, trajectories of passive scalars in 2D steady flows will have the same topology.

In order of increasing complexity follows the class of periodically time-dependent 2D flows.[1, 4, 7, 8] Here the stream function becomes time-dependent and the system is equivalent to a non-autonomous Hamiltonian one. In this case, integrable systems for which the phase space is completely foliated into invariant tori are very rare. The KAM theorem however, states that for small departures from integrability, the invariant tori form a finite measure subset of the phase space. For the passive scalars these invariant surfaces are barriers between which chaotically moving particles are trapped in the real space.

3D steady flows are somehow equivalent to the case described in the previous paragraph. Arnold[9] has shown that the motion of passive scalars is integrable if the corresponding fluid flow is inviscid ($\mathbf{u}$ satisfies an Euler equation) and if $\boldsymbol{\omega} \equiv \nabla \times \mathbf{u} \neq \lambda \mathbf{u}$ almost everywhere. At the other extreme, when the Beltrami condition is satisfied, $\boldsymbol{\omega} = \mathbf{u}$, it was found[10] that trajectories which are chaotic coexist with regular ones. The regular trajectories cover two-dimensional surfaces which are invariant under the time evolution of the passive scalar. In turn, the invariant surfaces separate the available space into disconnected regions to which the chaotic trajectories are confined.

The next step in complexity is given by the case of 3D flows with periodic time dependence. Stroboscopic maps of such systems are then three dimensional and in virtue of Eq. (1), volume preserving. In the following we will refer to this maps as 3D-Liouvillian[11] or in short, 3DLM. The purpose of this work is to study the invariant structures displayed by these maps and to determine whether or not such structures prevent chaotic passive scalars trajectories to reach all the space. In other words, we shall investigate the conditions under which unbounded *deterministic* diffusion is allowed in almost integrable 3DLM. This issue is central to the theory of mixing in fluids flows[1, 4] because of its implications for the mixing efficiency.

The study of 3DLM is also interesting from the dynamical systems theoretic point of view. The interest originates in the fact that Hamiltonian systems induce volume preserving maps on even-dimensional spaces. N degrees of freedom Hamiltonians produce Poincaré maps in $2(N-1)$ dimensions. Thus, to $N=2$ and $N=3$ correspond respectively two and four dimensional volume preserving maps. There is a very important difference between these two types of Hamiltonian systems, namely the existence of Arnold diffusion[12, 13] which has an intimate dimensional origin. For $N=2$ Hamiltonians (2D maps) the invariant structures preserved via KAM theorem completely separate the chaotic regions generated by the non integrable perturbations of each other. However, for $N=3$ the analogous invariant structures do not have enough dimensions to produce such a separation. As a consequence, a diffusive motion connecting all the chaotic regions and visiting any arbitrarily small neighborhood of each point in the phase space can in principle take place. The important characteristic of this diffusion is that it can occur at arbitrarily small values of the nonlinearity parameter ε although the diffusion rate decreases faster than exponentially as $\varepsilon \to 0$.[12,14] Since 3DLM lie (dimensionally) between the two qualitatively different Hamiltonian systems, it is a legitimate question weather or not a manifestation of the diffusive phenomenon described by Arnold is present also in this type of systems. It has been shown[15] that there is one class of 3DLM whose nearly integrable behavior is similar to the one found in $N=2$ Hamiltonians and another class for which unbounded diffusive motion takes place at arbitrarily small ε.

In Sec. 2 we briefly introduce the integrable structure of 3DLM. In Sec. 3 we review the behavior of the non-diffusive class. In Sec. 4 the origin and characteristic of diffusion in 3DLM are discussed in detail. Finally in Sec. 5 we point out several problems which are still unsolved.

2. Three dimensional maps and their invariant structures.

We concentrate our attention on volume preserving maps $L:T^3 \to T^3$ on the three dimensional torus of the form

$$(x',y',z')=(x+u(x,z),y+v(x',z),z+w(x',y'))\tag{4}$$

where u, v, and w are doubly-periodic functions. In particular, we shall investigate the following truncation of the Fourier expansions for these functions:

$$x'=x+A_1\sin z +C_2\cos y \quad (mod.\,2\pi)$$
$$y'=y+B_1\sin x'+A_2\cos z \quad (mod.\,2\pi)\tag{5}$$
$$z'=z+C_1 f(y')+B_2 g(x') \quad (mod.\,2\pi)$$

where the functions f and g are smooth on the circle.

In general we call *integrable* a map L_k^0 for which a set of variables $I \in R^k$, $\theta \in T^{3-k}$ can be found such that

$$I'=I\tag{6}$$
$$\theta'=\theta+\omega(I)$$

Clearly, the variables I which we call *actions* by analogy with Hamiltonian dynamics, parametrize a family of $(3-k)$-dimensional geometrical objects that are invariant under the dynamics. In turn, the θ variables specify the position of a point on a given invariant object. At each iteration the angular variables rotate an angle given by the corresponding component of the $(3-k)$-dimensional vector ω. This angle might in general vary from one invariant set to another but remains constant for all the points on each of such sets. Let us describe the four possible cases:

a) $k=0$. Since there are no actions, the frequency vector ω is constant and this integrable case corresponds to a uniform rotation on T^3.

b) $k=1$. The motion takes place on two-dimensional tori defined by $I=const.$ and a point on one of these tori is specified by the values of two angles θ_1, θ_2. Therefore, the motion on these two-tori is a uniform rotation for which the frequency only depends on the value of the action.

c) $k=2$. In this case, there are two actions I_1, I_2 which parametrize a family of invariant circles (one-dimensional tori). The angular variable rotates with a frequency $\omega(I_1,I_2)$.

d) $k=3$. In this case, the map (6) is the identity and each point of the space (parametrized by I_1, I_2 and I_3) is invariant. There are no angles in this problem.

For case a) the results of Ruelle and Takens[16, 17] indicate that small perturbations of the integrable system will produce completely chaotic maps. Although the study of the implication of volume preservation on these results is a very interesting problem, it lies outside the scope of this work which aims to the understanding of lower dimensional invariant objects. At the opposite extreme, $k = 3$, the nearly integrable systems are small perturbations of the identity $I' = I + \varepsilon F(I)$ or $\Delta I = \varepsilon F(I)$. In the $\varepsilon \to 0$ limit, the dynamics of this system can be described by a set of three autonomous differential equations after an appropriate rescaling of the time. Also this case lies outside the interest of the present paper. Therefore, we shall consider only small perturbations L_1 and L_2 around the respective integrable maps L_1^0 and L_2^0.

3. KAM surfaces in one-action three-dimensional Liouvillian maps.

An example of nearly integrable maps with a single action is the map L_1 given by the following restriction of Eq. (5):

$$x' = x + A \sin z + \varepsilon \alpha_C \cos y$$
$$y' = y + \varepsilon \alpha_B \sin x + A \cos y' \tag{7}$$
$$z' = z + \varepsilon \alpha_C \sin y' + \varepsilon \alpha_C \cos x'$$

At $\varepsilon = 0$ this map is integrable with only one action variable z. The motion in this limit takes place on the surfaces $z = const.$ and the angular variables x and y rotate at each iteration a constant angle $\omega_x = A \sin z$ and $\omega_y = A \cos z$ respectively. The map in Eq. (7) corresponds to a discretization of the ABC model proposed by Arnold[9] as an example of chaotic streamlines in a stationary 3-dimensional flow and extensively studied by Dombre et al.[10] The discretization is equivalent to the addition of periodic time dependence to the flow.

In order to investigate the effect of a finite but small value of the nonintegrable perturbation amplitude ε on the form of the invariant surfaces, we implement a perturbative scheme. The condition for a surface defined by the equation $\mathbf{r} = \mathbf{r}(t, s)$ (where $\mathbf{r} = (x, y, z)$ and t, s are parameters) to be invariant is

$$L_1(\mathbf{r}(t, s)) = \mathbf{r}(t', s') \tag{8}$$

We start with the invariant surface $z = z_0$ of the integrable case and suppose that the perturbed surface can be written in the form $z = z_0 + \sum \varepsilon^n H_n(x, y)$. Here the parameters t and s are identified with the coordinates x and y. Inserting this expression into the invariance condition we obtain an infinite system of linear functional equations for the unknown functions H_n that in principle can be solved order by order. The first step in this expansion leads to

$$z' = z_0 + \varepsilon H_1(x', y') + O(\varepsilon^2) = z_0 + \varepsilon H_1(x + A \sin z_0, y + A \cos z_0) + O(\varepsilon^2) \tag{9}$$
$$= z_0 + \varepsilon H_1(x, y) + \varepsilon \sin(y + A \cos z_0) + \varepsilon \alpha_B \cos(x + A \sin z_0) + O(\varepsilon^2)$$

Consequently, $H_1(x, y)$ satisfies the linear functional equation

$$H_1(x + A \sin z_0, y + A \cos z_0) = H_1(x, y) + \sin(y + A \cos z_0) + \alpha_B \cos(x + A \sin z_0) \tag{10}$$

The previous equation can be easily solved by expanding $H_1(x, y)$ in a double Fourier series. Setting $H_1(x, y) = \sum a_{mn} e^{i(mx + ny)}$ in (10) we obtain

$$a_{mn}[e^{i(mA \sin z_0 + nA \cos z_0)} - 1] = \frac{i}{2}[\delta_{-1,n} e^{-iA \cos z_0} - \delta_{1,n} e^{iA \cos z_0}] \delta_{0,m} \tag{11}$$

$$+ \frac{\alpha_B}{2}[\delta_{1,m} e^{iA \sin z_0} + \delta_{-1,m} e^{-iA \sin z_0}] \delta_{0,n}$$

Notice that the coefficients a_{mn} in Eq. (11) remain undefined if the resonance condition

$$m A \sin z_0 + n A \cos z_0 = 2\pi k \qquad (12)$$

is satisfied for $(m,n,k) \neq (\pm 1,0,k), (0,\pm 1,k)$. However, they can be uniquely determined by requiring that a_{mn}-s be continuous functions of z. Thus, the only four non-zero coefficients in $H_1(x,y)$ are $a_{0,\pm 1}$ and $a_{\pm 1,0}$. This is however, a peculiarity emerging from the form of the nonlinear terms in the equation for the action z. If higher order Fourier components were considered in the original map, then more non-zero coefficients would appear in the expression for H_1. Also in our case, when the expansion is carried over to higher orders in ε the corrections H_n will contain correspondingly higher order Fourier coefficients.

Eq. (12) has stronger consequences when it holds for $(m,n,k) = (\pm 1,0,k)$ and $(m,n,k) = (0,\pm 1,k)$. Namely, the values of $a_{0,\pm 1}$ and $a_{\pm 1,0}$ respectively are diverging. This occurs, of course only for few special values of z_0. The invariant surfaces where the condition (13) is satisfied, are called *resonant*. It is clear from the perturbative arguments that these resonant surfaces will exhibit a strongly singular behavior at finite ε. To understand this special behavior notice first that integrable motion on a resonant surface is such that each individual trajectory is not dense on the surface but rather fills an invariant curve contained in it. For $\varepsilon = 0$ there is a continuous family of such invariant lines covering the entire surface. At $\varepsilon \neq 0$ however, only a finite (and even) number of such lines survives the presence of the nonlinear perturbation. To illustrate this process we can implement a perturbative scheme for invariant lines which is similar to the one already used for surfaces. For the sake of clarity we restrict ourselves to the $(\pm 1,0,0)$ resonant surfaces. In the integrable case this resonance occurs if $\omega_x = A \sin z_0 = 0$ i.e., at $z_0 = z_i$ with $z_1 = 0$ and $z_2 = \pi$. It is clear that the trajectories on this resonant surfaces lie on lines parallel to the y-axis. Each initial condition x_0 on the resonant surface corresponds to a different line which is consequently defined by $x = x_0$ and $z = z_i$. We want to understand what happens with these lines when $\varepsilon \neq 0$. Therefore, we will look for perturbed lines of the form

$$x = x_0 + \varepsilon X(y) + O(\varepsilon^2) \quad ; \quad z = z_i + \varepsilon Z(y) + O(\varepsilon^2) \qquad (13)$$

and confine our calculations to order ε. By requiring invariance of (13) under the iteration of the map we find two functional equations, for $X(y)$ and $Z(y)$

$$X(y + A \cos z_i) = X(y) + A Z(y) \cos z_i + \cos y \qquad (14a)$$

$$Z(y + A \cos z_i) = Z(y) + \sin(y + A \cos z_i) + \alpha_B \cos x_0 \qquad (14b)$$

As before, expanding the two functions in Fourier series, $X(y) = \sum a_n{}^x e^{iny}$ and $Z(y) = \sum a_n{}^z e^{iny}$ we obtain for the coefficients

$$a_n{}^x (e^{inA \cos z_i} - 1) = A \cos z_i \, a_n{}^z + \frac{1}{2} (\delta_{1,n} + \delta_{-1,n}) \qquad (15a)$$

$$a_n{}^z (e^{inA \cos z_i} - 1) = \frac{i}{2} [\delta_{-1,n} e^{-iA \cos z_i} - \delta_{1,n} e^{iA \cos z_i}] + \delta_{0,n} \, \alpha_B \cos x_0 \qquad (15b)$$

Notice that when $n = 0$ the factor $(e^{inA \cos z_i} - 1)$ vanishes whereas the r.h.s of Eq. (15b) does not. Therefore, the first order correction for the invariant lines is finite only if $\cos x_0 = 0$. This implies that out of the infinity of invariant lines corresponding to the $(1,0,0)$ resonant surfaces $z_0 = z_i$, only those defined by the (x_0, z_i) pairs $(\pi/2, \pi)$, $(3\pi/2, 0)$, $(\pi/2, 0)$ and $(3\pi/2, \pi)$ survive at first order in ε. Numerical computations indicate that half of those lines are dynamically stable and the other half are unstable. This result is a remarkable manifestation in 3DLM of a scenario similar to the one which, for Hamiltonian systems, is predicted by the Poincaré-Birkhoff theorem.[13] While around stable lines, a family of elliptic invariant tubes is formed, associated with the unstable lines an H-shaped chaotic slab emerges. This scheme is repeated for higher order resonances at the corresponding order of the perturbation expansion. The location of the lowest order elliptic lines is indicated schematically in Fig. 1.

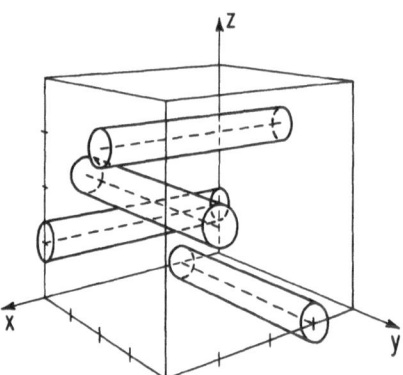

Fig. 1. Schematic representation of the lowest order *elliptic* resonances. The dashed lines parallel to the x and y-axis represent the stable invariant lines of the $(0,1,0)$ and $(1,0,0)$ resonant surfaces respectively, which survive when $\varepsilon \neq 0$ in Eq. (7). The cylinders sketch the tubular invariant surfaces that appear around these line at finite ε.

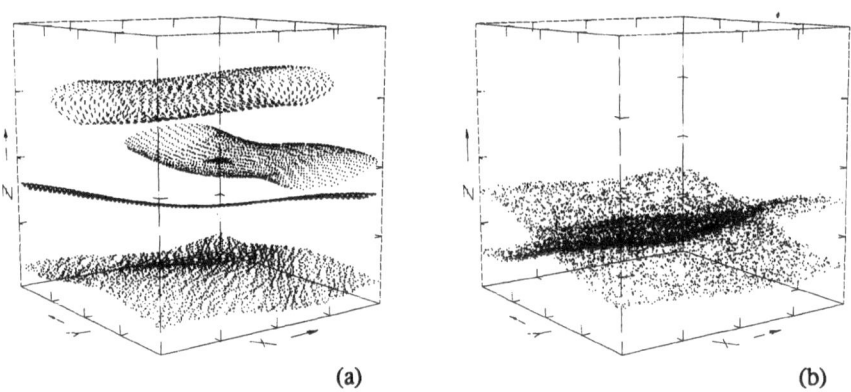

(a)   (b)

Fig. 2. Numerically obtained invariant surfaces and chaotic volumes. The parameter values in Eq. (7) are $A = 1.5$, $\alpha_B = 1$ $\alpha_C = 2$ and $\varepsilon = .1$. a) In order of increasing z: non-resonant surface $(x_0 = 0, \ y_0 = .56, \ z_0 = .11)$ and tubular surfaces around the $(1, -1, 0)$, $(1,0,0)$ and $(0,1,0)$ resonances -$(x_0 = 0, \ y_0 = .9, \ z_0 = .395)$, $(x_0 = .75, \ y_0 = 0, \ z_0 = .4626)$ and $(x_0 = 0, \ y_0 = .5, \ z_0 = .828)$ respectively-. b) H-shaped chaotic volume associated with the *hyperbolic* line of the $(0,1,0)$ resonance $(x_0 = 0, \ y_0 = .5$ and $z_0 = .24)$. All the initial conditions are indicated in fractions of 2π and the box represents the $[0, 2\pi]^3$ region.

For the surfaces where the frequencies ω are far from satisfying the resonance condition, we might expect that some version of the KAM theorem will hold and that slightly deformed invariant surfaces will persist if ε is not too big. Actually, there is extensive numerical evidence in favor of such a conjecture. In Fig. 2a we show one of those perturbed invariant surfaces obtained numerically. In the same figure we illustrate the existence of the tubular invariant surfaces surrounding the invariant lines to which the lowest order resonant surfaces collapsed.

An invariant tube of more complicated topology corresponding to second order resonances is also shown. Finally, a chaotic *slab* associated with one of the first order *hyperbolic* invariant line is shown in Fig. 2b. The reader can easily recognize that the preserved invariant surfaces are barriers through which a chaotically moving particle cannot penetrate. In other words, diffusion of an individual trajectory throughout all the space is not allowed. In the next section we show that the situation for L_2 maps is entirely different.

4. Diffusion in two-action maps.

Let us now consider the maps close to integrable cases with two almost conserved quantities, L_2:

$$I_1' = I_1 + \varepsilon P_1(I_2, \theta) \; ; \; I_2' = I_2 + \varepsilon P_2(I_1', \theta)$$
$$\theta' = \theta + \omega(I_1', I_2') \tag{16}$$

At $\varepsilon = 0$ the motion takes place on lines $I = const$. We could examine the perturbative behavior of those invariant lines with the hope of finding some sort of KAM result for them. However, by a mechanism similar to the one responsible for breaking the resonant surfaces in L_1 maps, all the invariant lines are destroyed to first order in ε. Later, we shall illustrate this process with a particular example. In fact, we can understand the origin of the singularity of this integrable case by means of the following argument based on an adiabatic approximation. Suppose that $\omega(I_1, I_2)$ is irrational for some given values of the arguments. In the limit of $\varepsilon \to 0$, we can assume that before I changes significantly, the angle θ covers uniformly the entire $(0, 2\pi)$ interval. Under these circumstances, the variation of I can only be sensitive to averages of $P = (P_1, P_2)$ over all the possible values of θ. Therefore,

$$\Delta I = \varepsilon <P(I, \theta)>_\theta \equiv \varepsilon \overline{P}(I) \tag{17}$$

where $< >_\theta$ stands for θ-average. Thus, the dynamics of the action variables decouples from θ for non-resonant ω. Eq. (17) leads in the limit $\varepsilon \to 0$ to a system of two ordinary differential equations

$$\frac{dI}{dt} = \overline{P}(I) \tag{18}$$

were the identification $\varepsilon \equiv \Delta t$ was made. This system can be easily integrated by variable separation. Its trajectories $I(t)$ satisfy

$$\int_0^{I_2} \overline{P}_1(I) dI - \int_0^{I_1} \overline{P}_2(I) dI = W(I_1, I_2) = const. \tag{19}$$

In other words, in the $\varepsilon \to 0$ limit, the action variables slowly evolve along the curves defined by Eq. (19). Including the fast motion in the θ direction, we infer that the originally invariant lines parallel to the θ axis coalesce in invariant surfaces Σ_β defined by the condition $W(I_1, I_2) = \beta$. A typical trajectory will cover densely such surfaces rather than move on an invariant curve. The adiabatic approximation is exact in the limit $\varepsilon \to 0$. One is then tempted to conclude that the situation is similar to the one described for one action maps in the sense that the adiabatic invariant surfaces can survive when finite nonlinearities are present. However, a new and very interesting phenomenon appears in L_2 maps. We first notice that the adiabatic approximation is bound to fail whenever the resonance condition $\omega(I_1, I_2) = 2\pi k/n$ is satisfied. Moreover, this condition defines a family of surfaces which in a generic case does not coincide with the family of invariant surfaces. As a consequence, each invariant surface will intersect at least one resonance sheet. At the intersections, the adiabatic approximation is spoiled and so is the smoothness of the invariant surfaces. Far from the intersection, however, one expects that trajectories evolve on surfaces which are slight deformations of the ones given by Eq. (19). In order to exemplify the characteristic behavior of L_2 maps, we use an appropriate restriction of the family of maps defined in Eq. (5):

$$x' = x + \varepsilon \alpha_{A_1} \sin z + \varepsilon \alpha_{C_2} \cos y$$

$$y' = y + \varepsilon \alpha_{B_1} \sin x' + \varepsilon \alpha_{A_2} \cos z \qquad (20)$$

$$z' = z + C_1 f(y') + B_2 g(x')$$

For $\varepsilon = 0$ the map in Eq (20) is integrable and has the form of Eq. (16). In this case the lines corresponding to constant values of x and y are invariant. A search for invariant lines of the form $(x, y) = (x_0, y_0) + \sum (X_n(z), Y_n(z)) \varepsilon^n$, in the nearly integrable case can be performed perturbatively. The order ε calculation leads to the following couple of functional equations

$$X[z + C_1 f(y_0) + B_2 g(x_0)] = X(z) + \alpha_{A_1} \sin z + \alpha_{C_2} \cos y_0 \qquad (21a)$$

$$Y[z + C_1 f(y_0) + B_2 g(x_0)] = Y(z) + \alpha_{B_1} \sin x_0 + \alpha_{A_2} \cos z \qquad (21b)$$

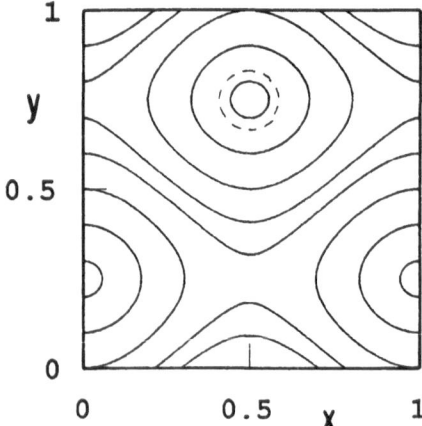

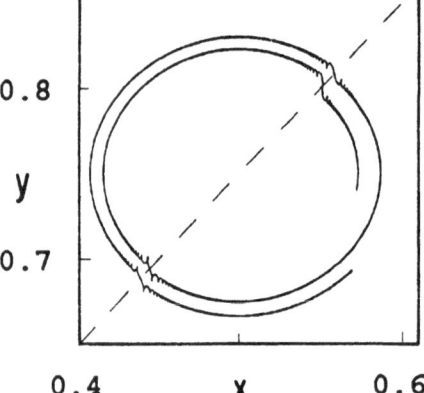

Fig. 3. Few members of the family of surfaces defined by Eq. (23) for $\alpha_{B_1} = 1.5$ and $\alpha_{C_2} = 2$ projected down the $z = 0$ plane. For comparison, the dashed curve is the first turn of the trajectory depicted in Fig. 4.

Fig. 4. One trajectory of the map in Eq. (20) for $\alpha_{A_1} = 1$, $\alpha_{A_2} = 2.5$, $C_1 = B_2 = 4$ and $\varepsilon = .001$, $f \equiv \cos$ and $g \equiv \sin$. α_{B_1} and α_{C_2} are the same as in Fig. 3. The dashed line indicates the location of the lowest order resonance.

Clearly, the constant terms containing x_0 and y_0 in Eq. (21) produce the same effect as the analogous terms containing x_0 in the Eq. (14b), i.e. the divergence of the zero order Fourier coefficient of the $X(z)$ and $Y(z)$ functions. Since $<\sin z>_z = <\cos z>_z = 0$ the adiabatic invariant surfaces Σ_β of Eq. (19) become

$$W_0(x,y) = \alpha_{C_2}\sin y + \alpha_{B_1}\cos x = \beta \qquad (22)$$

Fig. 3 shows some of these surfaces projected down the $z = 0$ plane. To illustrate the effect of the resonances we plotted in Fig. 4 a trajectory of the map in Eq. (20) for a finite but very small value of ε. The location of the lowest order resonance $\omega = 0$ for a particular election of the functions f and g is indicated by the dashed line. Close to this line, the trajectory oscillates wildly and jumps from one adiabatic surface to another.

To depict this behavior in a different way we show in Fig. 5 the time evolution of the value of W_0 which would be constant if the adiabatic approximation were exact. One can easily recognize intervals where W_0 is almost constant followed by relatively short times of oscillatory behavior. Naturally, this oscillations occur whenever the trajectory crosses the first order resonance. Notice that as a consequence of these oscillations W_0 randomly jumps from one asymptotically constant value to another corresponding to two different adiabatic surfaces.

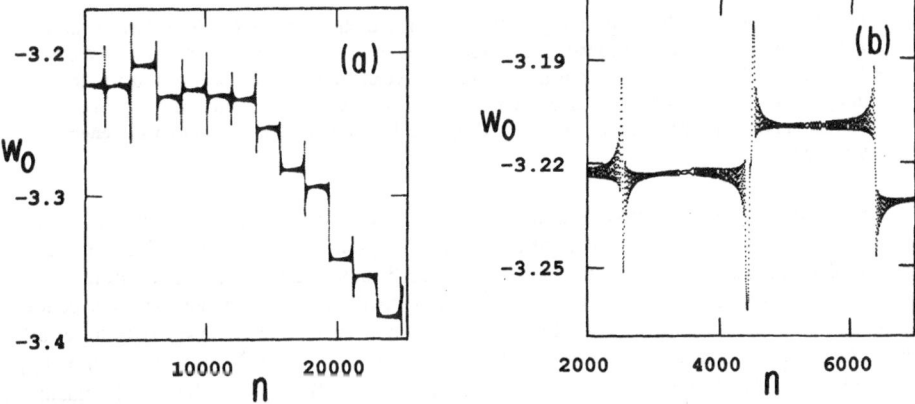

Fig. 5. a) $W_0(x_n, y_n)$ (Eq (23)) vs. n. b) Fine structure of this time evolution. The map and parameter values are the same as in Fig. 4.

One striking consequence of this dynamical behavior is that a single trajectory can in principle visit the entire region of space where the adiabatic surfaces which intersect with the first order resonance reside. The size of this region can be controlled by choosing the functional form of the frequency ω. In Fig. 6. we show two nearly extreme cases. In the first one (Fig. 6a), a case where the resonance condition almost coincides with one of the adiabatic invariant surfaces is shown. Several initial conditions have been used to generate this picture. Notice that most of the trajectories evolve on smooth surfaces which are roughly the same as those shown in Fig. 3. In addition, one can observe a small region of chaotic trajectories associated with the invariant surfaces intersecting the first order resonance condition. Of course, higher order resonances have similar effects in other regions of the space, but these are not evident on the time scale of the picture. On the other hand, Fig. 6b shows the opposite extreme. Here, the first order resonance indicated by the dashed line intersects almost all the surfaces of Eq. (22). In this figure, the iterations of only one trajectory are shown. It is now apparent that this single trajectory visits all the available space.

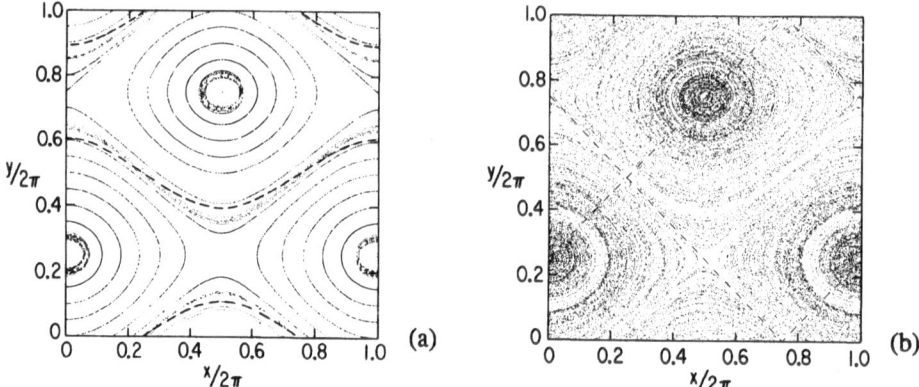

Fig. 6. Iterations of the map in Eq. (20) laying in the slice $0 \leq z \leq .01$. The dashed lines indicate the location of the lowest order resonances. a) $C_1 = 2.5$, $B_2 = 4$, $f \equiv \sin$, $g \equiv \cos$ and the remaining parameters are as in Fig. 4. Several initial condition distributed along the y-axis and the $x = \pi$ line have been necessary to obtain this picture. b) f, g and all the parameters are the same as in Fig. 4. Only one initial condition is required to generate this picture.

Since the adiabatic approximation described above is only valid to lowest order in ε it can be considered the first step of a perturbative expansion in powers of ε. The small amplitude oscillations still present in the intervals where β should remain constant could be attributed to higher order terms of this expansion. It is instructive to calculate the ε-correction to the adiabatic invariant in Eq. (22). Let us suppose that at a finite ε there exists a perturbed invariant surface which can be described by the equation $W(x,y) = \sum W_n \varepsilon^n = \beta$ where W_0 is given by Eq. (22). The condition of invariance reads now $W(x',y') = W(x,y)$. Replacing x' and y' from Eq. (20), expanding both sides in powers of ε and comparing $O(\varepsilon)$ terms we find a linear functional equation for W_1

$$W_1[x,y,z + C_1 f(y) + B_2 g(x)] - W_1(x,y,z) = \qquad (23)$$

$$\alpha_{A_1} \alpha_{B_1} \sin x \sin z - \alpha_{A_2} \alpha_{C_2} \cos y \cos z$$

If $W_1(x,y,z) = \sum a_n(x,y) e^{inz}$ is substituted in Eq. (24) we obtain

$$a_n(x,y) \{ e^{in[C_1 f(y) + B_2 g(x)]} - 1 \} = \delta_{1,n} \left(\frac{-i \alpha_{A_1} \alpha_{B_1} \sin x - \alpha_{A_2} \alpha_{C_2} \cos y}{2} \right) \qquad (24)$$

$$+ \delta_{-1,n} \left(\frac{i \alpha_{A_1} \alpha_{B_1} \sin x - \alpha_{A_2} \alpha_{C_2} \cos y}{2} \right)$$

Once again, we can define $a_n \equiv 0$ for $n > 1$ by continuity. However, $a_1(x,y)$ will necessarily be singular whenever the resonance condition $[C_1 f(x) + B_2 g(x)] = 2\pi(k/n)$ (with k and n integers) is satisfied for $n = 1$. The location of the $n = 1$ singularities coincides with the breakdown of the zero order adiabatic approximation. It is interesting to notice however, that when the calculation is carried out up to order ε^n the same type of singularities appear in a_n at the n-th order resonances. Therefore, the behavior described above for the first order resonances is reproduced for the n-th order ones on time scales of $O(\varepsilon^{-n})$.

Using Eq. (25) we can now compute $W(x,y)$ up to order ε. Fig. 7a shows the evolution in time of the quantity $W_\varepsilon = W_0 + \varepsilon W_1$ for the same part of the trajectory as in Fig. 5b. The divergence of W_ε at the resonances can be clearly appreciated in comparison with the bounded (although wildly oscillatory) behavior of W_0. On the other hand, the fine structure of W_0 in the intervals between the $n = 1$ resonances is smoothed out by the correction. Therefore, in these intervals W_ε is more accurately conserved than W_0. Notice however, that in the central part of these intervals W_t oscillates and jumps in a similar way as W_0 does close to the first order resonances but at a much smaller scale (Fig 7b). These oscillations are due to the $(k,n) = (1,2)$ resonance where, in turn, $O(\varepsilon^2)$ corrections diverge.

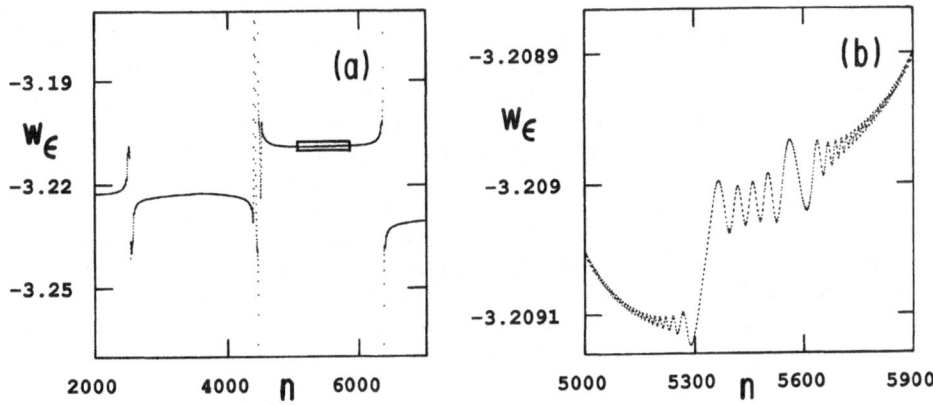

Fig. 7. a) Evolution of the adiabatic invariant $W_\varepsilon = W_0 + \varepsilon W_1$ along the same part of the trajectory shown in Fig. 5b and for the same map and parameter values. b) Enlargement of the small box indicated in a).

We can now discuss the relation between the diffusive motion in L_2 maps and the well known Arnold diffusion which appears in Hamiltonian systems with more than two degrees of freedom. First of all, both types of motion occur at arbitrarily small values of the nonlinearity parameter ε. However, the diffusion rate in the two processes has a different dependence on ε. While the Nekhoroshev estimate[14,18] provides a bound of $O(\exp(-1/\sqrt\varepsilon))$ on the Arnold diffusion rate, in L_2 maps we find a power law dependence. This difference originates in the mechanisms that lead to diffusion in each case. For nearly integrable Hamiltonian maps, as a consequence of the canonical structure, the evolution of each action variable does not depend explicitly on the other. On the other hand, $O(\varepsilon)$ sized non-integrable perturbations while preserving the invariance of the actions in the non-resonant regions of the *action space*, produce a motion of these variables along the complementary regions where the action dependent frequencies resonate. In general for these systems, the resonances are located on a interconnected network which is dense in the action space. Thus, an individual trajectory can in principle visit the neighborhood of any point in the space with a finite recurrence time. Moreover, this diffusive motion is allowed at arbitrarily small values of the perturbation. However, due the tenuousness of the resonance network the time scale on which Arnold diffusion is manifested, grows exponentially as $\varepsilon \to 0$. Calculations which involve the estimation of the size of high order resonances as a function of epsilon lead to the Nekhoroshev bound on the diffusion constant.

For L_2-type three-dimensional Liouvillian maps a different picture emerges. Here, there is no reason to consider the dynamics of the action variables independent of each other. In fact, a generic perturbation of the integrable case L_2^0 will involve some explicit coupling between the actions. Precisely this coupling is the one responsible for the adiabatic motion described above which unlike in Hamiltonian systems is in general transversal to the resonances.

Nevertheless, a contact between the two different behaviors can be established if only special nonintegrable perturbations that leave the actions explicitly uncoupled are considered. In the example of Eq. (20) this corresponds to setting $\alpha_{C_2} = \alpha_{C_1} = 0$. The absence of coupling has remarkable consequences. First of all, the perturbation expansion for the invariant lines is now well defined up to order ε except for a family of resonant lines where $\omega = 2\pi k$. One might expect that the perturbation expansion should be well defined to all orders at least for invariant curves located far enough from the resonances. However this is not the case. Already in the second order correction terms with non vanishing mean value (e.g. $\propto \sin^2 z$) show up. These have the same effect on the $O(\varepsilon^2)$ correction as the terms which couple the actions in Eq. (20) had on the $O(\varepsilon)$ one. Therefore, a massive breakdown of invariant lines with the same characteristics as in the coupled case occurs now at second order in ε. Once again, this singular behavior is associated with an adiabatic drift of the trajectories along invariant surfaces. To understand the origin and characteristics of this adiabatic drift, we consider a trajectory for which the initial value of x_0 and y_0 are close to a (k,n) resonance. Under these conditions, the n-th iterate will be necessarily close to the initial one. Defining $\Delta_n x = x_n - x_0$, $\Delta_n y = y_n - y_0$ and $\Delta_n z = z_n - z_0$ we find after lengthy but straightforward algebra that up to order ε^2

$$\Delta_n x = \frac{1}{2}\varepsilon^2 \alpha_{A_1}\alpha_{A_2}\frac{\partial\omega}{\partial y}\frac{n}{\sin^2(\omega/2)}\sin^2(z+\phi_1)+F_x(x,y,z) \tag{25a}$$

$$\Delta_n y = \frac{1}{2}\varepsilon^2 \alpha_{A_1}\alpha_{A_2}\frac{\partial\omega}{\partial x}\frac{n}{\sin^2(\omega/2)}\sin^2(z+\phi_2)+F_y(x,y,z) \tag{25b}$$

Where ϕ_1 and ϕ_2 are constants and $<F_x>_z = <F_y>_z = 0$. Since close to the resonance $\Delta_n z$ changes much faster $(O(\varepsilon))$ than $\Delta_n x$ and $\Delta_n y$ do $(O(\varepsilon^2))$, the averaging procedure is justified and we find

$$\Delta_n y = \frac{1}{4}\varepsilon_R^2 \alpha_{A_1}\alpha_{A_2}\frac{\partial\omega}{\partial y}\frac{n}{\sin^2(\omega/2)} \tag{26a}$$

$$\Delta_n y = \frac{1}{4}\varepsilon_R^2 \alpha_{A_1}\alpha_{A_2}\frac{\partial\omega}{\partial x}\frac{n}{\sin^2(\omega/2)} \tag{26b}$$

Proceeding as in the case of Eq. (19) we conclude that an $O(\varepsilon^2)$ drift of x and y along the curves $\omega(x,y)=const$ takes place. In other words, here the adiabatic invariant surfaces turn out to be parallel to the resonances. Therefore, the trajectories drift along the resonances in the action plane. This behavior resembles the one in Hamiltonian systems except for two facts.

First, in our type of systems the resonances do not intersect in the action space. Consequently, this drift alone does not allow a trajectory to visit the entire space. Secondly, the average drift per iteration Δ_n / n is proportional to ε^2 for all frequency values except at the $n = 1$ resonance. There, the expressions (25) and (26) are not valid any more because of non-vanishing $O(\varepsilon)$ terms. Notice the divergence of the r.h.s. of these equations at $\omega = 2\pi k$. This is consistent with the observation that around the first order resonance the average drift is $O(\varepsilon)$. It is also convenient to remark that the drift rate depends smoothly on the frequency. A more detailed description of the characteristics of this motion is given in Ref. 19.

Therefore, we can tentatively consider as a first approximation that the diffusive motion displayed by the generic L_2 maps where the actions are coupled, is the result of the combination of two effects: the adiabatic motion generated by the coupling and the drift along the resonances which becomes evident in the uncoupled case. Although this picture seems to be an oversimplification when high order resonances are involved, it gives a correct estimate for the diffusion rate in the regions dominated by the first order one. Since the time scale of the adiabatic motion is ε and the width of the first order resonance is $O(\sqrt{\varepsilon})$ the size of the jumps in the action plane as the trajectory crosses the $n = 1$ resonance is $O(\sqrt{\varepsilon})$. By assuming that these jumps are uncorrelated, we can consider the motion of this action variable as a random walk with $O(\varepsilon^{-1})$ time step. Therefore, the diffusion rate will be proportional to ε^2. In the Fig. 8 the agreement between this estimate and the results obtained from numerical simulations is stressed.

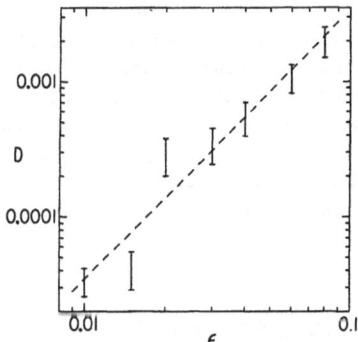

Fig. 8. Diffusion constant vs. ε. The map used here is obtained from Eq. (20) with $g(x) \equiv x$, $C_1 = 0$ and $B_2 = 1$ and the parameters α being the same as in the previous pictures. Error bars are obtained numerically. The best fit (dashed line) for $D \propto \varepsilon^\gamma$ gives $\gamma = 1.99 \pm 0.3$

5. Concluding remarks.

Three dimensional Liouvillian maps present an extremely rich variety of dynamical phenomena. Many of these phenomena show analogy in the nearly integrable limit, with the behavior of Hamiltonian systems of either two or three degrees of freedom which generate simplectic maps in two and four dimensions respectively. The similarity with one or the other is governed by the number of invariant quantities of the corresponding integrable system.

However, the detailed characteristics of the dynamics of three-dimensional maps differs from the Hamiltonian systems both in qualitative and quantitative aspects. In fact, the full understanding of many of the features of Liouvillian maps would probably require the development of appropriate techniques to deal with non-canonical systems.

Meanwhile, we shall indicate some interesting problems which are still open and whose solutions probably rely on more or less straightforward extension of the approaches used in other dynamical systems.

In the case of maps with one action, the smooth two-dimensional KAM surfaces break at a critical value of the nonlinearity parameter. The question of how these surfaces behave at the breakdown point arises as a natural extension of similar studies in two dimensional area preserving maps. However, in our case the dynamics on the invariant surface is a composition of two independent rotations and this problem becomes a version of the still unsolved transition to chaos in three-frequency systems.

A related problem is the study of the dynamics of extended objects, e.g. the evolution of one dimensional arrays of points in space. It is suggested by our previous findings that the invariant curves in 3D maps play a role similar to fixed points in lower dimensional systems as far as the scaling at the transition to chaos is concerned. Besides, in some physical applications, the quantities passively transported are intrinsically associated with this type of objects. This is the case, for example, for magnetic fields in highly conducting fluids or vortex lines in some special conditions.

On the other hand, the ergodic properties of L_2 maps should be extensively investigated. We were able to obtain good estimates for the diffusion rate in the regions dominated by the first order resonances. However, one can imagine a situation where a fraction of the adiabatic surfaces do not cross any first order resonance. In these regions the local diffusion rate is determined by the resonances with $n \geq 2$. A better understanding of the interaction between the adiabatic and resonant motions is required to obtain a similarly reliable estimate for this case. It is important to remark that in the applications to hydrodynamics, the dynamical variables of the Liouvillian maps are just the spatial coordinates of the passive particles. Thus, the diffusive motion described in this paper has a direct realization in the transport of those particles throughout the space.

Acknowledgements.

We thank U. Frisch, G. Gunaratne, S.A. Orzag, Y. Pomeau, I. Procaccia, P. G. Saffman, E. Spiegel and M. Tabor for useful discussions. This work has been supported in part by NSF-DMR under grant number 85-19460. M.F. acknowledges the support of a Dr. Chaim Weizmann Post-Doctoral Fellowship. O.P. acknowledges support of a fellowship from the Consejo Nacional de Investigaciones Científicas y Técnicas-Argentina.

References

1. H. Aref, *J. Fluid Mech.*, **143**, 1 (1984).
2. D.J. Tritton, *Physical Fluid Dynamics*, (Van Nostrand, Berlin, 1977).
3. C. Normand and Y. Pomeau, *Rev. Mod. Phys.*, **49**, 581 (1977).
4. W. L. Chien, H. Rising, and J. M. Ottino, *J. Fluid Mech.*, **170**, 355 (1986).
5. Y. B. Zeldovich and A. A. Ruzmaikin, "Nonlinear Problems of Turbulent Dynamo," in *Nonlinear Phenomena in Plasma Physics and Hydrodynamics*, (ed. R. Z. Sagadeev, p.

31, MIR Publishers, Moscow, 1986).

6. S. Childress, *J. Math. Phys.*, **11**, 3063 (1970).

7. J. Chaiken, R. Chevray, M. Tabor, and Q. M. Tan, *Proc. R. Soc. A*, **408**, 165 (1986).

8. T. H. Solomon and J. P. Gollub, in *Proceedings of the Fritz Haber International Symposium*, (ed. I. Procaccia, Plenum, New York, 1987).

9. V. I. Arnold, *C. R. Acad. Sci. Paris*, **261**, 17 (1965).

10. T. Dombre, U. Frish, J. M. Green, M. Henon, A. Mehr, and A. M. Soward, *J. Fluid Mech.*, **167**, 353 (1986).

11. E. Spiegel, private communication.

12. V. I. Arnold, *Mathematical methods in classical mechanics*, (Springer, New York, 1978).

13. A. J. Lichtenberg and M. A. Liberman, *Regular and Stochastic Motion*, (Springer Verlarg, Berlin, Heidelberg, New York, 1983).

14. N. N. Nekhoroshev, *Russ. Math. Surv.*, **32**, 1 (1977).

15. M. Feingold, L. P. Kadanoff, and O. Piro, *J. Stat. Phys.*, **50**, 529 (1988).

16. D. Ruelle and F. Takens, *Comm. Math. Phys.*, **20**, 167 (1971).

17. S. Newhouse, D. Ruelle, and F. Takens, *Comm. Math. Phys.*, **64**, 35 (1978).

18. G. Benettin and G. Gallavotti, *J. Stat. Mech.*, **44**, 293 (1986).

19. M. Feingold, L. P. Kadanoff, and O. Piro, to be published.

PERIODIC ORBITS OF MAPS ON THE DISK WITH ZERO ENTROPY

J.M. Gambaudo[1] and C. Tresser[2]

Let M be a set and f a map from M into itself. In this quite general setting, one can already define the **orbit** of a point x in M as the set $\{x, f(x), f^2(x), \cdots\}$ where f^n stands for $f \circ f^{n-1}$, f^0 being the identity map Id_M. Assume that for some x_0 in M, there is a $n > 0$ such that $f^n(x_0) = x_0$. Then there is a smaller such n, say $p = \min\{n > 0$ such that $f^n(x_0) = x_0\}$, and we will say that x_0 is a **periodic point** (of period p) and that the set $\{x, f(x), f^2(x), \cdots, f^{p-1}(x_0)\}$ is a **periodic orbit**.

We began by introducing M as a set and f as a map, but the more structure one puts on M the more questions one can ask about the orbit structure, and the more constraints on f one puts, the less freedom remains for this orbit structure. We will consider for instance the case when M is the 2 dimensional disk D^2. With no structure imposed on D^2, it would be natural to let f be any self map of D^2, and there would be few hope toward a global understanding of all possible orbit structures in this case. It occurs that a classical problem in dynamics consist in considering D^2 as a smooth manifold (or if you prefer a ball in R^2 equiped with its usual topology), and choosing f as a smooth orientation preserving embedding of D^2 (i.e. a map which has an inverse and which is differentiable as well as its inverse). For reasons which will be clear later, we will actually assume that f and its inverse are C^2 (i.e. twice diferentiable with second derivatives continuous). We will refer to this problem as the **main problem**.

But before dealing with this main problem, we will first consider a situation a bit simpler:

It is the case when M stands for D^1 which can be thought of as being the unit interval $I = [0,1]$. Choosing f as a diffeomorphism leads to an unexiting dynamics: for instance all the periodic points can only have periods 1 or 2!!!

[1]Depto. de Matemáticas y Ciencias de la Computación, Universidad de Chile, Casilla 170/3-correo 3, Santiago, Chile. Permanent addres: Laborartoire de Mathématiques, Parc Valrose 06034 Nice Cedex, France.
[2]Dept. of Mathematics, The University of Arizona, Tucson, Arizona 85721, U.S.A. Permanent addres: Laboratoire de Physique Théorique, Parc Valrose 06034 Nice Cedex, France.

E. Tirapegui and D. Villarroel (eds.), Instabilities and Nonequilibrium Structures II, 53–65.
© *1989 by Kluwer Academic Publishers.*

However an interesting problem arises when one endowes I with its usual order and chooses f as an endomorphism for this structure, i.e. simply a continuous map from I into itself. This can be considered as a special limiting case of the main problem, when an embedding of D^2 contracts D^2 on a tiny neighborhood of a segment.

Another interesting one dimensional problem concerns endomorphisms of the circle T^1 and we will also say some words about it.

It is a pleasure to one of the authors (J.M.G.) to thank FUNDACION ANDES for having supported his stay in Chile where part of this work has been done.

A. ONE DIMENSIONAL DYNAMICS.

I. First look on the interval.

THEOREM 1 (SARKOVSKII). *Let $<<$ stands for the following order on the set of the natural integers:*

$$3 << 5 << 7 << 9 << \cdots << 2.3 << 2.5 << 2.7 << 2.9 << \cdots 2^2.3 <<$$
$$2^2.5 << 2^2.7 << 2^2.9 << \cdots 2^n.3 << 2^n.5 << 2^n.7 << 2^n.9 << \cdots 2^n <<$$
$$2^{n-1} \cdots << 2^3 << 2^2 << 2 << 1.,$$

and let p stands fore any integer and assume that some continuous map f from the unit interval into itself has a periodic orbit with period p, then for all q such that $p << q$, f has a periodic orbit with period q.

This amazing resul goes back to 1964 [13] and fearly simple proofs have been given since (see e.g. [2]). This theorem is also relevant in the understanding of the appearance of **chaos** when varying the parameters defining families of maps on I. For us, chaos means **positive topological entropy**, and instead of giving a general definition (which can be found together with more background for instance in [4] or [7]), let us quote another nice theorem:

THEOREM 2 (MISIUREWICZ [10]). *Let f be an endomorphism of the unit interval I, then f has positive topological entropy if and only if there exist:*
- *a positive integer n,*
- *a subinterval J of I with non empty interior*
- *and two subintervals J_1 and J_2 of J such that booth $f^n(J_1)$ and $f^n(J_2)$ contain J. (see fig. 1).*

The same result holds true for endomorphisms of the circle, the subintervals being, in this case, arcs of the circle.

Definition. An f such that some of its iterates act as described above in some subintervals is said to have a **horseshoe**, and we say a n-**horseshoe** when n is explicitly needed. We let to the reader the pleasure to prove the

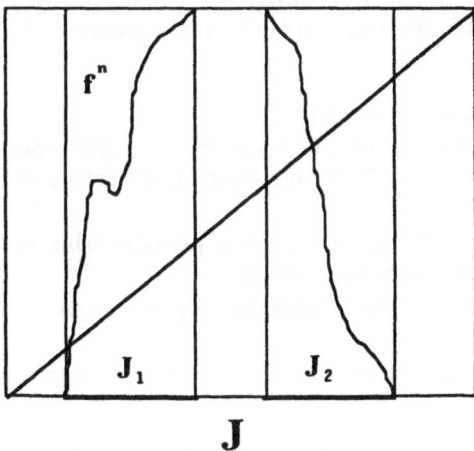

Figure 1.

PROPOSITION 1. *If f has a n-horseshoe then f^n has periodic orbits of all the periods;*

and let us check now the following.

PROPOSITION 2. *An endomorphism of the interval which possesses a periodic orbit with period 3, has a 2-horseshoe.*

The proof uses the following trivial but fundamental fact: "if f is an endomorphism of I and $p < q$ are 2 points of I then $f([p,q])$ contains the closed interval which has $f(p)$ and $f(q)$ as endpoints". This fact is a direct consequence of the mean value theorem. Then the proof of proposition 2 goes as follows:

Remark that in order for 3 points on I to be permuted, one of the extrema (call it **1**) has to be sent on the other extremum (call it **2**), which is sent itself on the middle point (call it **3**) which is in turn sent back to **1**. Without lose of generality, let us assume $2 < 3 < 1$. Since $f(3) = 1 > 3$ and $f(1) = 2 < 1$, there must be a point x in $]3,1[$ such that $f(x) = x$. Now, let us prove that $J = [2,x]$, $J_1 = [2,3]$ and $J_2 = [3,x]$ are good candidates. Using our "fundamental fact" it falls (see fig. 2):

$$f^2[2,3] \supset [1,2] \supset [2,x] \quad \text{and} \quad f^2[3,x] \supset [2,x] \qquad \text{Q.E.D.}$$

Proposition 2 is more than an example since combining it with Sarkovskii theorem, it gives:

THEOREM 3 (BOWEN AND FRANKS [3]). *If an endomorphism of the interval has a periodic orbit whose period p is not a power of 2, then it has positive topological entropy.*

The argument is the following:

i) Since p is not a power of 2, one can write $p = q.2^n$ where q is an odd number greater than 1, then using Sarkovskii theorem one gets that f has a periodic orbit with period 3.2^{n+1}.

ii) Hence the 2^{n+1}th iterate of f has a periodic orbit with period 3 and thus, thanks to proposition 2, a 2-horseshoe.

iii) It follows that f has a horseshoe and consequently positive topological entropy.   **Q.E.D.**

Collecting all the result we have gotten so far, we reach the main result of this section.

THEOREM 4 (MISIUREVICZ [11]). *Let f be an endomorphism of the interval, then f has zero topological entropy if and only if the set of periods of its periodic orbits is in one of the following two forms:*

$$P(f) = \cup_{0 \le i \le n}\{2^i\} \quad or \quad P(f) = \cup_{0 \le i \le \infty}\{2^i\}.$$

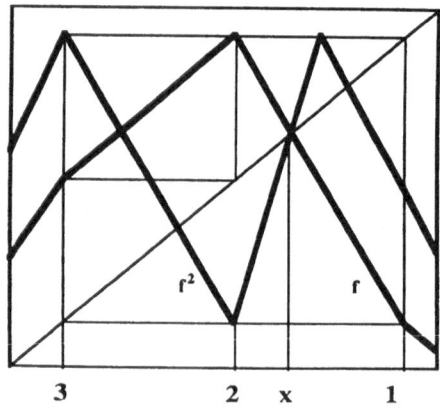

Figure 2.

II. Quick "petit tour" on the circle.

Theorem 4 is obviously "very false" for endomorphisms of the circle in that sense that any periodic orbit can occur without positive entropy and even without other periods that the period 1. This can be seen just by considering the rigid rotation with angle $2\pi \cdot p/q$, for some p coprime with q. We do not develop here the theory for the circle which parallels the one we have presented for the interval (see e.g. [9] and notice that theorem 2 and Proposition 1 still work) but we give directly the result which correspond to Theorem 4:

THEOREM 5. *Let f be an endomorphism of the circle, then f has zero topological entropy if and only if the set of periods of its periodic orbits is in one of the following three forms:*

i) $P(f) = \cup_{0 \leq i \leq n}\{q.2^i\}$ *for some q and n*
ii) $P(f) = \cup_{0 \leq i \leq \infty}\{q.2^i\}$ *for some q*
iii) $P(f)$ *is empty.*

This result can roughly be understands as follows:

i) First remark that two orbits cannot turn at different speed around the circle without creating entropy: just think about an arc joining them, as time goes on, the image of this arc would cover many times the circle yielding an obvious horseshoe effect.

ii) Hence, with zero entropy, everybody "turns at the same speed", if you turn yourself with all these people, what you see is a dynamics which looks like the one of endomorphisms of the interval. If the speed is a rational p/q, the set of periods is of one of the first two forms, if the speed is irrational, there is no periodic orbit.

III. A last remark concerning one dimensional dynamics.

A first glance at the results we have given in this section shows that some periodic orbits force positive topological entropy and others do not. Even if this is perfectly true, we can precise a bit this analysis. In fact, some periodic behaviors imply the existence of a horseshoe since, by continuity, the dynamics on the complement of the periodic orbit lacks of freedom, But a periodic behavior is not only determined by the period of the orbit but also by the order in which the periodic orbit is followed when we consider it as an ordered subset of the interval. In the case of the interval, it occurs that whatever the order we choose for a periodic orbit whose period is not a power of 2, this orbit implies positive topological entropy. But before living forever the magy of power of two let us suggest to the reader the exercise which consists to detect the horseshoe forced by a periodic orbit with period 4, which permutes cyclically the four ordered points on the interval.

It is with this state of mind that we are going to begin the second part of this paper, dedicated to 2-dimensional dynamics. Of course, there is no natural

order on the disk, but topology will give us nice tools to describe periodic orbits
and we will use them extensively!!!

B. TWO DIMENSIONAL DYNAMICS.

What remains when we allow our maps to live in a two dimensioanl space? As
in the case of endomorphisms on the circle, Sarkovskii theorem is far to be true
for embeddings of the disk. We can check it out by looking at the same example,
i.e., a rigid rotation of the disk around its center with angle $2\pi p/q$. We find that
all the points of the disk belong to a periodic orbit with period q (except the
center which remains fixed).

But fortunately everything has not disappeared, it remains for instance....

I. The notion of chaos.

The topological entropy is a topological invariant which can be defined for any
continuous map on any compact manifold. However, there exists, in the case of
C^2 embeddings of the disk, a result similar to theorem 2 that we are going to
present now. **In all the rest of this paper, f will stand for an orientation
preserving C^2 embedding of the disk.**

a) <u>Hyperbolicity, Stable and unstable manifold.</u>

Let P be a fixed point of f, $f(P) = P$, we say that P is an **hyperbolic saddle
point** if and only if the derivative of f in P, $Df(P)$, has two real eigenvalues, λ_s
(the stable one) which is inside the unit disk and λ_u (the unstable one) which is
outside the unit disk. To an hyperbolic saddle point P we can associate two one
dimensional manifolds, **the stable manifold** denoted $W^s(P)$ and **the unstable
manifold** denoted $W^u(P)$ and defined by:

$$W^s(P) = \{Q \in D^2 | f^n(Q) \to P \quad \text{when } n \to +\infty\},$$

and

$$W^u(P) = \{Q \in D^2 | f^n(Q) \to P \quad \text{when } n \to -\infty\}.$$

The two sets are not necessarily submanifolds as we will see in a while. But the
points in $W^s(P)$ (resp. $W^u(P)$) whose forward (resp. backward) orbits remain
in small neighborhood of P form a 1-dimensional submanifold tangent to the
eigenspace corresponding to λ_s (resp. λ_u) at P. Of course the same kind of
analysis extends to periodic hyperbolic orbits of saddle type.

b) <u>Homoclinic points and horseshoe.</u>

A **transverse homoclinic point** Q (to the saddle periodic point P) is a point
where the stable and unstable manifolds $W^s(P)$ and $W^u(P)$ intersect transversally. They where first discovered by Poincaré [12] who realized they yield a
complicated dynamics. In particular, it is obvious to see that the forward and
backward orbit of Q tend to the orbit of P.

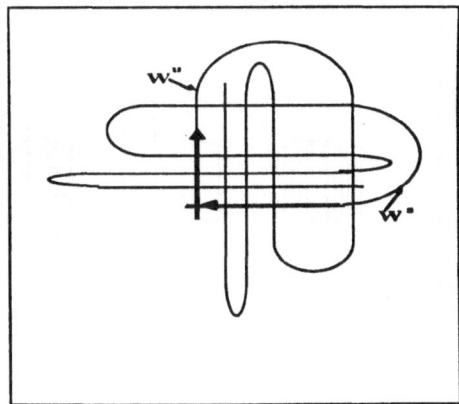

Figure 3.

In figure 3 we suggest the type of dynamics created by transverse homoclinic point.

That transverse homoclinic points imply infinitely many periodic orbits with arbitrarily large periods has been proved by Birkhoff. And Smale, trying to go deeper in Birkhoff results found his famous horseshoe (see [14] for the little story about this big work). Smale understood how the existence of transverse hommoclinic points determines the action of high iterates of the map on a small rectangle based on the stable manifold. This action can be decomposed in three essential steps described in figure 4:

   i) Contract the rectangle in one direction,
   ii) Expand the rectangle in the other direction,
   iii) Bend the rectangle and put it over the initial shape.

   c) Topological entropy.

Let us state the powerful

THEOREM 6 (KATOK [8]).
*Let M be a compact manifold and f a $C^{1+\epsilon}$ diffeomorphism * on M, then: f has positive topological entropy if and only if f has transverse homoclinic points to some periodic orbit.*

If we compare the description we gave of a Smale horseshoe with the definition we had of its one dimensional brother, it is easy to understand that Theorem 6 is actually a 2-dimensional version of Theorem 2. One important difference is

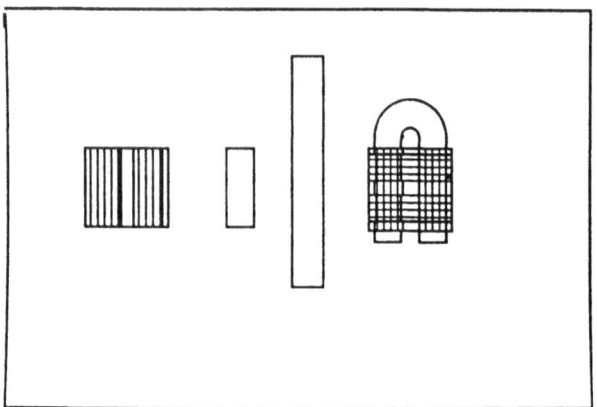

Figure 4.

that now we need more regularity for the map.

*The reader not familiar with this notation can read C^2.

II. Suspension and braids..

As we have seen at the beginning of this two-dimensional section, there is no hope that the period of a periodic orbit is enough to say something interesting about maps which possesse this periodic orbit. So we arrive to the problem of finding a better characterization of the periodic orbits. This can be done using **braid theory** that we are going to introduce now.

a) Suspension.

It is first instructive to have a closer look at the periodic orbits of rigid rational rotation of the disk. The rotation preserves circles surrounding the origin and the most fundamental fact concerning an orbit **O** on one of these circles is that two consecutive points on **O** are mapped onto two consecutive points of **O** with orientation preserved. It is then interesting to consider a rigid rotation with angle $2\pi \cdot p/q$ as the map we get by looking each time 1 at the effect of an actual revolution of the disk around its center with $2\pi p/q$ angular speed.

Let us consider the solid torus $T^2 = D^2 \times T^1$, and the field of lines $\{L_x\}_x$ in T^2 defined by

$$\{L_x = \{(R_{t \cdot 2\pi \cdot p/q}(x), t) \quad \text{for } t \in [0,1] \text{ and } x \text{ in } D^2\},$$

where $R_{t.2\pi.p/q}$ is the rigid rotation with angle $t.2\pi p/q$.

Each leaf of this foliation represents a solution of a differential equation (defined by a vector field) on T^2. Since the original disk $D^2 \times \{0\}$ cuts all these leaves transversally, it constitutes a **section** of the **flow** associated to this differential equation, the flow being the group of diffeomorphisms $\{R_{t \cdot 2\pi p/q}\}_{t \in \mathbb{R}}$. The rigid rotation $R_{2\pi \cdot p/q}$ turns to be the **Poincaré map** (or first return map) of this flow. More generally, for a flow with a section S, the first return map on S is defined by associating to any point P in S the next crossing with S of the leaf containing P.

What we did above is a simple example of a general procedure designed to associated a flow (or a vector field or a differential equation) to any diffeomorphism, and called **suspension**. It allows in fact to consider any diffeomorphism as the Poincaré map of some differential equation. In the context of our main problem, the construction is a bit more abstract that the one of our previous particular example, and we proceed as follows:

An orientation preserving embedding of the 2 disk, f, is isotopic to the identity: this means that we can find a continuous family $\{F_t\}_{t \in [0,1]}$ of embeddings of the disk such that F_0 is the Identity and $F_1 = f$. In the solid torus we construct a flow $\{\Phi_t\}_{t \in \mathbb{R}}$ satisfying:

$$\Phi_t(x,0) = (F_t(x), t) \quad \text{for } t \in [0,1] \text{ and } x \in D^2,$$

and

$$\Phi_t(x,0) = (F_{t-[t]}(\Phi_{[t]}(x,0)), t - [t]) \quad \text{for } t \in \mathbb{R}.$$

In this situation, all the orbits are not closed curves, the only one are these which contain a point $(x,0)$ where x is a periodic point of f. It is these closed curves in the solid torus that are going to give rise to the notion of braids.

b) Braids and knots.

Consider now the set E, of closed curves in the solid torus which are periodic orbits of a flow without singularities in the solid torus. In E we define an equivalence relation $C \approx C'$ if and only if there exists an homeomorphism isotopic to the Identity in the solid torus mapping C onto C'. An equivalence class is called a **braid**. Let $\mathbf{O_f}$ be a periodic orbit of an orientation preserving embedding of the disk f, and C_f the corresponding periodic orbit of one of its suspension. The **braid type** of $\mathbf{O_f}$ noted $\mathbf{B(O_f)}$ is the equivalence class of C_f.

Some remarks

1) The braid type is not uniquely determined, it depends on the suspension we choose. Nevertheless, the two different braids we can get by considering two different suspensions differ one from the other by a number of full twists, as shown in figure 5 which provides by the way an usefull representation of braids. These differences will not be relevant when we define some particular braids.

2) If two periodic orbits of two embeddings have the same braid type then they obviously have the same period, but the converse is not true: an example

Figure 5.

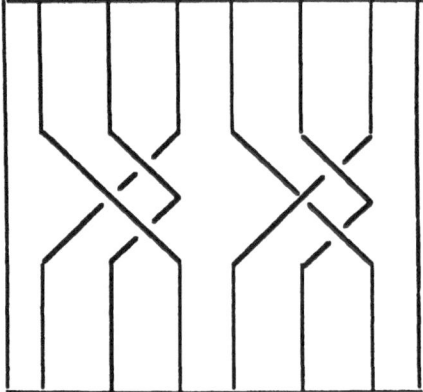

Figure 6.

is given in figure 6 where we can see two periodic orbits whith period 3 and different braid types.

3) A knot is the image by an homeomorphism of $\mathbf{R}^3$ or S^3 of a circle S^1. It can be proved that all knots can be mapped by an homoemorphism onto an element

in E. There exists another equivalence relation between knots defined by C $\sim C'$ if and only if there exists an homeomorphism isotopic to Identity of $\mathbf{R}^3$ mapping C onto C'. This equivalence relation is rougher that the previous one, for instance the two knots of figure 6 are $\sim$ equivalent but are not $\approx$ equivalent.

4) About braids, the classical references is the book of Birman [1].

c) Some particular braids.

Definition 1. We call rotation compatible braid the braid type of a periodic orbit of a rigid rotation of the disk.

Let us remark that we have the

PROPOSITION 1.

A braid B, is rotation compatible if there exist an element of B on the torus $S \times S^1$ in T^2, where S is a circle in D^2 centered in the origin.

Definition 2. Let $C \in E$, a torus **cabling** C is a torus $S \times C$ in D^2 where S is a small circle in D^2.

Definition 3. A braid B is an **iterated rotation compatible braid of the nth generation** if and only if there exists an element C in B which lies on a torus cabling an element C' of an iterated rotation compatible braid B' of the $n - 1^{th}$ generation.

Iterated rotation compatible braid of the first generation are rotation compatible braid.

The iterated rotation compatible braid of the generation zero is the equivalence class of he circle $\{0\} \times S^1$.

We let to the reader the pleasure to draw iterated rotation compatible braid of generation 2,3 and more...

III. Periodic orbits of embeddings and main results.

This little theory of braids we have developed in the previous section allows us to define some particular classes of periodic orbits for C^2 orientation preserving embeddings of the disk.

Definition 4. We say that a periodic orbit of a C^2 orientation preserving embedding of the disk is **rotation compatible** (resp. **hereditarily rotation compatible of the nth generation**) if its braid type is rotation compatible (resp. hereditarily rotation compatible of the nth generation).

Definition 5. Let O and O' be two periodic orbits of a C^2 orientation preserving embedding of the disk, f, and C and C' the two corresponding periodic orbits of a suspension of f. O is a father of O' (or O' a son of O) if and only if there exists an homeomorphism of the solid torus mapping C' on a torus cabling C.

More generally, **O** is an ancestor of **O'** if there exists a finite sequence of periodic orbits of f: $\mathbf{O}_1, \mathbf{O}_2, \cdots, \mathbf{O}_n$ such that $\mathbf{O}_1 = \mathbf{O}$, $\mathbf{O}_n = \mathbf{O}'$, and $\mathbf{O}_i$ is a father of $\mathbf{O}_{i+1}$ for $i = 1, 2, \cdots, n-1$.

Definition 6. Finally two periodic orbits of a C^2 orientation preserving embedding of the disk, f, **O** and **O'** are disjoint if and only if the two corresponding periodic orbits of a suspension of f, C and C' are such that C is homotopically trivial in $\mathbb{R}^3/C'$.

We are now able to state our main result:

THEOREM 7.

Let f be a C^2 orientation preserving embedding of the disk with entropy zero then:

1) *All periodic orbits of f are hereditarily rotation compatible,*
2) *If f possesses an hereditarily rotation compatible orbit **O** of the nth generation then f possesses a finite sequence $(\mathbf{O}_i)_{0 \leq i \leq n}$ of hereditarily rotation compatible orbits if the ith generation such that $\mathbf{O}_i$ is a father of $\mathbf{O}_{i+1}$ for $i = 0$ to $n-1$ and $\mathbf{O}_n = \mathbf{O}$.*
3) *If **O** and **O'** are periodic orbits of f then one of the following three situation occurs:*

   a) ***O** is an ancestor of **O'** or **O'** is an ancestor of **O**.*
   b) ***O** and **O'** have a common ancestor,*
   c) ***O** and **O'** are disjoint.*

The proof of this theorem is given in [6] and uses the work of Thurston on the classification of the diffeomorphisms of surfaces up to an isotopy [5]. A direct consequence of this theorem is that the periodic orbits with period 3 whose braid type is the one presented on the right side of figure 3 cannot occur without positive topological entropy.

REFERENCES

1. J.S. Birman, "Braids, Links and Mapping class groups," Ann. of Math. Studies no 82, Princeton University Press, Princeton, N. J., 1981.
2. L. Block, J. Guckenheimer, M. Misiurewicz and L.S. Young, *Periodic points and topological entropy of one dimensional maps*, Springer Lectures Notes in Math. **Vol. 819 (Springer, New York)** (1980).
3. R. Bowen and J. Franks, *The periodic points of maps of the disk and the interval*, Topology **15** (1976), 337–342.
4. J.P. Eckamnn and D. Ruelle, *Ergodic Theory of Chaos and strange Attractors*, Rev. Mod. Phys. **57** (1985), 617–656.

5. A. Fathi, F. Laudenbach and V. Poenaru, "Travaux de Thurston sur les Surfaces," As-
   terisque, 1979, pp. 66–67.

6. J.M. Gambaudo, s. Van strien and C. Tresser, *The periodic structure of orientation preserving embeddings of the Disk with entropy zero*, Submitted to Annales de L'I.H.P. Physique
   Théorique.

7. J.M. Gambaudo et C. Tresser, *Application du cercle de degré 1*, To appear in a monograph
   on chaos édited by P. Bergé, Masson.

8. A. Katok, *Lyapunov exponents, entropy and periodic orbits for diffeomorphisms*, Pub. Math.
   I.H.E.S. 51 (1980), 137–174.

9. R. Mackay and C. Tresser, *Transition to topological chaos for circle maps*, Physica 19D (1986),
   206–237.

10. M. Misiurewicz, *Horseshoes for mappings of the interval*, Bull. Acad. Pol. Ser. Sci. Math. 27
    (1979), 167–169.

11. M. Misiurewicz, *Structure of mappings of an interval with zero entropy*, Publ. Math. I.H.E.S.
    53 (1981), 5–16.

12. H. Poincaré, "Les méthodes nouvelles de la mécanique céleste III," Gauthier -
    Villars, Paris, 1951.

13. A. Sarkovskii, *Coexistence of cycles of a continuous map of the line into itself (in russian)*, Ukr.
    Mat. Z. 16 (1964), 61–71.

14. S. Smale, "The mathematics of time," Springer, New-York, 1980.

PERIODICALLY AND RANDOMLY MODULATED NON LINEAR PROCESSES

MIGUEL KIWI
Facultad de Física
Universidad Católica de Chile
Casilla 6177, Santiago 22, Chile
BENNO HESS and MARIO MARKUS
Max-Planck Institut für Ernärungsphysiologie
Rheinlandamm 201, D-4600 Dortmund 1, FRG
JAIME RÖSSLER
Departamento de Física, Facultad de Ciencias
Universidad de Chile
Casilla 653, Santiago, Chile

ABSTRACT. Many natural phenomena are governed by nonlinear recursive relations of the type $x_{t+1} = f(x_t)$, where f does depend on t. We focus our interest on the particularly simple case $x_{t+1} = r_t x_t(1-x_t)$, where r_t adopts either periodically or at random, the values A and B. Graphical representations of the Lyapunov exponent on the AB-plane show unexpected features, like self-similarity and early chaos (i.e. chaos for very low parameter values). The latter constitutes a novel mechanism to induce chaotic behavior. The meaning of the Lyapunov exponent for random processes is examined.

Introduction

Nonlinear processes governed by a relation of the type $x_{t+1} = f(x_t)$ have been extensively studied. In particular, the logistic equation

$$x_{t+1} = rx_t(1 - x_t), \tag{1}$$

has received much attention during the last years (see e.g. [1]). Its systematic study has led to the understanding of a wealth of nonlinear phenomena, since many of its properties are universal. However, a great variety of natural phenomena are described, at least qualitatively, by recursive relations of the type $x_{t+1} = f_t(x_t)$, where the application f_t also depends on the discrete variable t, which represents time or site. This kind of equations govern the behavior of e.g. one-dimensional alloys[2], quasi-one-dimensional ribbons[3] and biological populations with and without generational overlap.[4,5]

Formulation of the Problem

In this paper we investigate the particularly simple equation

$$x_{t+1} = r_t x_t(1 - x_t), \tag{2}$$

67

E. Tirapegui and D. Villarroel (eds.), Instabilities and Nonequilibrium Structures II, 67–74.
© 1989 by Kluwer Academic Publishers.

where r_t adopts, either in a periodic or random sequence one of only two values, which we denote by A and B. (In ecosystems r_t describes the changing environment). Our specific choice of f_t is not very restrictive, since two maps which are "topologically conjugated" exhibit[6] analogous features, as far as the stable or chaotic nature of trajectories on these maps is concerned. It should be mentioned that the work of Chang et al.[6] can be viewed as a special case of ours; their quartic map corresponds to the product $f_A[f_B(x)] \equiv f_{AB}(x)$, where r_t has period 2, $r_{2t} = B, r_{2t+1} = A$. Furthermore, Crutchfield et al.[7] did investigate the superposition of white noise onto the parameter r of Eq.(1), thus making it de facto t-dependent.

A process of the type $x_{t+1} = f_t(x_t)$ is fully specified by the sequence $S = \{r_1, r_2, r_3, \ldots r_t, \ldots r_N\}$ and the initial condition, or seed, $x_0 = x_{t=0}$. The successive values of x_t generate the history of the system, denoted by $X = \{x_1, x_2, x_3, \ldots x_t, \ldots x_N\}$. The sequence S may be random or periodic and is characterized by the concentration $p_\alpha \equiv N_\alpha/N, \alpha = A, B$, and the pair probability $P_{\alpha|\beta}(j) \equiv N_{\alpha\beta}(j)/N_\alpha = p_\beta + \gamma_j(\delta_{\alpha\beta} - p_\beta)$; here N_α is the number of $r_t = \alpha$ and $N_{\alpha\beta}(j)$ is the number of $\alpha\beta$-pairs j-steps apart, and γ_j is the Cowley parameter.[2] In particular, a Markovian sequence is specified by just two independent parameters, for example p_A and $P_{B|A}(1) = (1 - \gamma_1)p_A$. If $\gamma_1 \to 1$ then $S = \{\ldots AAAAAABBBBBB\ldots\}$, while $\gamma_1 = 0$ describes a truly random sequence. Values of $\gamma_1 < 0$ imply a tendency towards nearest neighbor AB pairs. The sequence becomes periodic $(BABABA\ldots)$ in the special case $p_A = p_B = 0.5$ and $\gamma_1 = -1$.

To study the dynamics of such systems it is convenient to introduce the density $D_\alpha(x) \equiv dN_\alpha/(Ndx)$, where dN_α is the number of $x_j \in [x, x + dx]$ with $r_{j-1} = \alpha$, $j = 1, 2, \ldots N$ and $\alpha = A, B$, with $N \to \infty$. The total density is $D(x) \equiv D_A(x) + D_B(x)$. For a random process $D_\alpha(x)$ is stationary, and independent of x_0 and S.

We define the Lyapunov exponent λ as a measure of the sensitiveness of the system to a change in initial conditions, for a fixed sequence S

$$\lambda \equiv \lim_{N \to \infty} \frac{1}{N} \ln \left| \frac{\partial x_N(S, x_0)}{\partial x_0} \right|$$
$$= \sum_{\alpha,\beta=A,B} \int_0^1 P_{\alpha|\beta} D_\alpha(x) \ln \left| \frac{\partial f_\beta}{\partial x} \right| dx , \tag{3}$$

where the latter equality is valid only for a Markov process.

Results

When we proceed to evaluate the magnitude and sign of λ obtained by alternating, either at random or periodically, two different values of r a great variety of results is obtained.[4,5] For example, with values of A and B both leading to non-chaotic behavior of Eq.(1) one may obtain chaos when a random or periodic succession of them is used in Eq.(2). Conversely, A and B values of the chaotic region of Eq.(1) may yield non-chaotic behavior[4] of Eq.(2). This led us to undertake a systematic evaluation of λ as a function of A and B, for several periodic and random sequences. In Fig.1 we display the Lyapunov exponent on the AB-plane, for different periodic sequences S. Apart from their aesthetic interest two features of these maps are quite apparent: self-similarity and multivaluedness. Self-similarity is generated by the repeated intersection of cascades of superstable lines[6] (defined by $\lambda \to -\infty$ and represented as dark lines on yellow regions of Fig.1); these cascades are continuations into the AB-plane of the superstable points present between two bifurcations of the map defined by Eq.(1). The asymmetry of $\lambda(A, B)$ in Fig.1a is related to multivaluedness under the $A \rightleftharpoons B$ exchange, for a fixed seed x_0. In particular, bi-valuedness can be detected visually in Fig.1a as a crossover of two branches with $\lambda < 0$. This bi-valuedness can be traced back to the fact that the function $f_{AB} \equiv f_A[f_B(x)]$ has several basins of attraction in some regions of the AB-plane. For example, if $f_{AB}(x) = x$ has three solutions, with x^* being the unstable one, then, if $1 - x^* < f_{AB}(1/2)$ and $x^* < f_{AB}(A/4) = f_A[f_{BA}(1/2)]$, at least two attraction basins do exist: $R_1 = [1 - x^*, x^*]$ and $R_2 = [x^*, A/4]$ with $f_{AB}(R_j) \subseteq R_j, j = 1, 2$. For a fixed seed x_0 the $A \rightleftharpoons B$ exchange induces a transition from one attraction basin to another; each basin is related to a particular branch of $\lambda(A, B)$. Inspection of Figs.1 reveals tri-, penta- and hepta-valuedness (crossover of three, five and seven branches with $\lambda < 0$).

In Fig.1d we display another causal non-Markovian case: the period 8 cycle defined by $S = (A^7B \ A^7B \ A^7B \ldots)$. Here, $\gamma_8 = +1$, $p_B = 1/8$. Its most striking feature is the onset of chaos for values of A and B both well below the threshold for chaos[1,7] of Eq.(1), which is $r_0 = 3.56994567\ldots$; for example, for $A = 3.36$ and $B = 2.25$. We call this phenomenon *early chaos*.[5] The multivaluedness does disappear when a small amount of randomness is superimposed on r_t; for example, for $\gamma_1 = -0.98$, as illustrated in Fig.2, full mirror symmetry around $r_t = A = B$ is recovered. In Fig.3 we display $\lambda(A, B)$ for a random sequence $\gamma_j = 0$ and $p_A = 1/2$; here early chaos is also present, e.g. for $A \cong 3.52$ and $B \cong 3.06$, but the superstable lines are more diffuse than in Figs.1 and 2.

We also find the inverse of the early chaos phenomenon. In fact, there are points (A, B) that have a negative Lyapunov exponent, but for which A alone, and B alone lead to chaos. For example, order is detected in Fig.1a. in the neighborhood of $A = 3.58$, $B = 3.98$.

Within our systematic exploration of the AB-plane we found early chaos quite generally, e.g. when $A = 3.34$ is combined with the surprisingly low value of $B = 1.199$, in the period 22 non-markovian sequence $S = (A^{21}B \ A^{21}B \ldots)$. This rather unexpected result can be understood with the aid of Fig.4. To be specific we start the above process with x_0 slightly larger than the unstable fixed point $x^* = 1 - 1/r$ of

Eq.(1) with $r = A = 3.34$, which corresponds to $x^* = 1 - 1/A \cong 0.7$. The 21 successive applications of f_A which follow yield in general, a value of $x(t = 21)$ very nearly equal to $x^- \cong 0.468$, the lowest of the two values of the period 2 regime of Eq.(1). Since this value is in the neighborhood of the maximum of $f_A(x)$, located at $x = 1/2$, the convergence of x_{2t+1} to $x^-(A)$ is very fast. The value of B is selected by imposing $f_B[x^-(A)] \cong 1 - x^* = 1/A$; this way almost always $x_t = f_A(x_{t-1})$ yields $x_t \cong x^*$ for $t = 23$ (recall that $f_A(x) = f_A(1 - x)$). Thus, the role of the application f_B is to reset x_t to the unstable point x^*, in order to start the process all over again. This way, chaotic behavior is induced by predominantly sampling the vicinity of the unstable fixed point and at the same time, keeping the system in a state of eternal transient. (A similar explanation holds for early chaos in Fig.1d). Two important features of the mechanism are quite apparent: the initial value of x is irrelevant and it is not very critical that the sequence S be periodic, which we also verified numerically. In fact, the argument only depends decisively on applying f_B on the lower point of the bi-cycle; i.e. the application has to have the "right phase". Thus, with great generality, chaotic behavior can be induced by interspersing, periodically or at random, a few events of the right strength in a basically stable system of period two. Early chaos may explain the fact that parameters leading to chaotic behavior in ecosystems are higher than those observed in nature.[8]

The processes discussed above may well be susceptible to experimental verification in electronic and mechanical devices, subject to periodic pulses. A suitable experimental setup may be the Josephson junctions implemented by Octavio *et al.*[9]

We now return to the discussion on the meaning of the Lyapunov exponent. The definition of λ given in (3) measures the sensitivity of x_N to a change in the seed x_0, when a fixed sequence S is considered. But, in the case of a random sequence there is no a priori knowledge of S. For example, an ecologist cannot foresee a randomly changing environment. This severely limits the feasibility of predicting the value of x_N, even when $\lambda < 0$. Consequently, the whole notion of a Lyapunov exponent may seem inappropriate to deal with random processes. However, there is an alternative interpretation of λ, well suited for experimental testing: assume that several isolated replicas of the system under study are available in different initial states, all of them subject to the same time sequence S. (As an ecological example one could consider a set of "islands" subject to the same ambient fluctuations, but different initial conditions). One expects that in a nonchaotic process all replicas with nearly the same initial conditions will converge to the same final state. Conversely, when $\lambda > 0$, one expects the final states to be unpredictable.

We have tested these ideas numerically calculating the time evolution of L systems, subject to the same S, but starting from L different seeds $\{x_0|\nu = 1, 2, \ldots L\}$, where ν labels a particular system. We then evaluated the distribution $D_{t,\alpha}(x)$ associated with $\{x_{t,\nu}\}$, where $\alpha = r_{t-1}$. While $D_{t,\alpha}(x)$ is generated on the basis of a seed average, $D_\alpha(x)$ was obtained for a fixed seed and t-average. We analyzed and compared the properties of these two distributions to conclude the following:

i) If the distribution $D_\alpha(x)$ has a compact support then $D_{t,\alpha}(x)$ is independent of the choice of $\{x_{0,\nu}\}$. Also, if $\lambda < 0$ and ignoring transients, $D_{t,\alpha}(x) = \delta[x - x(S_t)]$, where $S_t = \{r_1, r_2, \ldots r_{t-1}\}$. Instead, if $\lambda > 0$ then $D_{t,\alpha}(x)$ has a certain width.

ii) If the support of $D_\alpha(x)$ breaks into allowed and forbidden bands, then when $\lambda < 0$

the different replicas converge to a few final states, which are equal in number to the bands of $D_\alpha(x)$; the relative distance between replicas within the same band vanishes as $t \to \infty$. Conversely, if the process is chaotic, the final relative distance cannot be evaluated. However, if $0 \leq \lambda \ll 1$ (weak chaos) all replicas belonging to the same band will be close packed.

iii) In the chaotic regime we also compared the distributions $D_\alpha(x)$ and $D_{t,\alpha}(x)$, to find that they are different. Moreover, $D_{t,\alpha}(x)$ does depend on t, and is much more concentrated around a few peaks, than $D_\alpha(x)$. Thus, the value of λ plus the shape of $D_\alpha(x)$ yield useful and consistent information on the relative distance between final states, even when S is not known *a priori*.

Acknowledgments

This work was partially supported by DIB (JR), DIUC (MK) and FONDECYT (MK and JR). MM thanks FONDECYT and DAAD for travel funds.

References

1. P. Collet and J.P. Eckmann: *Iterated maps on the interval as Dynamical Systems,* Progress on Physics, Vol. I. (Birkhauser, Boston, 1980).
2. J. Rössler, G. Martínez and M. Kiwi, *Solid State Comm.* **61**, 395 (1987); P. Murilo Oliveira, M.A. Continentino and E.V. Anda, *Phys. Rev.* **B29**, 2808 (1984).
3. Y. Liu and K. Chao, *Phys. Rev.* **B33**, 1010 (1986).
4. M. Markus, B. Hess, J. Rössler and M. Kiwi, in *Chaos in Biological Systems*, ed. by H. Degn, A.V. Holden and L.F. Olsen (Plenum Publishing Co., NY, 1987), pp.267-277.
5. J. Rössler, M. Kiwi and M. Markus, in *From Chemical to Biological Organization*, ed. by M. Markus, S.C. Müller and G. Nicolis (Springer Verlag, Heidelberg, 1988), pp. 319-330.
6. S.J. Chang, M. Wortis and J.A. Wright, *Phys. Rev.* **A24**, 2669 (1981).
7. J.P. Crutchfield, J.D. Farmer and B. Huberman, *Phys. Repts.* **92**, 45 (1980).
8. W.M. Schaffer and M. Kot, *Bioscience* **35**, 342 (1985).
9. M. Octavio, *Phys. Rev.* **B29**, 1231 (1984); M. Octavio and C.R. Nasser, *Phys. Rev.* **B30**, 1586 (1984).

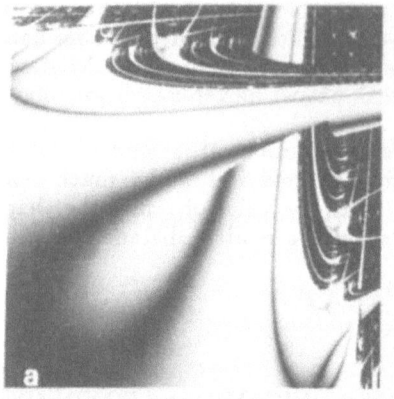
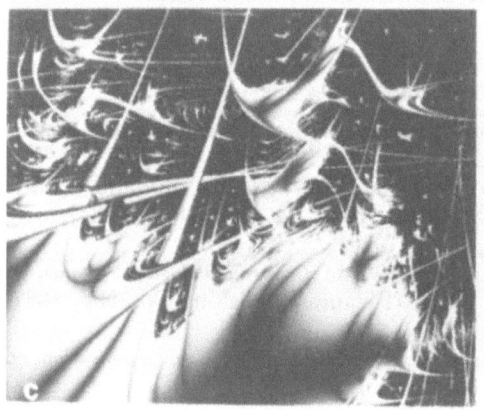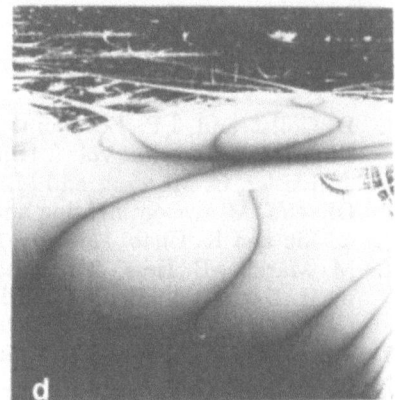

Figure 1. Lyapunov exponent λ for periodic sequences S:
(a) BA BA BA ......; (b) $B(BA)^2$ $B(BA)^2$......; (c) $B(BA)^3$ $B(BA)^3$......;
(d) A^7B A^7B ......   The shading changes from black to white as λ
increases from negative values to zero.   The black sea in the upper
right indicates $\lambda > 0$. Abscissas:B,   ordinates:A. The seed $x_0 = 0.5$.
Ranges:
(a) $2 \leq A \leq 4$, $2 \leq B \leq 4$;
(b) $3.543 \leq A \leq 3.667$, $3.452 \leq B \leq 3.608$;
(c) $3.212 \leq A \leq 4$, $2.759 \leq B \leq 3.744$;
(d) $2 \leq A \leq 4$, $2 \leq B \leq 4$.

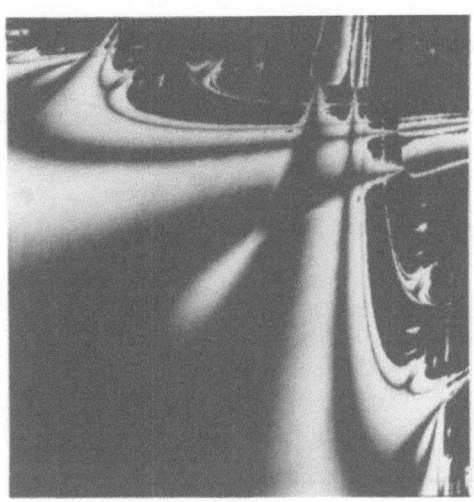

Figure 2. Lyapunov exponent for a partially random sequence. $p_A = p_B = 0.5$, $\gamma_1 = -0.98$. Abscissa: $2 \leq B \leq 4$, ordinate $2 \leq A \leq 4$. The shading changes from black to white as λ increases from negative values to zero. The black sea in the upper right corner indicates $\lambda > 0$.

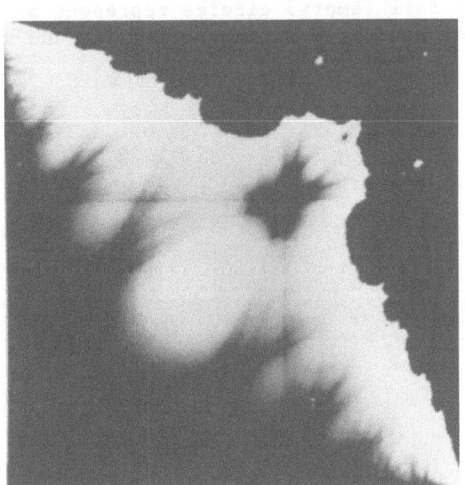

Figure 3. Lyapunov exponent for a random sequence S. $p_A = p_B = 0.5$, $\gamma_1 = 0$. Abscissa: $2 \leq B \leq 4$, ordinate $2 \leq A \leq 4$. The shading changes from black to white as λ increases from negative values to zero. The black wing shaped region in the upper right indicates $\lambda > 0$.

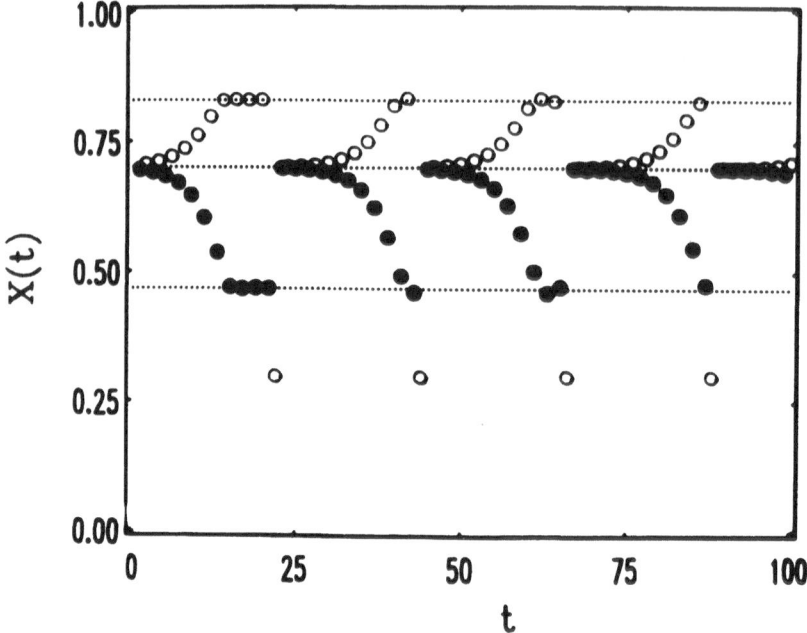

Figure 4. x(t) vs t for the sequence $A^{21}B\ A^{21}B\ \ldots$ with A = 3.34 and B = 1.199. The full (empty) circles represent odd (even) t-values. The dotted lines indicate the positions of x^+, x^* and x^- of Eq.(1) with r = A - 3.34.

THE GAP ROAD TO CHAOS AND ITS MAIN CHARACTERISTICS

M. C. de Sousa Vieira and C. Tsallis
Centro Brasileiro de Pesquisas Físicas
Rua Dr. Xavier Sigaud, 150
22290 Rio de Janeiro, RJ
Brazil

ABSTRACT. We study numerically the three types of asymmetry associated with the map $x'=1-\epsilon_i-a_i|x|^{z_i}$ ($i=1,2$ respectively correspond to $x>0$ and $x\leq0$). The first case is the amplitude asymmetry ($a_1 \neq a_2$), the second case is the exponent asymmetry ($z_1 \neq z_2$) and the last one is a discontinuous map ($\epsilon_1 \neq \epsilon_2$). In the two first cases the period-doubling road to chaos is topologically unmodified. In the last case the road to chaos is completely new ("gap road"). Chaos now is attained through sequences of inverse cascades. Various new features are observed, concerning the phase diagram, kneading sequences, Liapunov and uncertainty exponents, number of attractors, multifractality, among others. We also study the crossover between the discontinuous map and the continuous one.

1. INTRODUCTION

The evolution of a dynamical system governed by nonlinear one-dimensional difference equation presents a very rich structure[1]. Universal relations in difference equations presenting a single extremum of the $|x|^z$ class ($z>1$) were found by Grossmann and Thomae, Feigenbaum, and Coullet and Tresser[2]. The equation they considered was of the type

$$x_{t+1} = f(x_t) \equiv 1 - a|x_t|^z \tag{1}$$

The interesting behavior appears for $x_t \in [-1,1]$ and a $\in [0,2]$. For $z = 2$, Eq. (1) is equivalent to the logistic map [$x_{t+1} = 4\mu x_t(1-x_t)$]. When a increses from 0 to $a^*(z)$ [$a^*(2) = 1.401155...$; see ref. 3 for $a^*(z)$] the attractor (or long time solution) of the map (1) exhibits a sequence of periodic orbits with periods 2^k ($k=0,1,2,...$). The sequence $\{a_k\}$ where bifurcations occur, converges geometrically with a rate $\delta_k(z) \equiv (a_k - a_{k-1})/(a_{k+1} - a_k)$, which for large values of k, approaches $\delta(z)$ [$\delta(2) = 4.6692...$; see ref. 3 for other values of z]. Above a^* the chaotic regime appears, where aperiodic attractors are present, as well as an infinite number of periodic windows with period-doubling bifurcations.

After Feigenbaum's work, a great amount of theoretical as well as

E. Tirapegui and D. Villarroel (eds.), Instabilities and Nonequilibrium Structures II, 75–88.

experimental efforts have been dedicated to study the standard routes to chaos associated with continuous differentiable maps, namely, period-doubling, intermittency and quasiperiodicity. Nevertheless, little effort has been devoted to maps with an asymmetry at the extremum[4-10]. Experimental systems which can be described in terms of asymmetric maps are now appearing (see ref.8 for laser cavity and ref. 9 for nonlinear oscillators). When the singularity is a discontinuity, a new universal scenario to chaos appears[7,10].

The aim of the present paper is to study the following asymmetric map:

$$x_{t+1} = f(x_t) \equiv \begin{cases} 1 - \varepsilon_1 - a_1|x_t|^{z_1} , & \text{if } x_t > 0 \\ 1 - \varepsilon_2 - a_2|x_t|^{z_2} , & \text{if } x_t \leq 0 \end{cases} \tag{2}$$

The well known continuous symmetric case is recovered for $z_1 = z_2 \equiv z$, $a_1 = a_2 \equiv a$ and $\varepsilon_1 = \varepsilon_2 = 0$. Three types of asymmetry are studied, namely, case I ($a_1 \neq a_2$, $z_1 = z_2 \equiv z$ and $\varepsilon_1 = \varepsilon_2 = 0$), case II ($a_1 = a_2 \equiv a$, $z_1 \neq z_2$ and $\varepsilon_1 = \varepsilon_2 = 0$), and case III ($a_1 = a_2 \equiv a$, $z_1 = z_2 \equiv z$ and $\varepsilon_1 \neq \varepsilon_2$). Some features of these maps such as phase diagrams, attractors, Liapunov and uncertainty exponents, multifractality and others will be studied in the following sections.

2. AMPLITUDE ASYMMETRY (CASE I)

The route to chaos in this case is via period-doubling bifurcations. However, the set $\{\delta_k\}$ asymptotically ($k \to \infty$) presents an oscillatory behavior between two fixed values. Let us now mention at this point that preliminary numerical work[7] suggested that δ_k approaches a single value, namely that of the symmetric case. The present high accuracy calculations show that this is not so, but rather it exhibit the oscillatory behavior, first studied by Arneodo et al[4]. We have represented in Fig. 1(a) for $z_1 = z_2 = 2$ and $\varepsilon_1 = \varepsilon_2 = 0$, the critical lines [in (a_1, a_2) space] which generalize a^* (the first entrance into chaos) and a^M (value of a above which finite attractors disappear). In Fig. 1(b) we show the limiting values δ_∞ between which $\{\delta_k\}$ oscillates for k large enough; these limiting values are shown as function of a_1 along the critical line $a_2^*(a_1)$ of Fig. 1(a).

3. EXPONENT ASYMMETRY (CASE II)

The route to chaos in this case oncemore is via period-doubling bifurcation. However, a different behavior appears in the set $\{\delta_k\}$: the δ_k's do not converge for increasing k, but proceed in oscillatory fashion between two asymptotic lines (and not limiting values), one of them being divergent[6,7,9]. In Fig. 2 we have presented our results as well as those of ref. 6. This behavior has been recently exhibited experimentally[9]. Above a^* (chaotic region), the relative sizes of the various periodic windows are quite different from those of the $z_1 = z_2$ prototype. However, as in case I, the sequence of high-order windows is the same of the symmetric case, since this map satisfy the conditions required in ref. 11.

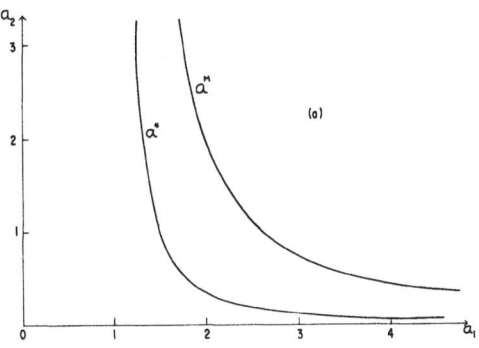

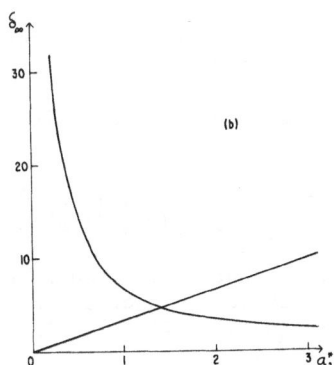

Fig. 1. (a) Special cuts of the "first entrance into chaos" and "finite-attractor-disappearance" hypersurfaces in the (a_1, a_2) space for $z_1 = z_2 = 2$ and $\varepsilon_1 = \varepsilon_2 = 0$; (b) values of δ_k for $k \gg 1$ as function of a_2 along the critical line $a_2^*(a_1)$ of (a). The numerical results for the asymptotic values are roughly reproduced by $\delta_\infty \cong 3.3 a_2^*$ and $\delta_\infty \cong 7/a_2^*$.

4. DISCONTINUOUS MAPS (CASE III)

Maps with a discontinuity at the extremum can be generated, for instance, by appropriate Poincaré sections in flows where trajectories on or near the attractor pass close to a saddle (or hyperbolic) point. In this situation the evolution of the dynamical variable depends on the side with respect to the saddle point, on which the preimages are localized. The standard example of such systems is the Lorenz model, the origin of which

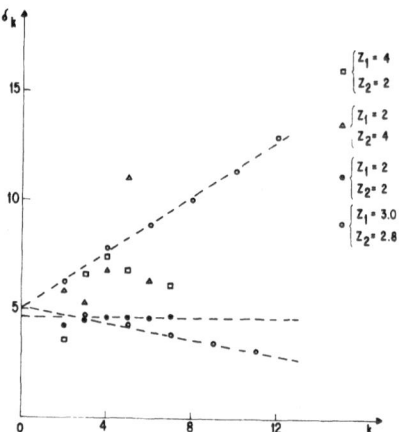

Fig. 2.   Evolution of the successive ratios $\{\delta_k\}$ for the $z_1 = z_2 = 2$ prototype and for the case II ($z_1 \neq z_2$).

is a saddle point. A typical map generated on this model is the case III, namely

$$x_{t+1} = f(x_t) \equiv \begin{cases} 1 - \varepsilon_1 - a|x_t|^z & , \text{ if } x_t > 0 \\ 1 - \varepsilon_2 - a|x_t|^z & , \text{ if } x_t \leq 0 \end{cases} \tag{3}$$

where $\varepsilon_1 \neq \varepsilon_2$. In ref. 7, it has been indicated that $f(0) = 1 - (\varepsilon_1 + \varepsilon_2)/2$, which is not convenient; this choice in fact alters the sequence of inverse cascades. The value of $f(0)$ actually used in the numerical calculations of ref. 7 was $f(0) = 1 - \varepsilon_2$.

4.1. Evolution of the Attractor

A very rich structure is present in the evolution of the attractor. The Fig. 3 shows the a-dependence of the attractor for a typical case, namely $(\varepsilon_1, \varepsilon_2) = (0, 0.1)$. We observe the appearance of sequences of inverse cascades in arithmetic progression ("inverse" refers the the fact that a has to decrease in order to approach the accumulation point associated with each cascade) initially mixed with pitchfork bifurcations. The first cascade we observe for increasing a is ...14→12→10→8→6→4, which accumulates on a=1. Immediately above this cascade we observe a couple of standard pitchfork bifurcations. Further on, a new inverse cascade appears as follows: ...21→17→13→9, and then again a pitchfork bifurcation into period 18. Then another inverse cascade appears as follows: ...76→58→40→22. After this cascade, no other standard pitchfork bifurcations are observed (until the entrance into chaos), but instead new inverse cascades are present: ... 70→48→26, and then ... 108→82→56, and then ...142→86→30, etc. A rule is observed: Within each cascade, the periods grow arithmetically by adding the first element immediately below its accumulation point. In

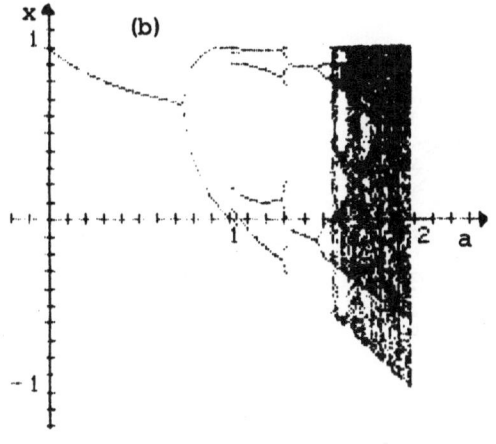

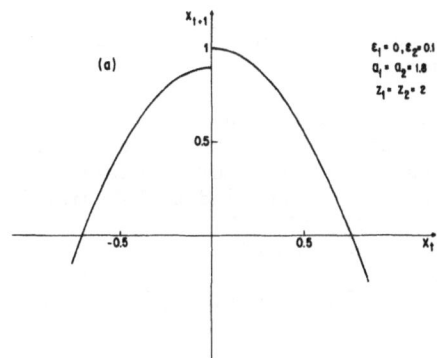

Fig. 3. (a) Discontinuous map and (b) a-evolution of the attractor for $(\varepsilon_1,\varepsilon_2)=(0,0.1)$, $z_1=z_2=2$ and $x_0=0.5$.

fact, we have a very fine structure. We observe that between any two consecutive elements of a cascade there is another inverse cascade whose periods grow with the same rule mentioned above. For example, between the
elements 40 and 22 of the third cascade, the cascade $...102\to62\to22$ exists.
Between the elements 102 and 62 of this cascade, we have the following
one $...266\to164\to62$, and so on. The elements of these cascade appear discon-
tinuously like tangent bifurcations. However, they do not present intermittency, since the iterated function $f(f(...f(x)))$ presents square cor-
ners which cross the $f(x)=x$ bissectrix.

In Fig. 4 we have represented, for a typical case, the "phase diagram" in the space of the size of the gap and of a. Such phase diagram
will be refered hereafter as <u>bunch of bananas</u>. Initially let us fix ε_1
and vary a. We see the behavior described above: inverse cascades of at-

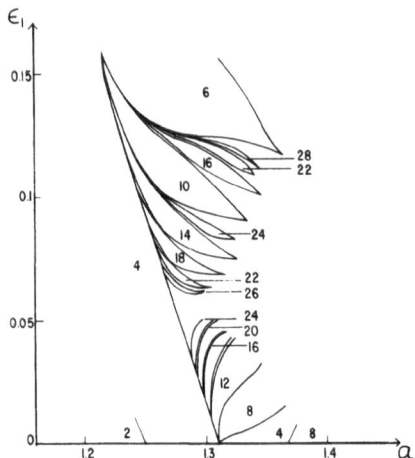

Fig. 4. Phase diagram for $z_1 = z_2 = 2$, $\varepsilon_2 = 0$ and $x_0 = 0.5$. The number indicates the period of the attractor. For $\varepsilon_1 = 0$ we recover the well known period-doubling sequence.

tractors whose periods grow arithmetically and accumulate on values of a, immediately below which appear cycles whose periods precisely are the corresponding adding constants. Furthermore, between any two bananas we always have another banana. The same kind of behavior is observed   by fixing a and varying ε_1 (or ε_2, of both, with $\varepsilon_1 \neq \varepsilon_2$). The accumulation points of the cascades in turn accumulate (for increasing a if $(\varepsilon_1, \varepsilon_2)$ are fixed) on a point which is the entrance into chaos. In other words, we have a (presumably) infinite number of accumulation points where there is no chaos (negative Liapunov exponents), as this only emerges at the accumulation point of the accumulation points.

    For fixed $(\varepsilon_1, \varepsilon_2)$, a given banana exists between a minimal value a_k^m and a maximal value a_k^M. Within a given cascade of bananas (whose sequence is noted with $k = 1, 2, 3 \ldots$), we verify

$$\left| a_k^m - a_{k+1}^m \right| \sim \left| a_{k-1}^m - a_k^m \right|^{z_1} \tag{4.a}$$

as well as

$$\left| a_k^m - a_\infty^m \right| \sim \left| a_{k-1}^m - a_\infty^m \right|^{z_1} \tag{4.b}$$

for k large enough. The same laws hold for $\{a_k^M\}$, for all values of $(\varepsilon_1, \varepsilon_2)$, such that $\varepsilon_1 \neq \varepsilon_2$, in the presence or absence of higher order terms in Eq. (3). Similar features are observed if we fix a and vary $(\varepsilon_1, \varepsilon_2)$.

4.2. Kneading Sequence

In the windows of the chaotic region for maps governed by (1) there are different kinds of stable periods of the same length. This differentiation is characterized by the order in which the points are visited. For every periodic orbit, there is one value of the control parameter for which the orbit includes the "critical point" (extremum) of the map. For this value of the control parameter a it is possible to form a word of finite length by noting whether each of the subsequent iterates in the orbit was less than (to the left: L) or greater than (to the right: R) of the critical point. Thus, a period-3 orbit $x_{crit} \rightarrow R \rightarrow L \rightarrow x_{crit}$, would be uniquely defined by the notation RL which we call the visitation pattern (or kneading sequence). Metropolis, Stein and Stein[11] (see also Derrida et al.[12]) discovered that the order in the arrangement of these words is independent of the unimodal map studied. In particular, if the word P corresponds to a period which actually occurs, we can construct another word $H(P)=P\mu P$, where $\mu=R$ if there is an even number of R's in P and $\mu=L$ otherwise. H is called the harmonic of P and represents the doubled-period adjacent to P.

For the discontinuous maps governed by Eq. (3) the inverse cascades which are observed depend on the size of the gap, as is shown in Fig. 4. For a fixed gap, the construction of the kneading sequences, for a given cascade, obeys the following rule: If P_n is the pattern of the n-cycle that exists immediately below the accumulation point of the cascade and P_k is the pattern of the k-cycle of the cascade, then the pattern of the m-cycle that results form the addition of the n-cycle and of the k-cycle is $P_m=P_k\mu P_n$, where $\mu=R(L)$ if the last bifurcation below that inverse cascade reaches the x=0 axis by $0^+(0^-)$. For instance, in the case $(\varepsilon_1,\varepsilon_2) = (0.1,0)$ when one branch of the attractor with period-4 reaches the x=0$^-$ axis the inverse cascade ...18→14→10 appears (see Fig. 4). The patterns associated with these periods are $P_4=RLR$, $P_{10}=RLRRRLRLR$, $P_{14}=RLRRRLRLRLRLR$ $=P_{10}LP_4$, $P_{18}=RLRRRLRLRLRLRLRLR=P_{14}LP_4$, etc. Between the elements 14 and 10 the cascade ...38→24→10 is present. The pattern associated with its elements are $P_{24}=P_{10}LP_{14}$, $P_{38}=P_{24}LP_{14}$, etc.

4.3. Crossover to the Period-doubling Scenario

The number of pitchfork bifurcations in the discontinuous maps is a function of the size of the gap. It increases when the size of the gap decreases, and diverges when the gap vanishes. For example, for $(\varepsilon_1,\varepsilon_2) = (0,0.0001)$ we observe six pitchfork bifurcation (mixtured with inverse cascades), whereas for $(\varepsilon_1,\varepsilon_2) = (0,0.1)$ there are only three pitchfork bifurcations. In the a-evolution of the attractor, a cycle which results from a pitchfork bifurcation can reappear further on: See cycle of size two in Fig. 3. This cycle disappears at the value $a_d = (1-\varepsilon_1)^{1-z_1}$ and reappears at $a_r = (1-\varepsilon_1)/(1-\varepsilon_2)^{z_1}$. To be more precise, these values are slightly modified according to the attractor towards which the system evolves, which in turn depends on the initial condition x_0, as discussed in section 4.5. The reappearance of the cycle with period two can even happen above the first entrance into chaos. For example, the map with $(\varepsilon_1,\varepsilon_2)=(-0.1,0.1)$ first enters into chaos at $a^* \cong 1.23$, whereas $a_r=1.358$. When $\varepsilon_1=\varepsilon_2 \equiv \varepsilon$ (continuous map), we have $a_d = a_r = (1-\varepsilon_1)^{1-z_1}$, a fact which clearly illustrates how the crossover to the period-doubling scenario occurs. A similar study can be made for cycles with period 4,8, etc.

4.4. Liapunov Exponent

The chaotic regime is characterized by a high sensitivity on the initial condition (value of x_0). The Liapunov exponent λ provides a quantitative measure of this dependence ($\lambda<0$ and $\lambda>0$ respectively correspond to periodic orbits and to chaotic motion). The Liapunov exponent is defined through

$$\lambda \equiv \lim_{N\to\infty} \frac{1}{N} \sum_{t=0}^{N-1} \ln|f'(x)|_{x=x_t} \tag{5}$$

In Fig. 5 we present the a-evolution of the Liapunov exponent for the typical case $(\varepsilon_1,\varepsilon_2) = (0,0.1)$. The first entrance into chaos occurs in this case at $a^* \cong 1.5447414$. We see in Fig. 5 remarkable features: (i) the structure is roughly self-similar; (ii) the "fingers" corresponding to high (low) periods are narrow (large); for a given cascade, they monotonously become narrower when the periods increase and shift towards negative values of λ, thus exhibiting (presumably) <u>infinitely large periods with no chaos</u>; the higest and largest finger of each cascade corresponds to the lowest period of that cascade; if we consider increasingly large lowest periods, the tops of the fingers approach $\lambda=0$ and drive the system into chaos; (iii) changements of periods occur for $\lambda\to-\infty$, in remarkable contrast with changement of periods in the period-doubling road, which occur at $\lambda=0$.

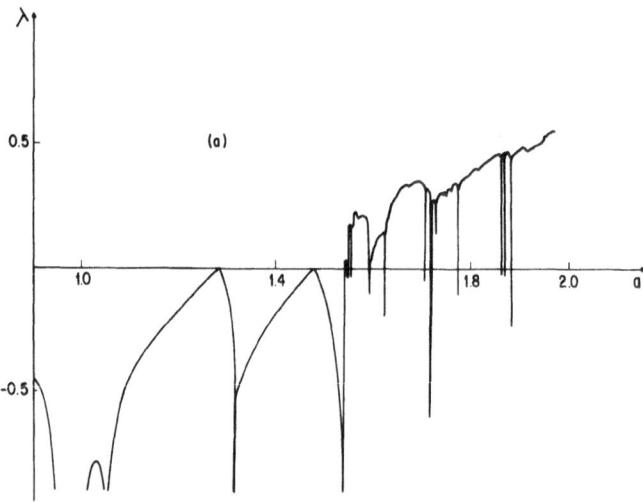

Fig. 5. Evolution of the Liapunov exponent as function of a for $(\varepsilon_1,\varepsilon_2)=$ (0,0.1), $z_1=z_2=2$ and $x_0=0.5$. The numbers in the fingers indicate the period of the attractor (c) is the amplification of the small rectangle in (b), which in turn is an amplification of (a).

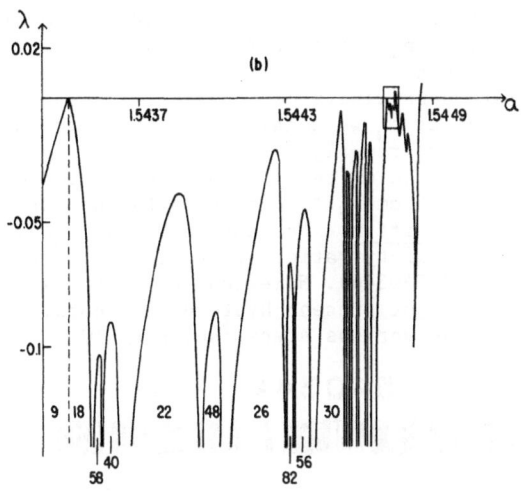

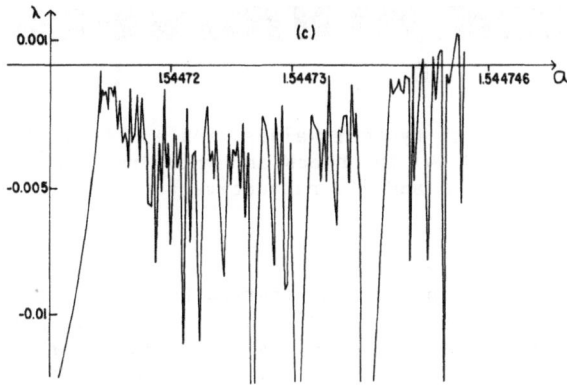

Fig. 5. Continuation.

4.5. Uncertainty Exponent

Continuous one-dimensional maps presenting an unique extremum have at most one finite attractor, which is independent on the initial conditions. In maps with a gap at the extremum we verify that this picture is modified. In such cases, more than one finite attractor (typically two attractors) appear when we cross from one banana (see Fig. 4) to a neighboring one (we made this observation in several crossings, it probably happens in all of them). The attractor which is chosen depends on the

initial value x_0. Two examples are presented in Fig. 6 for a=1.3 and
a=1.540344; the black and white regions are euclidean (dimensionality
D=1) whereas the border-set between them is a fractal with capacity dimension d. The uncertainty exponent[13] α_u is given by α_u=D-d. The system
is said to present <u>final-state</u> sensitivity or non-sensitivity according
to be $0 \leq \alpha_u < 1$ or $\alpha_u = 1$. To calculate α_u we consider, in the interval of x_0
corresponding to finite attractor (roughly [-1,1])N randomly chosen val-
ues (typically $N=10^4$). We then chose ε (say 10^{-3} and below) and check
whether both attractors starting from $x_0 \pm \varepsilon$ coincide with that of x_0; if
not, that value of x_0 is said <u>uncertain</u>. We denote N_u the total number
of uncertain points. The uncertainty ratio N_u/N varies as ε^{α_u}. We find
$\alpha_u \cong 0.85$ ($\alpha_u \cong 0.22$) for a=1.3 (a=1.540344). Numerical experiments based on
forth and back variations of a might present hysteresis according to the
initial value x_0 retained for the various steps (see ref. 10).

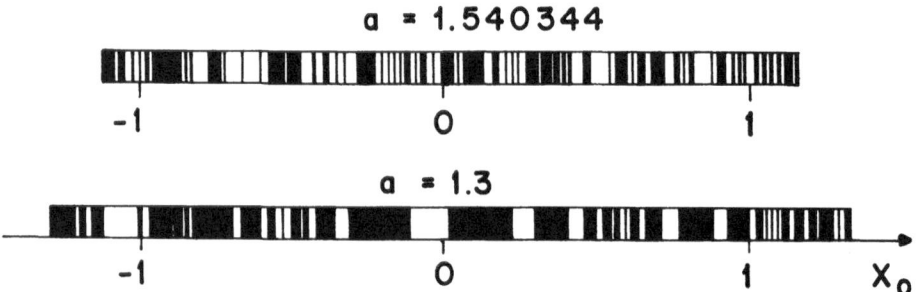

Fig. 6.   Basins of attraction for typical values of a and $(\varepsilon_1,\varepsilon_2)=(0,0.1)$
and $z_1=z_2=2$. For a=1.3 (a=1.540344) the black and white regions correspond
to cycles with period 8 and 2 (25 and 21) respectively.

4.6. Multifractality

The attractor of discontinuous maps at the first entrance into chaos is[10]
a multifractal[14]. The formalism used to study multifractals consists in
covering the attractor with boxes , indexed by i, of size l_i and assume
that the probability density scales like $p_i \propto l_i^\alpha$, in the limit $l_i \to 0$. The
characterization of a multifractal is through the function $f(\alpha)$, which
is the dimension of the set of boxes which share a given index α. Through
a Legendre transformation, $f(\alpha)$ is related to the generalized dimension
D_q[15]. The minimal and maximal values of α respectively coincide with D_∞
and $D_{-\infty}$; the maximal value of $f(\alpha)$ coincides with the Hausdorff dimension
D_0. In Fig. 7 we illustrate $f(\alpha)$ for the case $(\varepsilon_1,\varepsilon_2) = (0,0.1)$. Its
shape is different (more square-like) from that obtained without gap (period-doubling road to chaos), and the values we obtain are $D_0 \cong 0.95$, $D_\infty \cong 0.45$
and $D_{-\infty} \cong 5.7$. Notice that the period-doubling relation $D_{-\infty}=zD_\infty$ fails in
the gap case.

4.7. Discontinuous Map as a Limit Case

Since physical maps presumably do not exhibit a (sharp) discontinuity,

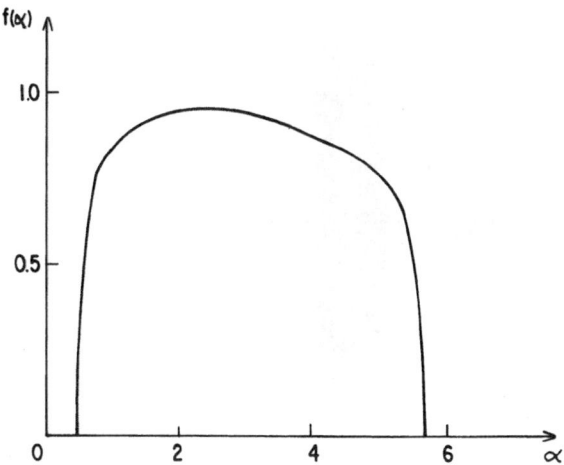

Fig. 7. Multifractal funtion $f(\alpha)$ for $(\varepsilon_1,\varepsilon_2)=(0,0.1)$, $z_1=z_2=2$ and $x_0=0.5$ (chaos appears at $a*\cong1.5447414$).

we consider in this section the following continuous map[16]

$$x_{t+1} = f(x_t) \equiv 1 + \varepsilon|x|^w sgn(x) - a|x|^z \qquad (6)$$

with $w<1<z$. When $w<<1$ this map has a general shape very similar to the map (1), but without discontinuities. In the case $w=0$, the rapidly changing part of $f(x)$ around $x=0$ is replaced by a jump, and we recover the discontinuous map (with $\varepsilon_1=\varepsilon_2\equiv\varepsilon$). Notice that the map (6) has a Schwarzian deriv̱ative which is positive for an interval of x around $x=0$. In Fig. 8 we show the a-evolution of the attractor for a typical case. This picture   is roughly similar to Fig. 3. Near $a=1$ we observe the cycles with periods $2\to6\to4$, a fact which is a reminiscence of the inverse cascade $2...10\to8\to6\to4$. When w increases the map becomes more and more similar to the logistic map (Eq. (1)), therefore a crossover to the period-doubling scenario is expected. In Fig. 9 we show this crossover for a particular range of a. Initially, we observe that both extrema correponding to changement of periods $2\to6$ and $6\to4$ become chaotic and then merge in one extremum and then becomes chaotic again.

For $\varepsilon>0$ we found, in all cases studied, a sudden entrance into chaos when the external parameter a is varied. For $\varepsilon<0$ we found, before the first entrance into chaos, a sequence of period doubling bifurcations with the convergence ratio of the set $\{\delta_k\}$ being approximately the same of the logistic equation. Therefore we observe indications for a route to chaos via period-doubling in maps with positive Schwarzian derivative for an interval of x around $x=0$. This situation is unusual and   seems to us an interesting question, worthy to be studied, since the maps studied until now that present period-doubling route to chaos have nega-tive Schwarzian derivative for all finite values of the dynamical variable.

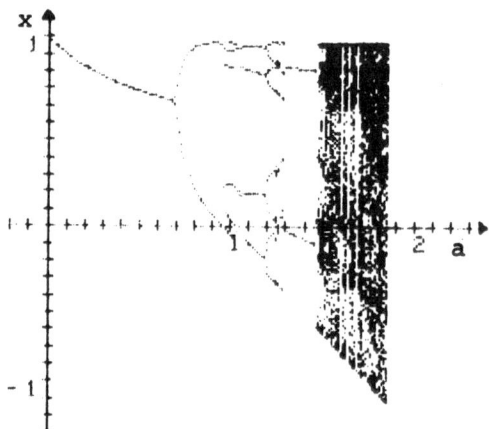

Fig. 8.   Evolution of the attractor as a function of a for the equation
(6) with ε=0.1, w=0.1 , z=2 and x_0=0.5.

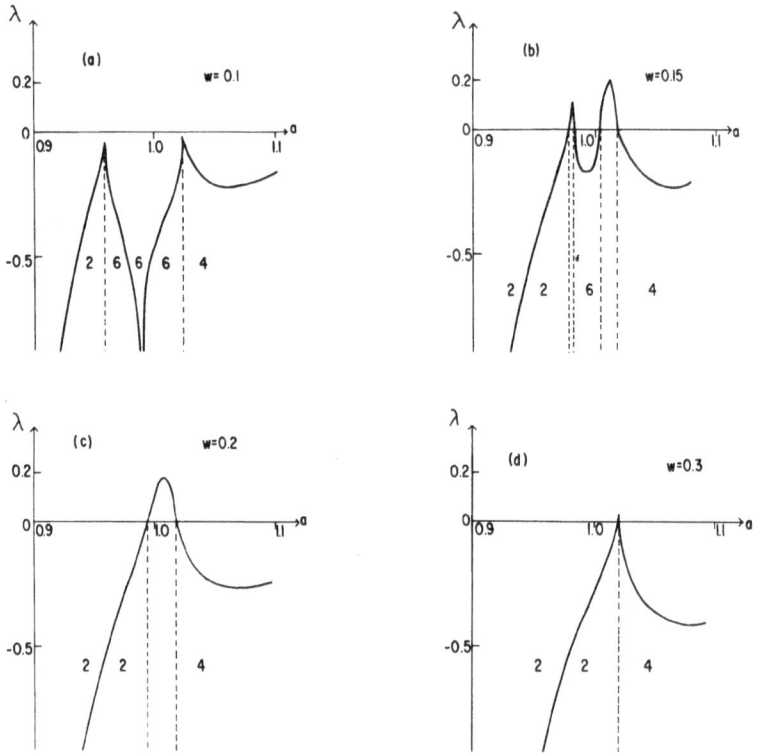

Fig. 9.   Liapunov exponent as a function of a for ε=0.1, z=2, x_0=0.5 and
w=0.1,0.15, 0.2, 0.3. The numbers indicate the period of the attractor.

5. CONCLUSIONS

We have shown that an asymmetry introduced in the logistic map can alter some of its basic features. Amplitude asymmetry ($a_1 \neq a_2$) and exponent asymmetry ($z_1 \neq z_2$) do not alter the bifurcation sequence, but the unique tendency associated with the set $\{\delta_k\}$ disappears. In the discontinuous map ($\varepsilon_1 \neq \varepsilon_2$) the route to chaos is completely modified. Sequences of inverse cascades in arithmetic progression are observed in the evolution of the attractor. There is no chaos at the accumulation point of these cascades, which appears only at the accumulation point of the accumulation points. Several other unusual features were found at the phase diagram, Liapunov and uncertainty exponents, multifractality, among others.

We acknowledge with pleasure very fruitful suggestions by H.W. Capel, M. Napiórkowski, as well as interesting remarks by A. Coniglio, E.M.F. Curado, H.J. Herrmann, J.P. van der Weele and Ph. Nozières. We are indebted to P. Coullet and C. Tresser for calling our attention on ref. 4.

REFERENCES

1. R.M. May, Nature **261**, 459 (1976).
2. S. Grossmann and S. Thomae, Z. Naturf. **32A**, 1353 (1977); M.J. Feigenbaum, J. Stat. Phys. **19**, 25 (1978); P. Coullet and C. Tresser, J. Phys. (Paris) Colloq. **5**, C25 (1978).
3. P.R. Hauser, C. Tsallis and E.M.F. Curado, Phys. Rev. A **30**, 2074 (1984). Bambi Hu and Indubala I. Satija, Phys. Lett. **98A**, 143 (1983); J. P. Eckmann and P. Wittwer, Computer Methods and Borel Summability Applied to Feigenbaum's Equation, Lectures Notes in Physics, vol. 227 (Springer, Berlin, 1985); J.P. van der Weele, H.W. Capel and R. Kluiving, Phys. Lett. **119A**, 15 (1986); J.K. Bhattacharjee and K. Banerjee, J. Phys. A **20**, L269 (1987); M.O. Magnasco and D.L. Gonzalez, private communication; M.C. de Sousa Vieira, unpublished.
4. A. Arneodo, P. Coullet and C. Tresser, Phys. Lett. **70A**, 74 (1979).
5. P. Szépfalusy and T. Tél, Physica **16D**, 252 (1985); J.M. Gambaudo, I. Procaccia, S. Thomae and C. Tresser, Phys. Rev. Lett. **57**, 925 (1986).
6. R.V. Jensen and L.K.H. Ma, Phys. Rev. A **31**, 3993 (1985).
7. M.C. de Sousa Vieira, E. Lazo and C. Tsallis, Phys. Rev. A **35**, 945 (1987).
8. A.A. Hnilo, Optics Commun. **53**, 194 (1985); A.A. Hnilo and M.C. de Sousa Vieira, J. Opt. Soc. Am., in press.
9. M. Octavio, A, Da Costa and J. Aponte, Phys. Rev. A **34**, 1512 (1986).
10. M.C. de Sousa Vieira and C. Tsallis, unpublished; M.C. de Sousa Vieira and C. Tsallis, to appear in Disordered Systems in Biological Models, eds. L. Peliti and S.A. Solla (World Scentific, 1988); M.C. de Sousa Vieira and C. Tsallis, to appear in Universalities in Condensed Matter, eds. R. Jullien, L. Peliti, R. Rammal and N. Boccara (Springer Proc. Phys.,1988)
11. M. Metropolis, M.L. Stein and P.R. Stein, J. Combinatorial Theory A **15**, 25 (1973).
12. B. Derrida, A. Gervois and Y. Pomeau, J. Phys. A **12**, 269 (1979).

13. C. Grebogi, S.W. McDonald, E. Ott and J.A. Yorke, Phys. Lett. **99A**, 415 (1983); M. Napiórkowski, Phys. Lett. **113A**, 111 (1985).
14. T.C. Halsey, M.H. Jensen, L.P. Kadanoff, I. Procaccia, and B.I. Shraiman, Phys. Rev. A **33**, 1141 (1986).
15. H.G.E. Hentschel and I. Procaccia, Physica **8D**, 435 (1983).
16. Z. Kaufmann, P. Szépfalusy and T. Tél, unpublished.

POTTS MODEL AND AUTOMATA NETWORKS

E. Goles
Departamento de Matemáticas Aplicadas
Facultad de Ciencias Físicas y Matemáticas
Universidad de Chile
Casilla 170-3, Correo 3, Santiago
Chile

ABSTRACT. In this paper we study Automata Networks whose local rules are "compatibles" with the Potts Hamiltonian. We prove that there exists compatible local rules with a very complex global behaviour (automata configurations may code any logic function) and that Maximal Local Rules admit a Lyapunov functional derived from the Potts Hamiltonian. Furthermore, in the last case, the steady-state behaviour is very simple : fixed points and/or two cycles.

1. Introduction

The study of automata networks has become very related to several physical models: hidrodynamics, percolation, spin glasses, etc.$|1,2|$. Introduced by Von Neumann $|3|$ Automata Networks can be defined as interconnected cells. Each cell i has internal states $0,1,...,q-1$ and interacts in discrete time steps with its neighbour cells. In turn, it updates its own state according to a local rule:

$$x_i^{t+1} = f_i(x_j^t; j \in V_i) \quad t = 0,1,2,.. \tag{1.1}$$

where V_i is the set of neighbours of site i.

The update rule is synchronous; i.e. all the cells update their state simultaneously.

A particular case of automata networks are Cellullar Automata, (CA) where the cells are located in a regular array with short-ranged interactions and the local transition rule is uniform; i.e. it is the same for any site in the array.

In this paper we shall study Automata Networks (not neccesarily in a regular array) such that local rules are "compatible" with the Hamiltonian introduced in the Potts model; i.e.

$$H(x) = - \sum_{<i,j>} \delta(x_i,x_j) \tag{1.2}$$

89

E. Tirapegui and D. Villarroel (eds.), Instabilities and Nonequilibrium Structures II, 89–96.
© *1989 by Kluwer Academic Publishers.*

where $x \in Q^n, Q = \{0,1,\ldots,q-1\}$ is the set of states and δ is a symmetric function. In the classical Potts model, δ is the Kronecker function: $\delta(a,b) = 1$ iff $a = b$ and 0 otherwise.

A local rule is "compatible" with (1.2) if its application on site i may decrease the Hamiltonian. An Automata Network with local compatible rules will be called a "Potts Automata" (PAN).

The compatibility is not enough to insure that the synchronous update of the PAN is driven by H. In fact, we give examples where the dynamic behaviour is extremely complex (in terms of computing capabilities of the PAN). Instead of this "negative result" we prove that a subclass of PAN is driven by a Lyapunov operator derived from H. Furthermore we prove in this last case that the periodic behaviour is very "simple" : fixed point or two-cycles. This particular class is important because it contains neural networks, majority networks and automata used in digital image processing $|4,5,6|$.

Other applications of PAN have been developped, as generalizations of Q2R, in $|7|$.

2. Potts Automata Networks (PAN)

Let $Q = \{0,\ldots,q-1\}$ be the set of states and $G = (N,E)$ a non-oriented graph, where N is the finite set of nodes or sites ($|N| = n$) and E is the set of links. The neighbourhood of a site is defined by:

$$V_i = \{i \in N; (i,j) \in E\} \quad \forall i \in N$$

Since $|N| = n$ the set of configurations on G is Q^n, where $|Q^n| = q^n$. Given $x \in Q^n$ we define the Hamiltonian

$$H(x,x) = -\sum_{\langle i,j \rangle} \delta_1(x_i,x_j) + \sum_{i=1}^{n} \delta_2^i(x_i,x_i) \qquad (2.1)$$

where δ_1 is a symmetric function from $Q \times Q$ into $\mathbb{R}$, $\{\delta_2^i\}$ is a set of symmetric functions from Q into $\mathbb{R}$, and $\langle , \rangle$ noted the interaction with the neighbours states. In the classical Potts model δ_1 is the Kronecker function and $\delta_2^i = 0$.

One may write (2.1) as:

$$H(x,x) = \sum_{i=1}^{n} H_i(x,x)$$

where

$$H_i(x,x) = -\sum_{j \in V_i} \delta_1(x_i,x_j) + \delta_2^i(x_i,x_i)$$

We shall say that a local rule f_i (on site i) is compatible with H iff:

$$H_i(\tilde{x},\tilde{x}) \leq H_i(x,x)$$

where $\tilde{x} = (x_1,\ldots,f_i(x),\ldots,x_n)$.

A set of local functions $\{f_i\}_{i=1}^n$ is compatible iff each f_i is compatible.

There are two major ways to introduce a dynamics on the previous model: the synchronous (or parallel) and the sequential update. The synchronous update is defined as follows:

$$x_i(t+1)=f_i(x_j(t); \ j \in V_i) \qquad \forall i=1,\ldots,n \qquad\qquad (2.2)$$

and the sequential update:

Given $x(t) \in Q^n$, and given $i \in \{1,\ldots,n\}$:

$$x_i(t+1)=f_i(x_j(t); j \in V_i)$$
$$x_k(t+1)=x_k(t) \qquad \forall k \neq i \qquad\qquad\qquad (2.3)$$

The choice of i in the sequential iteration can be done at random or cyclically; i.e. site i updates its state immediatly after site i-1 does.

Clearly, for the sequential iteration, a PAN is driven by the Hamiltonian H. In fact, let $k \in \{1,\ldots,n\}$ be the site to be updated, hence:

$$x_k(t+1) = f_k(x_j(t); j \in V_k)$$
$$x_j(t+1) = x_j(t) \text{ for any } j \neq k$$

Since f_k is compatible:

$$H_k(x(t+1),x(t+1)) \leq H_k(x(t), x(t))$$

Then the PAN evolution with sequential updating is very simple: if one has strictly compatibility (i.e. $H(x(t+1)x(t+1)) < H(x(t),x(t))$ if $x_k(t+1) \neq x_k(t)$) the network converges, for any initial condition, to a limit set of fixed points that are local minima of H.

Now we shall see that for synchronous update the dynamic behaviour may be very complex. For instance, let δ be the Kronecker function and the Hamiltonian $H(x) = - \sum_{<i,j>} \delta(x_i,x_j)$. Let $G = \mathbb{Z}_n$ (the one-dimensional n-torus), $Q = \{0,1,2,3\}$ be the state set and the neighbourhood $V_i = \{i-1,i+1\}$, $1 \leq i \leq n$, where indexes are taken mod n.

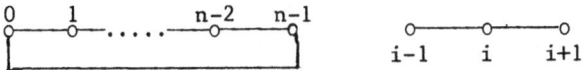

The local rule, f, is the following:

$$f(x_{i-1},x_{i+1}) = \begin{cases} x_i+1 \bmod 4 & \text{if } \mathrm{card}\{j \in V_i/x_j=x_i+1 \bmod 4\,\} \geq \\ & \qquad \geq \mathrm{card}\{j \in V_i/x_j=x_i\} \\ x_i & \text{otherwise} \end{cases} \qquad (2.4)$$

The global transition function is given by the synchronous application of f to any site, i.e:

$$F(x)(i) = f(x_{i-1},x_{i+1}) \text{ for any } i \in \{0,\ldots,n-1\}$$

Clearly f is a compatible rule. In fact, given $i \in \{0,\ldots,n-1\}$, let us suppose:

$$\mathrm{Card}\{j \in V_i/x_j=x_i+1 \bmod 4\} < \mathrm{Card}\{j \in V_i/x_j=x_i\}$$

hence, $f(x_{i-1},x_{i+1}) = x_i$, then $H_i(\tilde{x},\tilde{x}) = H_i(x,x)$ where
$\tilde{x}=(x_0,\ldots,f(x_{i-1},x_{i+1}),\ldots,x_n)$. Now, if $\mathrm{Card}\{j \in V_i/x_j=x_{i+1} \bmod 4\} \geq$
$\mathrm{Card}\{j \in V_i/x_j=x_i\}$ then $f(x_{i-1},x_{i+1}) = x_i+1 \bmod 4$.
Since $H_i(x,x) = - \sum\limits_{j \in V_i} \delta(x_i,x_j) = - \mathrm{Card}\{j \in V_i/x_i=x_j\}$ we have

$H_i(\tilde{x},\tilde{x}) \leq H_i(x,x)$.

The dynamics of this PAN is complex: there exist configurations breaking the space symmetry:

```
Gliders        3 2 1     or      1 2 3
               3 2 1             1 2 3
                     .         .
                      .       .
```

Also, the crash of gliders give the quiescent configuration:

```
3 2 1 0 1 2 3    or    , 3 2 1 0 0 1 2 3
  3 2 1 2 3               3 2 1 1 2 3
    3 2 3                   3 2 2 3
      3                       3 3
      0                       0 0
```

Meeting between gliders and stable configurations:

```
3 2 1 0 2 2 2        where 222 is a stable
  3 2 1 2 2 2        configuration
    3 2 2 2 2
      3 2 2 2
        3 2 2
          3 2
            3
```

By using previous configurations one may built any logic gate by coding

1's as gliders:

NOR- gate : $a \to \bar{a}$

3 2 1 0 0   — — —   — — —

                input cells   output cells

there the input is $1 \equiv 123$ or $0 \equiv 000$

OR- gate : $(a,b) \to a \vee b$

3 2 1 $\xleftrightarrow{\hspace{2cm}}$ X X X $\longleftrightarrow$ 2 2 $\xleftrightarrow{\hspace{2cm}}$ Y Y Y   - - -

        m cells   input   1 cells input   output

where $XXX \in \{000,321\}$; $YYY \in \{000,123\}$

AND- gate : $(a,b) \to a \wedge b$

3 2 1   X X X   Y Y Y - - 2 2   1 2 3 - - -

      $\xrightarrow{\hspace{4cm}}$ $\xrightarrow{\hspace{2cm}}$

       1 cells   1 cells   output

where $XXX,YYY \in \{000,321\}$

    For instance let $(a,b) = (1,1)$, hence:

```
321 --- 321 -- 321-22---------123 ----
321      321     32122         123
 321       321    3222         123   output
  321       321    322         123
   321       321    32        123
    321       321    2        123
     321       321          123
      321       321        123
       321       321    123
        321       321 123
         321      32123
          321      323
           321      3            the output is 321
            321
```

    As $\{NOR, OR, AND\}$ is an universal set of logical gates, with this automaton we may simulate, in 1-dimension, any logic function, then its dynamic behaviour is complex.

3. Maximal Local Rules

Now we study a class of local transition functions where the compatibility with the Hamiltonian H is enough to associate a Lyapunov functional to the synchronous update of the network.

    Let H be the Hamiltonian defined in (2.1) and let f_i be the following local rule:

$$f_i(x)=s \Longleftrightarrow \sum_{j \in V_i} \delta_1(s,x_j)-\delta_2^i(s,x_i) \geq \sum_{j \in V_i} \delta_1(r,x_j)-\delta_2^i(r,x_i)$$
$$\text{for } r=0,\ldots,q-1$$

In case of tie one takes the bigger state.

This class of local rules is called Maximal Local Rules and they are compatibles with H.

The synchronous update is given by:

$$x_i(t+1) = f_i(x(t)) \text{ for } t = 0,1,2,\ldots; \text{ and } i=1,\ldots,n$$

We associate to the previous network the quantity:

$$E(x(t+1)) = H(x(t+1),x(t))$$

$$= \sum_{<i,j>} \delta_1(x_i(t+1), x_j(t)) + \sum_i \delta_2^i(x_i(t+1),x_i(t))$$

Proposition 1. The operator E is a Lyapunov functional associated to the synchronous update of PAN.

Proof. Let us take $\Delta_t E = E(x(t+1)) - E(x(t)) =$

$$= - \sum_{<i,j>} \delta_1(x_i(t+1),x_j(t)) + \sum_i \delta_2^i(x_i(t+1),x_i(t))$$

$$+ \sum_{<i,j>} \delta_1(x_i(t),x_j(t-1)) - \sum_i \delta_2^i(x_i(t),x_i(t-1))$$

Since G is a non-oriented graph and the functions δ_1, δ_2^i are symmetric:

$$\Delta_t E = \sum_i \{ - \sum_{j \in V_i} \delta_1(x_i(t+1),x_j(t)) + \delta_2^i(x_i(t+1),x_i(t))$$

$$- (- \sum_{j \in V_i} \delta_1(x_i(t-1),x_j(t)) + \delta_2^i(x_i(t-1),x_i(t)) \}$$

From definition of f_i, each term of $\Delta_t E$ satisfies $(\Delta_t E)_i \le 0$ ☐

It is not difficult to see that in case of tie $\Delta_t E$ may vanish and $x(t+1) \ne x(t-1)$.

In order to avoid this fact one takes the perturbation of E:

$$E^*(x(t+1)) = (q+1) E(x(t+1),x(t)) - \sum_i (x_i(t+1) + x_i(t))$$

where $q = \max \{s; s \in Q\}$

Proposition 2. The operator E^* is a Lyapunov functional and it is strictly-decreasing in the transient phase.

Proof. Clearly,

$$\Delta_t E^* = (q+1) \sum_i (\Delta_t E)_i - \sum_i (x_i(t+1) - x_i(t-1))$$

hence, the i-th term of $\Delta_t E^*$ is the following:

$$(\Delta_t E^*)_i = (q+1)(\Delta_t E)_i - (x_i(t+1) - x_i(t-1))$$

If $x_i(t+1) \ne x_i(t-1)$ and $(\Delta_t E)_i = 0$ (the tie case) one has, by definition of f_i, $x_i(t+1) > x_i(t-1)$ then $(\Delta_t E^*)_i < 0$

Otherwise (i.e.: $x_i(t+1) \neq x_i(t-1)$ and $(\Delta_t E)_i > 0$) since δ_1 and δ_2^i are integral functions:

$$(\Delta_t E^*)_i \leq -(q+1) - (x_i(t+1) - x_i(t-1))$$

$$\leq -(q+1) + q \leq -1 < 0$$

then $\Delta_t E^* = \sum (\Delta_t E^*)_i$ is strictly decreasing. ☐

From previous proposition one may characterize the steady-state behaviour of PAN:

Corollary. The PAN admits, in the steady state, only fixed points and/or two cycles.

Proof. Since E* is a strictly-decreasing Lyapunov functional if $x(t+1) \neq x(t-1)$, we have that in the steady-state $x(t+1) = x(t-1)$ then there exists only fixed points and/or two cycles. ☐

As an example of maximal local rules we have the majority network

$$f_i(x) = s \Leftrightarrow \text{card}\{j \in V_i \ / \ x_j = s\} \geq \text{card}\{j \in V_i \ / \ x_j = r\}$$

$$\forall r \in Q$$

In case of tie, one takes the maximum value. Other examples are given by local average rules used in graph transformations $|5,6|$:

$$x_i(t+1) = x_i(t) + \begin{cases} -1 & \text{if} \quad \sum_{j \in v_i} x_j(t) - d_i x_i(t) < 0 \\ 0 & \text{if} \quad x_j(t) = x_i(t) \qquad \forall j \in V_i \\ +1 & \text{otherwise} \end{cases}$$

where $d_i = \text{card } (V_i)$, and $x(0) \in N^n$.

Asymptotic behaviour and bounds of the transient are given in $|5,6|$.

ACKNOWLEDGMENTS

This works was partially supported by Fondo Nacional de Ciencias 88 grant 554, TWAS and DTI, Universidad de Chile 88.

REFERENCES

|1| S. Wolfram, *Theory and Applications of Cellular Automata*, World
      Scientific, 1986.
|2| E. Bienenstock, F. Fogelman-Soulie, G. Weisbuch, 'Disordered
      Systems and Biological Organization', *NATO ASI Series F. Computer
      and Systems Sciences*, Vol.20, Springer-Verlag, 1986.
|3| A. Burks, *Essays on Cellular Automata*, University of Illinois
      Press, 1970.
|4| E. Goles, 'Dynamics of Positive Automata Networks', *Theor. Comp.
      Sci.* 41(1985) 19-32.
|5| E. Goles, 'Local Graph Transformations Driven by Lyapunov
      Functionals', Res. Rep. Dep. Mat., U. Ch. (1988), send to Complex
      Systems.
|6| E. Goles, A.M. Odlyzko, 'Decreasing Energy Functions and Lengths
      of Transients for some Cellular Automata', to appear in *Complex
      Systems* 2-(1988).
|7| Y. Pomeau, G. Vichniac, 'Personal Communication'.

CELLULAR-AUTOMATA FLUIDS

GERARD VICHNIAC
Plasma Fusion Center
Massachusetts Institute of Technology
Cambridge, MA 02139, USA

1 Definitions and motivations

Invented in 1948 by von Neumann and Ulam, cellular automata are fully discrete dynamical systems. The dynamical variables take their values in a finite set (typically a few bits), they are arranged on a lattice where they evolve in discrete time [8,14,22,53,48]. The global dynamics results from the parallel application, at each lattice site, of a local transition rule defining the new value at that site as a function of the current values in a small neighborhood around the site.

This paper presents an introduction to the applications to fluid dynamics of these mathematical objects, often called in this context *lattice-gas automata*. For more formal and extended treatments, see the reviews [17,54,21] and the conference proceedings[1,37,2]. It may appear surprising that fully discrete objects can simulate fluids at all. Feynman (cited in [15]) explained this as follows:

> The behavior of a fluid depends very little on the nature of the individual particles in that fluid. For example, the flow of sand is very similar to the flow of water or the flow of a pile of ball bearings. We have therefore taken advantage of this fact to invent a type of imaginary particle that is especially simple for us to simulate. This particle is a perfect ball bearing that can move in one of six directions. The flow of these particles on a large enough scale is very similar to the flow of natural fluids.

There has been an increased interest in cellular automata fluids during the past three years, the motivation of which stems from both physics and computer science.

1.1 EFFICIENT SIMULATIONS OF FLUID FLOWS

The automata method differs sharply from the standard numerical approaches to fluid dynamics. In the latter, the flow equations must undergo both truncation and roundoff before their computer treatment. In effect, one constructs a discrete model derived from

E. Tirapegui and D. Villarroel (eds.), Instabilities and Nonequilibrium Structures II, 97–116.
© *1989 by Kluwer Academic Publishers.*

the original differential equation, but which is clearly distinct from it in ways which are often uncontrollable and prone to numerical instabilities.

The cellular automata modeling is fully discrete from the start. It lends itself to *exact* simulation by "finitistic" means (as von Neumann called them[18]), and lets digital computers do what they do best; namely, iterative Boolean decisions rather than necessarily imprecise floating-point arithmetic. The attempt to reduce fluid dynamics to logical primitives thus offers the practical computational advantages of intrinsic stability and efficiency. But to be considered a viable alternative to the successful finite-difference and spectral methods of fluid dynamics, cellular automata must demonstrate both correctness and efficiency. Correctness implies that all the lattice artifacts can be removed and that lattice gases indeed represent solutions of the hydrodynamic equations; section 7 presents recent efforts in this direction. Efficiency means that the computational cost of low-noise simulations of flows with high values of the Reynolds number scales gracefully with that parameter. Zaleski performed a careful analysis of this question[57]. Eventually, what makes cellular automata most promising is the ease with which they treat problems generally considered to be extremely difficult: complex and moving boundaries, porous media, multiphase flows and combustion[1,50,45,51,12,41,42].

1.2 EFFICIENT USE OF MASSIVELY PARALLEL COMPUTING MACHINES

Cellular automata not only lend themselves to exact computer implementation, they also are naturally adapted to massively parallel implementation. From the vantage point of computer science, this remedies the vexing infancy of the art of parallel programming in the face of the availabilty of powerful parallel machines. Arrays of processing elements with local communications can, within the cellular automata approach, be used to their full potential for fluid dynamics, and, for that matter, for the simulation of a variety of distributed natural phenomena, most naturally those occurring in homogeneous physical systems with short-ranged microscopic interactions[49,46,22]. The possibility of simulating the potentially boundless complexity of hydrodynamics sidesteps the formidable problem of efficient universal parallel programming, and arguably contributes to it.

2 The HPP and FHP models

The first fully discrete model[19,20] for fluid dynamics was proposed in 1973 by Hardy, de Passis, and Pomeau (HPP). The HPP gas lives on a square lattice, where each site is endowed with four bits of information that code the particle occupation for each allowed direction. The evolution rule is very simple: particles travel straight at unit velocity (one site per time step) except when particles collide head-on, in which case they scatter at right angle (see Figure 1). This microdynamics obviously conserves the total number of particles as well as both components of the momentum. Collisions will not occur in this model if a third particle is also entering the site, so that the links will contain at most one particle. This arbitrary rule constitutes a Boolean exclusion principle and puts a bound on the local density (four particles per site). This makes simulations not only exact, but also *stable*: solutions cannot run away to unphysical regimes. (Note, however, that derived quantities such as the viscosity can diverge with the size of the 2D lattice, as expected[4,34,29].)

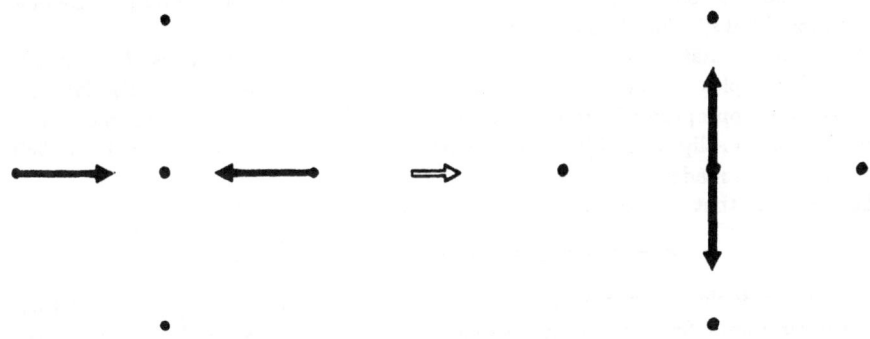

Figure 1: Collision in the HPP model.

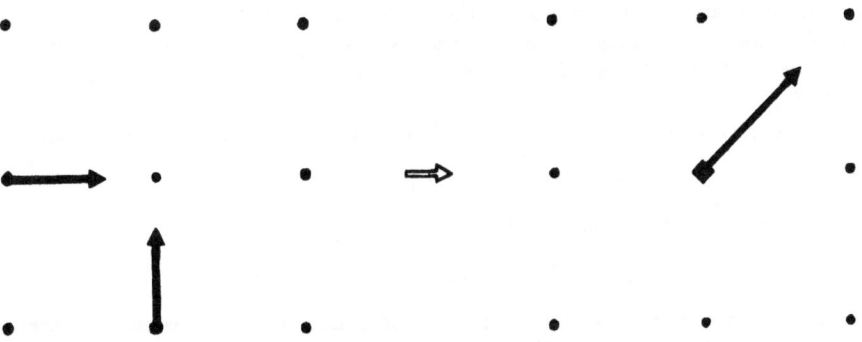

Figure 2: Collision with energy exchange in a multispeed model. Two particles of energy 1 interact and form a rest particle (represented with a diamond) and a particle of energy 2 (see section 7).

Figure 2 shows a collision in an extension of the HPP model that allows particle motion along diagonal links[24,25,11,36,35]; see section 7.

Lattice-gas automata rules can be easily altered to accommodate (possibly moving) obstacles with complex geometries. It suffices to "draw" such obstacles on the lattice and implement the appropriate boundary conditions on their surface. The amendments enabling specular, no-slip, and diffusive boundary conditions are straightforward and "left as an exercise to the reader."

The HPP rule that applies at each lattice site can be expressed as

$$n'_i = n_i + n_{i+1}n_{i+3}\bar{n}_i\bar{n}_{i+2} - n_i n_{i+2}\bar{n}_{i+1}\bar{n}_{i+3}, \tag{1}$$

where n'_i is the Boolean variable for a particle in direction i exiting the considered site, the unprimed variables refer to particles entering it, and $\bar{n} \equiv 1 - n$ is the Boolean complement of n. If a particle moving in the i direction is the only one reaching the site at time t, it will propagate undisturbed: $n'_i = n_i$. It will be deflected if it meets a particle moving in the $i + 2$ (mod 4) direction: hence the last term in (1). This term is an annihilation (or depletion, or loss) operator A_i for the i direction. Similarly, the second term in the r.h.s. of (1) is a creation (or populating, or gain) operator C_i. It shows that direction i becomes populated if a collision occurs ($n_{i+1} = n_{i+3} = 1$) and the Boolean exclusion principle holds ($n_i = n_{i+2} = 0$).

In 1985, Frisch, Hasslacher and Pomeau (FHP) proposed a very consequential modification of the HPP model. The change consists of replacing HPP's square lattice by a *hexagonal* lattice (strictly speaking, crystallography's *triangular* lattice, with triangular plaquettes and six nearest neighbors—hence the six directions referred to in Feynman's quote above). The rule is essentially the same: particles meeting head-on scatter at $\pm 60°$; there is also a three-body collision rule (see below). The FHP rule can be expressed under a form similar to (1).

The lattice-gas rules can in general take the form of a master equation (change=gain−loss)

$$n'_i - n_i = C_i - A_i. \tag{2}$$

Note that A_i and C_i contain imbedded projectors

$$\bar{n}_i C_i = C_i \qquad \text{and} \qquad n_i A_i = A_i. \tag{3}$$

The first relation expresses the Boolean exclusion principle. The second one says that to remove a particle, it must be there in the first place (exclusion principle for holes). Equation (2) reads, omitting the direction indices,

$$n' = n(1 - A) + C = n\bar{A} + C. \tag{4}$$

A most interesting aspect of the cellular automata approach is that the same equation can be written as a master equation (palatable to the physicist) or as a logical algorithm (palatable to the programmer or the hardware designer) that begs for direct parallel implementation. This is particularly conspicuous in more general models (see section (7)), where several types of particles and collision processes are defined. In such models, rules can be written formally as

$$n' = n \prod_p \bar{A}^p + \sum_p (C^p \prod_{p' \neq p} \bar{C}^{p'}), \tag{5}$$

where the sums and products are over all the defined processes p. Thanks to the exclusion principle as expressed in (3), different physical collision processes readily translate into mutually exclusive logical predicates. This allows one in many cases to evaluate with impunity both exclusive ($\oplus$) and inclusive ($\vee$) disjunctions as regular *arithmetic* sums: the l.h.s. of expressions such as (5) is guaranteed to be 0 or 1, a Boolean value, despite the complicated arithmetic expressions that can appear in the r.h.s.

3   The "unary" representation in cellular automata

Now that we are acquainted with the HPP and the FHP models, let us review how, in general, cellular automata can represent continuous quantities. Figure 3 (from[34]) shows the evolution of an HPP gas of half maximum density perturbed at $t = 0$ by a small square of maximum density. Configurations at time steps 30 and 90 display an outgoing circular wave. The figure was made using the CAM fast simulator[47,48]; the dots indicate sites with three or four particles. Figure 4 illustrates the original way whereby cellular automata represent gradually varying quantities. In standard computer arithmetic, a continuous function $f(x)$ such as the bell-shaped curve in Figure 4, is represented with a sequence of floating-point numbers, cf. the rectangles in the figure, much as a Gaussian is approximated by a binomial distribution. The cellular automata representation is radically different. It is reminiscent of the primitive "unary" system, where the number 4, for example, is represented as '1111.' In cellular automata, one has a ceiling of value 1 (actually, 4 in HPP) for the Boolean variable $n(x)$, and $f(x)$ can be reconstructed with a folding

$$f(x) = \int n(x')t(x - x')dx', \qquad (6)$$

where $t(x)$ is an appropriate (box-car or truncated Gaussian) convolutional filter. Looking again at Figure 3, we note that the eye automatically performs something like this filtering; it "computes" the local density of 1-bits at each point (the local darkness), and reconstructs the "height $f(x)$" of the wave. To be sure, the "unary"representation requires a finer mesh for a given level of noise. Moreover, its dynamic range is very limited. But it is precisely this limitation that guarantees stability. It reflects actual physics: hydrodynamical variables, such as the density, typically vary within one order of magnitude only (even in shock waves) and not between 10^{-30} and 10^{30}, as allowed in most floating-point representations. In cellular automata, there are no most nor least significant bits. This "bit democracy" (an expression popularized by Frisch) permits a more efficient use of computer memory. It also offers another way to understand the stability; numerical instabilities are indeed errors on the least significant bit that creep up during iterations all the way to the most significant bits.

Figures 3 and 4 call for three further observations:

i) the wave in Figure 3 is circular, despite the underlying square lattice,

ii) the particles exhibit a conspicuous *collective behavior*. Nothing in the rules says that the particles should form a growing circle or any particular pattern;

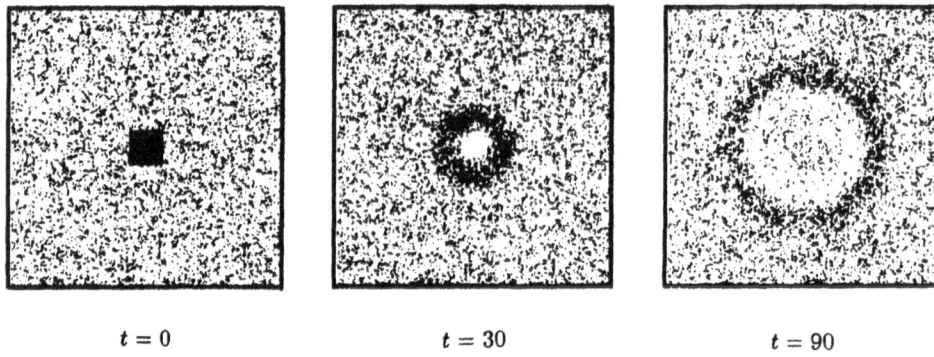

$t = 0$ $t = 30$ $t = 90$

Figure 3: Isotropic "sound" wave in the HPP model.

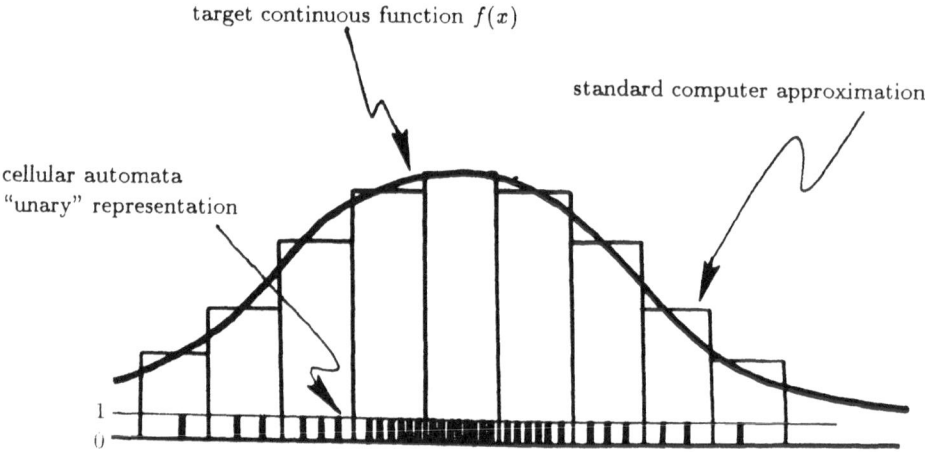

Figure 4: A continuous function $f(x)$ can be represented in cellular automata as the local density of 1-values of a Boolean function $n(x)$, via a folding $f(x) = \int n(x')t(x - x')dx'$, where $t(x)$ is an appropriate test function. Note on Figure 3 that the eye automatically performs this filtering operation.

the particles themselves are, just as the rule, "short-sighted." In fact, particles in the inner ridge of the circle are pushed *inwards* by the local density gradient; at $t = 90$, they have formed a secondary maximum at the center,

iii) a most schematic, naive, *discrete* model emulates wave motion—a paradigm of the mathematics of the *continuum*.

The *isotropy* (point i above) can be understood informally by viewing the central perturbation as scattered by the *random* background. Indeed, random walk on the square lattice is itself isotropic (its space properties involve coordinates x and y only in the combination $\sqrt{x^2 + y^2}$). Actually, the isotropy holds only in the linear (sound wave) limit; see section 6. The emergence of collective behavior out of simple elements (point ii) is pervasive in cellular automata[14,22,53,48]. Figure 3 is a simple case of *pattern formation*, such as realized in nature by the formation, for example, of complex snow-flakes by water molecules which of course do not have the "blueprint" of the pattern they make.

The third point is discussed in detail the next two sections.

4   From the microscopic to the macroscopic levels

4.1 LEVELS OF DESCRIPTION AND CHARACTERISTIC TIMES

The ability of lattice-gas automata to achieve wave motion is actually a (linear) fallout of their ability to simulate fluids. The fact that flow phenomena lend themselves to approximation by very naive microscopic interactions is not too surprising, considering Feynman's remarks above. The microscopic details average out in the macroscopic limit; a liquid and a gas under similar symmetries and identical values of dimensionless numbers are described by the *same* Navier-Stokes equation. The standard macroscopic derivation of this equation (see e.g., Ref. [31]) takes as ingredients nothing more than conservation laws and reasonable smoothness assumptions. One can thus hope that a microscopic interaction with the correct symmetries is bound to yield in the right limit the Navier-Stokes equations. As we shall see in section 6, incorporating the correct symmetries is not a trivial matter.

In a 1986 column entitled "On two levels" and dedicated to lattice-gas automata, Leo Kadanoff writes[28]:

> I find it fantastic and beautiful that the tiny, trivial world of the lattice gas can give rise to the intricate structures of fluid flow. The physical universe is also wonderfully simple at some levels, but overpowerfully rich in others ...,
> The exact connection between the two levels will be worked out in the next few years.

The current wisdom on how the connection should be achieved is to follow the manner[32, 39,6,9] in which continuum kinetic theory (CKT) arrives at the hydrodynamics equations .

For convenience, Kadanoff's two levels will be refined here into four. Each level L is characterized by relevant physical quantities (RPQ(L)), ruled by a specific law of evolution:

LEVEL 0: Exact microscopic. The RPQ are interacting particles, ruled by the Newton's laws in CKT and by transition rules such as (1) in cellular automata.

LEVEL 1: Statistical microscopic. The RPQ are the N-particle distribution functions, ruled by Liouville's equations or the complete BBGKY hierarchy.

LEVEL 2: Kinetic. The RPQ are the 1-particle distribution functions, ruled by the Boltzmann equation.

LEVEL 3: Hydrodynamic. The RPQ are the hydrodynamic variables, ruled by the Euler and Navier-Stokes equations.

One could place the separation between the microworld and the macroworld (Kadanoff's two levels) between the LEVELS 0 and 1 above, since it is where the continuum is introduced, or, perhaps better, between LEVELS 1 and 2, since that is where the passage from reversibility to irreversibilty occurs. (Note, however, that it can be very advantageous to introduce irreversibility at the lowest level by defining stochastic microscopic cellular automata rules—see section 7.)

At the lowest level, the cellular automata approach is simpler because it is based on discrete space-time. Also, it involves simpler interactions (binary rules) at the microscopic level rather than the more complex intermolecular potentials. A chief motivation of this field is showing that one is not penalized by these simplifications; in other words, the complexity attained at the macroscopic level with cellular automata matches that of the continuum modeling.

The exact and statistical microscopic levels reflect the same physics but under different descriptions. Besides the general hypothesis of ergodicity, the passage to LEVEL 1 does not invoke any particular assumption. The passage to the kinetic and hydrodynamic levels, on the other hand, implies that specific physical conditions hold. Stated somewhat formally, one assumes that for L=1 and 2, the time dependence of the RPQ(L) becomes fully carried by the RPQ(L+1) in a time shorter than the period under which the RPQ(L+1) vary significantly. Informally, this statement reads: *Who carries the time dependence and how quickly?* It is further decyphered by noting that at each level, the time evolution of a gas is dominated by a characteristic relaxation time during which the RPQ can vary notably. A collision time τ_{collis} pertains to the microscopic LEVELS 0 and 1. In cellular automata, one can set τ_{collis} of the order of τ, the elementary time step. The kinetic level is characterized by a mean free path λ_{mfp} and a mean free time τ_{mfp}. A macroscopic scale λ_{hydro} and time τ_{hydro} describe phenomena at the hydrodynamic level. In the most favorable cases, these time scales are well separated:

$$\tau_{\text{collis}} \ll \tau_{\text{mfp}} \ll \tau_{\text{hydro}}. \tag{7}$$

The separation between these time scales quantitatively measures the physical meaningfulness of distinguishing between the levels. In particular, the first inequality (7) implies that the Liouville N-body distribution functions decorrelate (more precisely, have their time dependence carried by Boltzmann 1-particle functions) faster than local equilibria are reached[3]. The second inequality (7) holds in those physical situations where the Boltzmann 1-particle distribution functions become *normal solutions*, i.e., have their time dependence carried by the hydrodynamical variables faster than the latter undergo significant changes—there are cases where this hypothesis does not hold, typically in dilute plasmas, where τ_{mfp} can be very large.

We shall now review the construction of the hydrodynamic description out of the cellular automata microworld. Section 5 shows in more detail how this works in HPP and FHP.

LEVEL 0 → LEVEL 1. While LEVEL 1 is obtained from LEVEL 0 by formal averages in CKT, it is reached in cellular automata by averages over space—cf. Figure 4. Time averages may also be necessary to remove instabilities pointed out in Refs. [16,55].

LEVEL 1 → LEVEL 2. This passage can be implemeted in cellular automata, just as in CKT, by assuming $\tau_{collis} \ll \tau_{mfp}$ and also by truncating the BBGKY hierarchy, neglecting correlations (the *Stosszahlansatz*). These two assumptions are generally verified at low densities, they generate irreversibility and yield the Boltzmann equation.

LEVEL 2 → LEVEL 3. As in CKT, the Euler and the Navier-Stokes equations arise from the zeroth and the first order of a Chapman-Enskog expansion (or of a multi-scale expansion[17], which addresses explicitly the discreteness of space in cellular automata). Such expansions assume $\tau_{mfp} \ll \tau_{hydro}$.

4.2 ALTERNATIVE PATHS

Let us mention for completeness that lattice-gas automata have also borrowed other standard roads to the flow equations. The Boltzmann equation (LEVEL 2) can be by-passed altogether and the Navier-Stokes equation (LEVEL 3) can be obtained directly from LEVEL 1, via the Kubo-Green autocorrelation formalism; see Rivet's application to cellular automata[40]. This formalism yields the transport coefficients; it is in principle also valid at higher densities, where the first inequality (7) does not necessarily hold.

The kinematic viscosity ν, for example, is expressed as the time integral of the velocity autocorrelation function $F(t)$. A reasoning due to Alder and Wainwright[4] shows that $F(t)$ should decay (for t large) as:

$$F(t) \sim t^{-D/2}, \qquad D \geq 2. \tag{8}$$

where D is the dimension of space. Measurements[29] of ν show F's logarithmic divergence in 2D for the FHP gas, in agreement with (8). Direct measurements[34] of $F(t)$, however, suggest an anomalous tail t^{-2} for FHP (see also the contributions by Frenkel and Ladd to Ref. [2]).

Finally, the flow equations and the transport coefficients can be reached from an arguably even higher level:

LEVEL 4: Stochastic, ruled by Langevin's equation.

Section 8.1 of Ref. [17] applies to cellular automata the "noisy hydrodynamics" developed in the last chapter of Landau and Lifshitz's classic treatise[31].

5 From kinetic theory to hydrodynamics

A wealth of numerical simulations of lattice gases indeed document that *kinetic* regimes, solutions of a Boltzmann equation (section 5.1) set the stage for *hydrodynamical* regimes where *local equilibria* establish themselves (section 5.2). The requirement of consistency in what Frisch *et al.*[17] call "gluing" together these local equilibria is most useful: it imposes constraints on observables that are nothing but the equations of hydrodynamics (section 5.3).

5.1 THE BOLTZMANN EQUATION

Let us see how all of this works in the HPP and FHP cases. First, we average both sides of (1) (LEVEL $0 \rightarrow$ LEVEL 1), denoting by N_i the average $\langle n_i \rangle$ of n_i. The passage LEVEL $1 \rightarrow$ LEVEL 2 is implemented (in the HPP case) by replacing in the r.h.s. each average $\langle n_i n_j \bar{n}_k \bar{n}_l \rangle$ of a product by a product $N_i N_j (1 - N_k)(1 - N_l)$ of averages, hence truncating a cluster expansion to its lowest term and neglecting correlations.

It has recently been proposed to take N_i instead of the Boolean n_i as the primitive computational objects (see Ref. [33] and the contribution of Succi, Higuera, and coworkers to Ref. [2]). This is a blatant breach of the "fully discrete" party line, since the time evolution of N_i, a number between 0 and 1, requires the use of inexact floating-point arithmetic. Early results indicate that at low Reynolds numbers, this "lattice Boltzmann" method is more cost effective than the lattice-gas one, at least in a general-purpose computing environment.

Let us come back now to the orthodox cellular automata approach, where the only objects one evolves in time are the Boolean n_i themselves. Again, the introduction of macroscopic quantities and their laws of evolution is achieved here automatically by the eye (see section 3) or by an explicit, independent postprocessing treatment along the lines of (6). After propagation (of duration τ) along the lattice link (of size l) the l.h.s. of (2) reads

$$n_i(\mathbf{r} + l\mathbf{e}_i, t + \tau) - n_i(\mathbf{r}, t) \tag{9}$$

where $\mathbf{e}_i$ is the unit vector in direction i (in HPP, one sets $l = \tau = 1$).

We now define the density of matter $\rho(\mathbf{r}, t)$ and the velocity field $\mathbf{u}(\mathbf{r}, t)$ using

$$\rho(\mathbf{r}, t) = \sum_i N_i \tag{10}$$

and

$$\rho(\mathbf{r}, t)\mathbf{u}(\mathbf{r}, t) = \sum_i N_i \mathbf{c}_i, \tag{11}$$

where $\mathbf{c}_i = (l/\tau)\mathbf{e}_i$ is the velocity in direction i. Replacing in (9) n_i with N_i and the finite difference by a material derivative (i.e., using the chain rule), we obtain an equation of the Boltzmann form:

$$\frac{\partial N_i}{\partial t} + \mathbf{u} \cdot \nabla N_i = J_i(\{N_j\}), \tag{12}$$

where J_i is a collision operator, a sum of products of 1-particle functions N_j and their complements $(1 - N_l)$.

5.2 LOCAL EQUILIBRIA

In CKT, the complete single particle distribution $f_c(\mathbf{r}, \mathbf{v}, t)$ characterizes the state of a gas at all times. But after several molecular collisions have occurred, the state is adequately described by the much simpler *normal* solution $f(\mathbf{r}, \mathbf{v}, \rho(\mathbf{r}, t), u(\mathbf{r}, t), T(\mathbf{r}, t))$, where the time dependence is carried by the hydrodynamical variables ρ, $\mathbf{u}$, and the temperature T. This results from the Hilbert contraction theorem, whose constructive proof gives an explicit expansion for f

$$f = f^{(0)} + \theta f^{(1)} + \theta^2 f^{(2)} + \cdots, \tag{13}$$

thus providing the basis for the Chapman-Enskog method[10]. (This method extends Hilbert's approach by also expanding the time-derivative in the Boltzmann equation.) The expansion parameter $\theta = \lambda_{\mathrm{mfp}} \nabla$ involves the product of the mean-free path λ_{mfp} with the gradients of the distribution functions and of the hydrodynamical variables. It is of the order of magnitude of the Knudsen number $\mathrm{Kn} = \lambda_{\mathrm{mfp}}/\lambda_{\mathrm{hydro}}$.

The physical meaning of Hilbert's theorem (see [13] for a lucid discussion) is that around any point $\mathbf{r}$, some quantities remain invariant even though the velocity $\mathbf{v}$, and thus the complete $f_c(\mathbf{r}, \mathbf{v}, t)$, vary violently with each collision. As the state of *local equilibrium* is rapidly reached (as strongly suggested by the spatially inhomogeneous H–theorem), the gas at $\mathbf{r}$ can be characterized by the local values of the hydrodynamical variables and by their gradients. (The hydrodynamical variables are nothing but the densities of the invariant quantities). It may be argued that the relevance of these variables is gained by construction at the expense of that of the microscopic velocity, since they are defined as the first *moments* of f_c, a procedure that makes $\mathbf{v}$ a dummy variable.

The equilibrium distribution function $f^{(0)}$ (so denoted because it correspond to the zeroth order of the Chapman-Enskog expansion) is obtained in cellular automata as in CKT by requiring that the r.h.s. (the collision term) of the Boltzmann equation vanish. This constraint expresses the logarithm of $f^{(0)}$ as the linear combination of all the additive (or "summational") collision invariants, yielding in CKT the celebrated Maxwell-Boltzmann form:

$$f^{(0)}(v) \propto \rho \exp\left(-\frac{m}{2kT}(\mathbf{v} - \mathbf{u})^2\right). \tag{14}$$

In cellular automata on the other hand, it is a Fermi-Dirac functional form which is attained at equilibrium

$$N_i^{(0)} = 1/(1 + \exp \Sigma_i), \tag{15}$$

where

$$\Sigma_i = \alpha + \gamma \cdot \mathbf{c}_i, \tag{16}$$

just as the exponent in (14), is a linear combination of collision invariants (in multispeed models, such as illustrated in Figure 2 (cf. section 7), the third invariant—energy, labelled by T—is present). The Fermi-Dirac form is not a ghostly influence of quantum mechanics, it simply results from the Boolean exclusion principle. It occurs because in the relation

$$\log \hat{N}_i^{(0)} = \Sigma_i, \tag{17}$$

of the form occurring in the standard kinetic derivation of (14), the argument of the log is not directly the 1-particle function $N_i^{(0)}$ (as the $f^{(0)}$ of CKT), but the ratio

$$\hat{N}_i^{(0)} = N_i^{(0)}/(1 - N_i^{(0)}) \tag{18}$$

because the exclusion principle yields additional $(1 - N_i)$ factors in the collision term of (12).

Here we have just hit upon a major technical difference between cellular automata and CKT. The Gaussian form (14) makes the integral counterpart of (10) and (11) straightforward, and greatly facilitates the interpretation of the weights α and γ of (16) in terms of ρ and u. An additional advantage of (14) is the manifestly Galileo invariant dependence $(u - v)^2$, which makes life in CKT particularly easy. One starts with the global equilibrium by setting mean velocity u identically to zero in (14). Deviations from the global equilibrium, characterized by nonzero values of u, are readily recovered with local Galileo transforms. In contrast, the more complicated form (15) makes the interpretation of the weights quite cumbersome (especially in FHP, where there are six unknown functions $N_i^{(0)}$), and one must resort to an additional approximation to connect equilibria with vanishing and nonvanishing values of u. The most natural approximation is expanding $N_i^{(0)}$ around $u \equiv |u| = 0$. (This low Mach number limit confines the validity of the theory to the neighborhood of the incompressible regime.) One expands the weights $\alpha = \alpha(\rho, u)$ and $\gamma = \gamma(\rho, u)$ and identifies the Taylor coefficients power by power, using the conservation laws. For example, (10) implies $N_i(\rho, u = 0) = \rho/b$, denoting by b the number of bits per site (4 in HPP, 6 in FHP). This gives us the leading order. (In more complex model, this identification is not unique, a fact that we shall use to an advantage in section 7.) One obtains here, up to the second order

$$N_i^{(0)}(\rho, u) = \frac{\rho}{b} + \frac{2\rho}{b}c_i \cdot u + \frac{2\rho}{b}g(\rho)(2(c_i \cdot u)^2 - u^2), \tag{19}$$

where the presence of the function

$$g(\rho) = \frac{1}{2}\frac{2\rho - b}{\rho - b} \tag{20}$$

is a lattice artifact that breaks the Galilean invariance of the advection (of second order in u) at the macroscopic level, as we shall see below.

5.3 THE HYDRODYNAMIC EQUATIONS

Just as in CKT, one gets the equations of evolution by multiplying both sides of the Boltzmann equation (12) by the collision invariants and by summing over the four (or six) velocities c_i. Due to the conservation laws, the r.h.s. contributes to zero. Taking for collision invariant the number 1 and the velocities c_i, one gets respectively the continuity equation

$$\frac{\partial}{\partial t}\rho + \nabla \cdot \rho u = 0 \tag{21}$$

and the generic equation for the momentum density

$$\frac{\partial}{\partial t}\rho u + \nabla \cdot \Pi = 0, \tag{22}$$

where Π is the momentum flux tensor, defined by

$$\Pi_{\alpha\beta} = \sum_i c_{i\alpha}c_{i\beta}N_i. \tag{23}$$

As it stands, (22) is not very useful because via (23), it still involves N_i, and a bona fide flow equation must involve hydrodynamical variables ρ, $\mathbf{u}$ only. The microscopic N_i's must be replaced in (23) by their values in terms of these quantities, as derived by a Chapman-Enskog or a multiscale time expansion[17]. The zeroth and first order of such treatment yield respectively the Euler and the Navier-Stokes equations.

5.3.1. *Zeroth Order: the Euler Equation.* The zeroth order of the Chapman-Enskog expansion (and, for that matter, the zeroth order of any reasonable expansion) takes for $f^{(0)}$ the local equilibrium distribution of the last section. It consists in lattice gases of taking $N_i^{(0)}$ for N_i in (23). This implies physically that the local equilibria vary slowly and their space interdependence need not be included in the hydrodynamic description. Hilbert expansion (13) shows that the lowest order in the Knudsen number will yield an equation with only first order derivatives (from (22)), i.e., the Euler equation. This equation indeed assumes closeness to global equilibrium since it excludes stress and heat transport. When one plugs $N_i^{(0)}$ above in the definition (23) of the momentum-flux tensor where the sum is taken over the four HPP directions, the quadratic term in c_i of (19) yields a contribution

$$\sum(c_i \cdot u)^2 c_{i\alpha}c_{i\beta} \tag{24}$$

that still depends on the c_i's, and thus on the orientation of the lattice. More formally, the quadratic term in the subsonic expansion (in u) of (23)

$$\Pi_{\alpha\beta} = \frac{1}{2}\rho\delta_{\alpha\beta} + T_{\alpha\beta\gamma\delta}u_\gamma u_\delta + O(u^4) \tag{25}$$

is not isotropic, and the resulting HPP flow equation—from (22)—will not be isotropic either. Note, however, that since $\delta_{\alpha\beta}$ *is* isotropic, the linear part of the HPP flow equation, together with (21), yield an isotropic propagation of sound. This is why the wave in Figure 3 is rotationally invariant. The scalar term in the r.h.s. of (25) gives a speed of sound

$$c_s = \left(\frac{\partial P}{\partial \rho}\right)^{\frac{1}{2}} = 1/\sqrt{2} \tag{26}$$

in units of l/τ. The reader is invited to verify this directly on Figure 3 (given that the snapshots were taken at $t = 0$, 30, and 90, and that the square frame is of size 256×256).

The lack of isotropy of the nonlinear terms casts a cloud on the ambition to simulate a variety of actual flow processes with cellular automata. This is where FHP's small modification plays a huge role. The magic of the hexagonal neighborhood is that the culprit term (24), when the sum applies now to the six directions of the lattice, yields a quantity $\frac{3}{4}u^2\delta_{\alpha\beta} + \frac{3}{2}u_\alpha u_\beta$ which is independent of the c_i's, allowing for genuine isotropy. Section 6 shows how this magic happens.

Thanks to isotropy, the generic equation (22) is in the FHP geometry {almost} the Euler equation, up to terms in $O(u^4)$. The almost aspect is indicated below by the terms in braces; without them, one would have a bona fide Euler equation for an ideal gas:

$$\frac{\partial}{\partial t}(\rho u_\alpha) + \frac{\partial}{\partial x_\beta}((\rho u_\alpha u_\beta)\{g(\rho)\}) = -\frac{\partial}{\partial x_\alpha}P \tag{27}$$

where the pressure P is given by the equation of state:

$$P = \frac{1}{2}\rho(1 - \{g(\rho)u^2\}). \tag{28}$$

The function $g(\rho)$ in front of the advective term breaks the Galilean invariance. To see this formally, just add a constant velocity V to u_α and u_β in (27)); in more physical terms, (27) advects vorticity with velocity $g(\rho)u$ instead of u. Fortunately, there are ways to force $g(\rho) \equiv 1$; see section 7.

5.3.2. First Order: the Navier-Stokes Equation.

As the hydrodynamical variables vary with r, a more refined description will also involve their gradients, in the spirit of (13). Including the first order in the Knudsen number permits a description further removed from global equilibrium, where heat flow and stress may occur. This term brings off-diagonal, second-order derivatives contributions to the pressure, weighted by shear (ν) and bulk (ν') viscosities. Adding to the r.h.s. of the Euler equation the terms

$$\rho\nu\nabla^2 u_\alpha + \rho\nu'\frac{\partial}{\partial x_\alpha}\nabla \cdot \mathbf{u} \tag{29}$$

gives the Navier-Stokes equation within the same approximations as in (27). The physical consistency condition in gluing neighboring local equilibria translates mathematically into a solubility constraint in the equation for $N^{(1)}$. This constraint turns out to be the Navier-Stokes equation.

It was found that the lattice induces a spurious viscosity[40,23]. Levermore traced this discretization artifact to the second order Taylor term in the expansion of the finite difference (9) (see e.g., Ref. [54], and also Ref. [5], which contains a suggestive analysis showing that the Chapman-Enskog theory on a lattice has the flavor of convergence proofs of finite-difference schemes). Consistency requires that the second Taylor term be taken into account to modify the coefficients of the second-order space derivatives (29), namely, the viscosities. This correction contributes to a *negative* viscosity; it has been called "propagation viscosity," since it exists even if no collision occurs. Using the H-theorem, Hénon proved that for rules obeying "semi-detailed balance" (see e.g., Ref. [17]), this negative contribution is always smaller, in absolute value, than the ordinary positive one[23]. But Rothman has shown that negative total viscosity can occur in a modification of the FHP rule that violates semi-detailed balance and uses *irreversible* collisions sensitive to momentum gradients[43]. Appropriate admixtures of Rothman's modification to the standard FHP rule could in principle result in arbitrary small positive viscosity and open the possibility of simulations of flows with very high values of the Reynolds numbers Re=$u\lambda_{\text{hydro}}/\nu$.

Simulations of the FHP (and of closely related) models have successfully reproduced, qualitatively at least, the whole menagerie of the classic hydrodynamic instabilities (Kelvin-Helmholtz, Rayleigh-Taylor, Rayleigh-Bénard), see Refs. [1,7,21,37,2].

The insensitivity to the details of the microscopic interactions and the scaling properties are reminiscent of the situation in statistical mechanics where universality allows one to focus one's efforts on the simplest model representative of an entire universality class. Similarity in lattice-gas hydrodynamics, much as universality in statistical mechanics, comes to the rescue of naive mechanistic modeling.

6   Symmetries

The previous remark points to the need to make certain that the schematic model with which one works possesses the correct symmetries. The invariance by translation is no major problem: for all lattices, the limit of vanishing mesh sizes eventually "fills the gaps" between the lattice points. The invariance by rotation is different; the above limit does not necessarily erase the lattice's orientation. In more technical terms, the rotation group is compact (the macroscopic limit does not involve large angles).

For the theory to be isotropic, the momentum flux tensor must reduce to a scalar. The sound wave in Figure 3 is rotationally invariant, but later configurations would show that the damping of the wave (the ring does not only expand, it also diffuses) will not be the same along the lattice axes and along 45° lines[56,33], because the viscosity is not isotropic. Damping, i.e., the nonlinear contribution of (25), will eventually betray the existence of the underlying lattice, and, more to the point, its orientation.

The observable anisotropy stems from properties of the fourth order tensor $T_{\alpha\beta\gamma\delta}$, the coefficient of the quadratic term of (25). The fact that isotropy can be saved follows from a feature of the triangular lattice and an elementary theorem of tensor calculus. The theorem states that fourth order tensors that are both isotropic and symmetric in its two first (or two last) indices must take the form

$$T_{\alpha\beta\gamma\delta} = 2\lambda\delta_{\alpha\beta} + \mu(\delta_{\alpha\beta}\delta_{\gamma\delta} + \delta_{\alpha\delta}\delta_{\beta\gamma}) \tag{30}$$

where λ and μ are nothing but the good old Lamé moduli of elasticity. The meaning of these coefficients is not as important here as the fact that there are only TWO of them. We are thus looking for a lattice that will reduce the D^4 components $T_{\alpha\beta\gamma\delta}$ in (30) to only two independent moduli. This requirement holds in any number of dimensions D.

In the hexagonal (FHP) symmetry, a rotation of $\pi/3$ changes the coordinates $\xi = x + iy$ and $\eta = x - iy$ into $\xi\exp(i\pi/3)$ and $\eta\exp(-i\pi/3)$. To remain unchanged under such a rotation, the moduli must be zero or possess a η for each ξ index (remember that $T_{\alpha\beta\gamma\delta}$ varies as the product $r_\alpha r_\beta r_\gamma r_\delta$ of coordinates of a vector, the position $\mathbf{r}$ for example). The only nonvanishing moduli are then: $T_{\xi\eta\xi\eta}$ and $T_{\xi\xi\eta\eta}$.

In the square (HPP) lattice, the rotations are of $\pi/2$. This introduces two new moduli: $T_{\xi\xi\xi\xi}$ and $T_{\eta\eta\eta\eta}$, giving altogether (coming back to the x, y coordinates)

$$T_{xxxx}, \; T_{xxyy}, \; T_{xyxy}. \tag{31}$$

We are thus eventually left with THREE independent components, one too many.

The two vs. three components argument shows in the most elementary way why the hexagonal neighborhood wins and the square one loses in their attempts to achieve isotropy in 2D, more specifically why one can detect the four basic directions of HPP but no evidence that the FHP model has only six directions. A very simple derivation of (30) can be found in the delightful 1931 monograph by Jeffreys[27], and the crystallographic argument is a 2D projection of a discussion given in Landau and Lifshitz's treatise on elasticity[30]. More general and elegant derivations are given in [17] and in [54], which gives a detailed group-theoretical treatment.

We found that the square lattice cannot reproduce isotropy. We could be legitimately surprised that the FHP hexagonal lattice, or any lattice, can achieve isotropy at all. The argument shown above sheds light on the fact that isotropy is obviously achieved only within the limits of the Navier-Stokes equation, that is, only insofar as one is not concerned with tensors of degree higher than four. One could easily imagine an experiment, involving for example polarized light, that would implicate higher rank tensors and thus reveal anisotropy of the FHP lattice. Note also that isotropy too confines the theory to the neighborhood of the incompressible regime (low Mach numbers). Higher values of u would require higher tensors in the subsonic expansion (25), eventually breaking the isotropy of *any* regular lattice.

The FHP model cures another disease of the HPP model: along a given line (or column), HPP collisions add or remove particles by pairs only. As a result, each line and column has its own private conserved momentum. Lack of lateral momentum transfer in effect makes the square HPP model one-dimensional, as verified by measuring the velocity autocorrelation function[34]. The latter indeed decays as $t^{-2/3}$, in agreement with Pomeau and Résibois's derivation[38] which extends (8) to $D=1$.

These spurious invariants are a serious flaw in the theory because they partition the energy surface. The system will be stuck in a small region of that surface, which must be metrically transitive for the ergodic theorem to hold. In other words, the randomization to equilibrium that we assumed in section 5 can be greatly inhibited by the existence of the spurious invariants. The problem is solved by the introduction of a 3-body collision rule in the FHP model, whereby three particles reaching symmetrically (i.e., with mutual 120° angles) a given site will bounce back, thus breaking the lines private momentum conservation.

When everyone believed that all was safe and that the FHP model was free of spurious invariants, McNamara and Zanetti discovered a family of them in this model (see d'Humières's contribution to Ref. ([37]). The effect of these invariants on the validity of FHP hydrodynamics is still under debate.

The triangular lattice achieves isotropy in 2D. But it turns out (by inspecting, e.g., section 10 of Ref. [30]) that none of the fourteen 3D Bravais lattices can produce pairwise symmetric fourth order tensor with only two independent moduli. However, d'Humières and coworkers have found that in 4D, the face-centered hypercubic lattice (FCHC) has the desired property, and that an appropriate projection into 3D can host an isotropic gas[24]. Rivet performed impressive simulations of this gas on a CRAY-2; see his contribution to Ref. [37].

7   Removing discreteness artifacts

This section presents efforts to eliminate the lattice artifacts indicated with braces in (27) and (28). First note that particles in the HPP and FHP all move at the same speed; this makes energy conservation equivalent to that of mass. Fluids in single speed models live at fixed temperature, they have equal specific heats, $C_P = C_V$, and equal compressibilities $\kappa_S = \kappa_T$. True thermal effects can be readily introduced by allowing particles with various speeds, thus lifting the mass-energy degeneracy. This can be done by introducing stopped particles[26] (with enegy 0), and by enlarging the elementary HPP neighborhood so as to include next nearest neighbors and particles with energy 2 (in units of $\frac{1}{2}(l/\tau)^2$), moving along diagonal links[25,24,11]. Figure 2 shows two particles of energy 1 interact and create a rest particle and a fast particle of energy 2. This process has an inverse where a fast particle hits a rest particle and decays. Weighting the direct and inverse processes with different probabilistic rates[7] achieves $g(\rho) \to 1$ for $\rho \to 0$ (instead of the unphysical $g(\rho) \to \frac{1}{2}$ implied by (20)), and thus restoring Galilean invariance in the limit of vanishing density. In the same vein, simulations have been performed[45] with values ρ^* such that $g(\rho^*) = 1$.

It has been recently shown that using a 3D model derived from a FCHC neighborhood enlarged to include energy 2 particles, one can force $g(\rho) \equiv 1$ and at the same time remove the spurious terms in ρu^2 from the definition (28) of the static pressure[35,36]. In a nutshell, the strategy relies on integrating, in a way that achieves both isotropy and Galilean invariance, the following three ingredients:

(i) particles with several velocities, allowing for true thermal effects,

(ii) different rates for direct and inverse processes,

(iii) use of underdeterminedness of the subsonic expansion for the energy.

With energy as a new collision invariant, (16) reads now

$$\Sigma_i = \alpha + \beta \epsilon_i + \boldsymbol{\gamma} \cdot \mathbf{c}_i, \tag{32}$$

where $\epsilon_i = 0, 1, 2$ are the particle energies, and we have a total energy

$$U = U_{\text{int}} + \frac{1}{2}\rho u^2 \tag{33}$$

which is the sum of an internal energy and a hydrodynamic flow energy. Equations (21) and (22) are augmented with an equation for the density of energy

$$\frac{\partial U}{\partial t} + \nabla \cdot Q = 0, \tag{34}$$

where $Q = \sum_i \mathbf{c}_i \epsilon_i N_i$ is the full energy flux. The g function (20) now depends on the energy through

$$g(\rho, U_{\text{int}}) = \frac{D}{D+2} \frac{\rho \langle \epsilon^2 \rangle}{U_{\text{int}}^2}, \tag{35}$$

where the term in brackets is the moment $\langle \epsilon^2 \rangle = \sum_i N_i \epsilon_i^2$.

The underdeterminedness of the subsonic expansion (point iii above) means here that the normalizations (21), (22), and (34) do not constrain all the Taylor coefficients in the expansion around $u = 0$ of $\alpha(u), \beta(u)$, and $\gamma(u)$, or equivalently of $\rho(u)$ and $U(u)$. Specifically, some freedom is left in distributing the first two Taylor terms of $U = U_0 + U_2 u^2$ over the two contributions of (33). This freedom, together with a proper choice for the relative rates of creation a decay of fast particles (point ii above) allows us to adjust the moment

$\langle\epsilon^2\rangle$ so as to force the g function (35) identically to 1 and at the same time cancel the spurious term in ρu^2 in the static pressure, which reads now

$$P = \frac{2}{D}U_{\text{int}} = \rho T_{\text{k}} \tag{36}$$

with the equipartition $U_{\text{int}} = \frac{D}{2}\rho T_{\text{k}}$. These equations define a kinetic temperature T_{k}, which in general differs from the thermodynamic definition

$$T_{\text{th}} = 1/\beta = \left(\frac{\partial S}{\partial U_{\text{int}}}\right) \tag{37}$$

where β is the same as in (32) and S is the standard statistical entropy for fermions[35].

It turns out that unlike the equation for the density of momentum, (34) obeys Galilean invariance up to first order in ρ only. Full invariance can be restored by adjusting the higher moment $\langle\epsilon^3\rangle$. This can be done by adding new particles types, or even "photons" (carrying momentum but no mass) with tunable decay rates[35].

The inclusion of the energy degree of freedom in the subsonic expansion, together with different rates for direct and inverse processes, yield realistic Euler dynamics and thermodynamics. The Navier-Stokes level of this model and the form of the transport coefficients are currently under study.

I am thankful to B. Chopard, E. Eisenberg, K. Molvig, and D. Rothman for enlightening discussions. This work was supported by a grant from AFOSR, No. 89–0119.

References

[1] *Complex Systems* 1 (1987) No. 4 pp. 545–851, (Proc. Workshop on Large Nonlinear Systems, Santa Fe, Oct. 1986).

[2] *Cellular Automata and the Modeling of Complex Physical Systems*, Proc. of the Feb.– Mar. 1989 workshop at les Houches, France. (Springer-Verlag, 1989).

[3] A. Akhiezer and S. Peletminski, *Methods of Statistical Physics* (Pergamon, 1981).

[4] B. Alder and T. Wainwright, *J. Phys. Soc. Jpn.* 26 (1968), and *Phys. Rev. A* 1 (1970) 18.

[5] B. M. Boghosian and C. David Levermore, in Ref.[37].

[6] J. P. Boon and S. Yip, *Molecular Hydrodynamics* (McGraw Hill, 1980).

[7] C. Burges and S. Zaleski, *Complex Systems* 1 (1987) 31.

[8] A. W. Burks, *Essays on Cellular Automata* (University of Illinois Press, 1970).

[9] C. Cercignani, *The Boltzmann Equation and its Applications* (Springer-Verlag, 1988).

[10] S. Chapman and T. G. Cowling, *The Mathematical Theory of Non-Uniform Gases* (Cambridge University Press, 1939).

[11] B. Chopard and M. Droz, *Phys. Lett.* A126 (1988) 476.

[12] P. Clavin, P. Lallemand, Y. Pomeau, and G. Searby *J. Fluid Mech.* 188 (1988) 437.

[13] E. G. D. Cohen, in *Transport Phenomena in Fluids*, H. J. M. Hanley, ed, (Marcel Dekker, 1969).

[14] D. Farmer, T. Toffoli, and S. Wolfram, eds. *Cellular Automata* (North-Holland, 1984).

[15] W. D. Hillis, *Physics Today*, Feb. 1989, p. 78.

[16] U. Frisch, B. Hasslacher, and Y. Pomeau, *Phys. Rev. Lett.* 56 (1986) 1505.

[17] U. Frisch, D. d'Humières, B. Hasslacher, P. Lallemand, Y. Pomeau and J. P. Rivet, *Complex Systems* 1 (1987) 649.

[18] H. M. Goldstine and J. von Neumann, *unpublished* (1946), reprinted in *J. von Neumann's Complete Works*, Vol. 5 (Pergamon, 1961), pp. 1–32.

[19] J. Hardy, Y. Pomeau, and O. de Pazzis, *J. Math. Phys.* 14 (1973) 1746.

[20] J. Hardy, O. de Pazzis, and Y. Pomeau, *Phys. Rev.* A13 (1976) 1949.

[21] B. Hasslacher, *Los Alamos Science*, Special Issue, no. 15, 1987.

[22] B. Hayes, *Scientific American*, 250:3 (March 1984) 12.

[23] M. Hénon, *Complex Systems*, 1 (1987) 475.

[24] D. d'Humières, P. Lallemand, and U. Frisch, *Europhys. Lett.*, 2 (1986) 291.

[25] D. d'Humières and P. Lallemand, *Helvetica Physica Acta* 59 (1986) 1231.

[26] D. d'Humières and P. Lallemand, *Complex Systems*, 1 (1987) 599.

[27] H. Jeffreys, *Cartesian Tensors* (Cambridge University Press, 1931).

[28] L.P. Kadanoff, *Physics Today*, Sept. 1986 p. 7.

[29] L.P. Kadanoff, G. R. McNamara and G. Zanetti, "From Automata to Fluid Flow: Comparisons of Simulation and Theory," *U. of Chicago preprint*, 1987.

[30] L.D. Landau and E. M. Lifshitz, *Elasticity* (Pergamon Press, 1959).

[31] L.D. Landau and E. M. Lifshitz, *Fluid Mechanics* (Pergamon Press, 1959).

[32] R. L. Liboff, *Introduction to The Theory of Kinetic Equations* (Wiley, 1969).

[33] G. R. McNamara and G. Zanetti, *Phys. Rev. Lett.* 61 (1988) 2332.

[34] N. Margolus, T. Toffoli, and G. Vichniac, *Phys. Rev. Lett.* 56 (1986) 1694.

[35] K. Molvig, P. Donis, J. Myczkowski and G. Vichniac, *Multi-Species Lattice Gas Hydrodynamics*, Plasma Fusion Center Report PFC/JA-88-23, M.I.T. (1988).

[36] K. Molvig, P. Donis, J. Myczkowski and G. Vichniac, in Ref. [37]

[37] R. Monaco, ed, *Discrete Kinetic Theory, Lattice Gas Dynamics and the Foundations of Hydrodynamics* (World Scientific, 1989).

[38] Y. Pomeau and P. Résibois, *Phys. Rep.* 19 (1975) 63.

[39] P. Résibois and M. DeLeener, *Classical Kinetic Theory of Fluids* (Wiley, 1977).

[40] J.-P. Rivet, *Complex Systems* 1 (1987) 838.

[41] D. H. Rothman, *Geophysics*, 53 (1988) 509.

[42] D. H. Rothman and J. M. Keller, *J. Stat. Phys.*, 45 (1988) 1119.

[43] D. H. Rothman, "Negative-viscosity lattice gases," *MIT-DEAPS preprint,* 1989, see also Ref. [2].

[44] A. Rucklidge and S. Zaleski, *J. Stat. Phys.*, 51 (1988) 299.

[45] D. d'Humières, P. Lallemand, and G. Searby, *Complex Systems* 1 (1987) 632.

[46] T. Toffoli, *Physica* 10D (1984) 117, reprinted in Ref. [14].

[47] T. Toffoli, *Physica* 10D (1984) 195, reprinted in Ref. [14].

[48] T. Toffoli and N. Margolus, *Cellular Automata Machines: A New Environment for Modeling* (MIT Press, 1987).

[49] G. Y. Vichniac, *Physica* 10D (1984) 96, reprinted in Ref. [14].

[50] G. Y. Vichniac, in *Disordered Systems and Biological Organization*, E. Bienenstock, F. Fogelman, and G. Weisbuch, eds., les Houches Winter School 1985, Proceedings (Springer-Verlag, 1986).

[51] G. Y. Vichniac, in *Chaos and Complexity*, R. Livi, S. Ruffo, S. Ciliberto. and M. Buatti, eds. (World Scientific, 1988).

[52] J. von Neumann, *Theory of Self-Reproducing Automata* (edited and completed by A. W. Burks), (University of Illinois Press, 1966).

[53] S. Wolfram, ed. *Theory and Applications of Cellular Automata* (World Scientific, 1986).

[54] S. Wolfram, *J. Stat. Phys.* 45 (1986) 471.

[55] V. Yakhot and S.Orszag, *Phys. Rev. Lett.* 56 (1986) 169.

[56] V. Yakhot, B. Bayly, and S.Orszag, *Phys. Fluids* 29 (1986) 2025.

[57] S. Zaleski, in Ref. [37].

CYLINDER DISTRIBUTION OF THERMODYNAMIC LIMIT ON BETHE LATTICES

S. Martínez
Departamento de Matemáticas Aplicadas
Facultad de Ciencias Físicas y Matemáticas
Universidad de Chile
Casilla 170-3, Correo 3, Santiago
Chile

ABSTRACT. We study the thermodynamic limit τ on Bethe lattices when $L_n \to L$, for any finite spin set S. We show the cylinder values $\tau\{\sigma \in \Omega: \sigma_{L_k} = r|k|\}$ are determined by vectors $\overline{\psi}$ which verify a period 2-equation: $\overline{\psi} = T*^2\overline{\psi}$. In the case q = 2 i.e. $L = \mathbb{Z}$ we arrive to a Perron – Frobenius equation, then to a unique distribution τ.

1. INTRODUCTION

Consider the Bethe lattice $L = \lim_{n \to \infty} L_n$ where the L_n are the set of points of L at distance $\leq n$ of the central point 0 (see below figure 1). For a finite spin set S we study the thermodynamic limits $\{\tau\}$ of the Gibbs ensembles ν_n defined in L_n^S.

We first show $\nu_n(r|k|)$ can be written as (see (3.33)):

$$\nu_n(r|k|) = \prod_{\ell=0}^{k} \prod_{(i_1 \ldots i_\ell) \in L_\ell - L_{\ell-1}} \alpha_{n-\ell}^{r(i_1 \ldots i_{\ell-1})}(r(i_1, \ldots, i_\ell))$$

where the vector $\alpha_{n-\ell}^{r(i_1 \ldots i_{\ell-1})}(\cdot)$, for $\ell \geq 1$, corresponds to one step transition probabilities, and the vector $\alpha_n^{r(i_1 \ldots i_{\ell-1})}$ for $\ell = 0$ (also noted α_n^{ϕ}) evaluates the one-site distribution. After we find an homeomorphism δ which changes the vector $\alpha_n(r|k|) = (\alpha_{n-\ell}^{r(i_1 \ldots i_{\ell-1})}(\cdot): \ell = 0 \ldots k)$ into a new vector $\psi_n(r|k|) = \delta(\alpha_n(r|k|))$, in such a way that the limit points $\overline{\psi}(r|k|)$ of the sequence $(\psi_n(r|k|))$ can be identified. We prove they verify the period 2 equation (see (3.37)):

117

E. Tirapegui and D. Villarroel (eds.), Instabilities and Nonequilibrium Structures II, 117–130.
© 1989 by Kluwer Academic Publishers.

$$\overline{\psi}(r|k|) = T*^2\overline{\psi}(r|k|)$$

for a well defined transformation T* (see (3.38)). Then we deduce
our main result, that the thermodynamic limit τ verifies:

$$\tau(r|k|) = \prod_{\ell=0}^{k} \prod_{(i_1\ldots i_\ell)\in L_\ell - L_{\ell-1}} \overline{\alpha}^{r(i_1\ldots i_{\ell-1})}(r(i_1\ldots i_\ell))$$

with $\overline{\alpha}(r|k|) = (\alpha^{r(i_1\ldots i_{\ell-1})} : \ell = 0\ldots k) = \delta^{-1}(\overline{\psi}(r|k|))$, where $\overline{\psi}(r|k|)$
verify the above period 2 equation.

When $q = 2$ i.e. $L = \mathbb{Z}$ the vectors $\alpha_n^{r(i_1\ldots i_{\ell-1})}$ do not depend on
n for $\ell \geq 1$, then the only dependence on n is found for $\ell = 0$, i.e. for
α_n^ϕ. Its transformed limit points $\overline{\psi}^\phi = \delta(\overline{\alpha}^\phi)$ verify a Perron
Frobenious equation: $\widehat{A\overline{\psi}^\phi} = \lambda\overline{\psi}^\phi$. So we get a unique thermodynamic limit
with cylinder values completely defined.   Deeper results concerning
the unicity of Gibbs states on Bethe lattice have been established in
|3| using entropy arguments. The $\mathbb{Z}$ case has been studied in |2| where
the Gibbs state is shown to verify a Perron-Frobenius equation by using
the transfer matrix technique. The limit set of the one-site
distribution sequence (α_n^ϕ) has been studied by the same techniques of
this work in |4,5|.

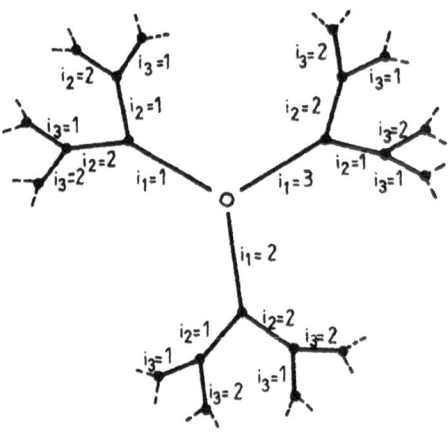

Figure 1. Bethe Lattice for $q = 3$.

2. THE BETTE LATTICE

The Bethe lattice of coordination number q is formed by the points
$i = (i_n)_{n \geq 1}$ such that $i_1 \in Q = \{0,1,\ldots,q\}$, $i_n \in Q = \{0,1,\ldots,q-1\}$ for
$n \geq 2$, $i_m = 0$ implies $i_n = 0$ for all $n \geq m$, and there exists some $m \geq 1$
such that $i_m = 0$ (see figure 1).

For any $i \in L$ we note $m(i) = \inf\{m \geq 1: i_m = 0\}$, we write
$0 = (0,0,\ldots)$ and for any $i \neq 0$ we identify i with $(i_1,\ldots,i_{m(i)})$.
Define $L_n = \{i \in L: m(i) \leq n\}$ and $L^*(i_1) = \{j \in L_n - \{0\}: j_1 = i_1\}$.

We write $i \to i'$ iff $m(i') = m(i) + 1$ and $i_n = i'_n$ for $n = 1.,,,n(i)$.
Two points $i,i' \in L$ are called neighbors iff $i \to i'$ or $i' \to i$, in this
case we write $(i,i') \in V$.

All the disjoint graphs $(L_n(i_1),V \cap (L_n(i_1) \times L_n(i_1)))_{i_1 \in Q_1}$ are

equivalent (the one-to-one correspondence $(i_1,i_2,\ldots,i_k) \to (i'_1,i_2,\ldots,i_k)$
preserves the neighbor relation), then we identify them and note
$L^*_n = L_n(i_1)$ with the common neighbor relation. With this notation we
have for any $0 < k < n$: $L_n - L_k = \bigcup_{i \in L_{k+1} - L_k} L^*_{n-k}(i)$, where the $L^*_{n-k}(i)$
of the union are mutually disjoint copies of L^*_{n-k}.

Let $S \subset \mathbb{R}^d$ be a finite set of spins endowed with the inner
product $\langle , \rangle$. Write $\Omega_n = L^S_n$, $\Omega^*_n(i) = L^*_n(i)^S$, $\Omega^*_n = L^*_n{}^S$.

For the configuration $\sigma \in \Omega_n$ its hamiltonian is (see ref $|1|$):

$$H_n(\sigma) = -\{K \sum_{(i,j) \in V \cap L_n \times L_n} \langle \sigma(i),\sigma(j) \rangle + \sum_{i \in L_n} \langle N,\sigma(i) \rangle\} \qquad (2.1)$$

where K is the interaction and N the external magnetic field.

Let us note by σ_A the restriction of the configuration σ to the
set A. For $\sigma \in \Omega_n$ note $\sigma^{(i_1)} = \sigma_{L^*_n(i_1)} \in \Omega^*_n(i_1)$ the restriction of σ
to $L^*_n(i_1)$. When we develop the Hamiltonian we find:

$$-H_n(\sigma) = \langle N,\sigma(0) \rangle + \sum_{i_1 \in Q_1} W_n(\sigma(0),\sigma^{(i_1)}) \qquad (2.2)$$

where the function W_n is defined by induction:

$$W_n(r,\sigma') = \langle Kr + N, \sigma'(j_1) \rangle + \sum_{j_2 \in Q} W_{n-1}(\sigma'(j_1), \sigma'^{(j_1 j_2)})$$

$$\text{(2.3)}$$

for $\sigma' \in \Omega_n^*(j_1)$

being $\sigma'^{(j_1 j_2)} = \sigma'_{L_{n-1}^*}(j_1 j_2)$.

If we put $H_k(\sigma) = H_n(\sigma_{L_k})$ for $k > 0$, we get:

$$-H_n(\sigma) = -H_k(\sigma) +$$

$$\text{(2.4)}$$

$$+ \sum_{i_1 \in Q_1} \sum_{i_2 \in Q} \cdots \sum_{i_{k+1} \in Q} W_{n-k}(\sigma(i_1 \ldots i_k), \sigma^{(i_1 \ldots i_{k+1})})$$

where $\sigma^{(i_1 \ldots i_{k+1})} = \sigma_{L_{n-k}^*}(i_1 \ldots i_{k+1})$. Define:

$$g_n(r) = \sum_{\sigma \in \Omega_n^*} \exp(W_n(r,\sigma)) \qquad \text{(2.5)}$$

It is easy to show that it verifies the recursive equation:

$$g_n(r) = \sum_{s \in S} (g_{n-1}(s))^{q-1} \exp(\langle N + Kr, s \rangle) \qquad \text{(2.6)}$$

We can express the partition function $Z_n = \sum_{\sigma \in \Omega_n} \exp(-H_n(\sigma))$ (the factor $\beta = \dfrac{1}{kT}$ has been included in the Hamiltonian) in terms of $g_n(r)$. In fact it can be shown that:

$$Z_n = \sum_{r \in S} (g_n(r))^q \exp(\langle N, r \rangle) \qquad \text{(2.7)}$$

The above expressions (1.2)-(1.6) are found by using the recursive method of reference |1|, which is similar to the transfer matrix method used in |2| to study the case $L = \mathbb{Z}$, i.e. $q = 2$.

Let n be fixed. The Gibbs ensemble v_n is the following probability measure on Ω_n:

$$v_n(\sigma) = Z_n^{-1} \exp(-H_n(\sigma)) \text{ for any } \sigma \in \Omega_n. \qquad \text{(2.8)}$$

We shall study the thermodynamic limit of the Gibbs ensemble ν_n, when $L_n \to L$, $n \to \infty$. First of all there exists a subsequence $n' \to \infty$ such that the following limits exist:

$$\lim_{n' \to \infty} \theta_{n,n'} \cdot \nu_{n'} = \tau_n \text{ for all } n \geq 1$$

(2.9)

where $\theta_{n,n'} : \Omega_{n'} \to \Omega_n$ is the coordinate projection

For this family (τ_n) there exists a unique probability measure τ defined on the configuration space $\Omega = L^S$ such that:

$$\theta_{n,\infty} \tau = \tau_n \text{ for all } n \geq 1$$

(2.10)

where $\theta_{n,\infty} : \Omega \to \Omega_n$ is the coordinate projection

The measure τ is a thermodynamic limit.

Recall τ is entirely determined if we know the values it takes on the finite cylinders. Take $r|k| \in \Omega_k$, from the definitions we have:

$$\tau\{\sigma \in \Omega: \sigma_{L_k} = r|k|\} = \lim_{n' \to \infty} \nu_{n'}\{\sigma \in \Omega: \sigma_{L_k} = r|k|\}$$

(2.11)

Then to determine the values $\tau(r|k|)$ we are lead to study the set of limit points of the sequence $(\nu_n(r|k|))$. In fact if we find that the set of limit points is a singleton, which will happen when $q = 2$ ($L = \mathbb{Z}$), then the set of thermodynamic limits $\{\tau\}$ (of $L_n \to L$) will also be a singleton. In section 3 we shall find the equations that verify the limit points of $(\nu_n(r|k|))$, then these equations must necessarily be verified by $\tau(r|k|)$.

3. LIMIT EQUATIONS

Now let ν_n be the Gibbs ensemble on Ω_n. The one-site Gibbs state on $i = 0$ is given by:

$$\xi_n^\circ(r) = \nu_n\{\sigma \in \Omega_n : \sigma(0) = r\}$$

(3.1)

From the definitions we find:

$$\xi_n^\circ(r) = z_n^{-1}(g_n(r))^q \exp(<N,r>)$$

(3.2)

Let us evaluate the distribution ν_n on the cylinders $r|k| \in \Omega_k$ for $k \leq n$. We have:

$$\xi_n^{L_k}(r|k|) = \nu_n\{\sigma \in \Omega_n : \sigma(i) = r(i) \text{ for } i \in L_k\} \qquad (3.3)$$

By definition:

$$\xi_n^{L_k}(r) = Z_n^{-1} \sum_{\{\sigma \in \Omega_n : \sigma(i) = r(i), i \in L_k\}} \exp(-H_n(\sigma)).$$

It is not hard to show that:

$$\xi_n^{L_k}(r|k|) = \qquad (3.4)$$

$$Z_n^{-1} \exp(-H_k(r|k|)) \prod_{(i_1 \ldots i_k) \in L_k - L_{k-1}} (g_{n-k}(r(i_1 \ldots i_k)))^{q-1}$$

Now suppose $k \geq 2$, so $k - 1 \geq 1$. We note $r|k-1| \in \Omega_{k-1}$ the restriction of $r|k|$ to L_{k-1}. Then:

$$\xi_n^{L_{k-1}}(r|k-1|) = Z_n^{-1} \exp(-H_{k-1}(r|k-1|)) \cdot \qquad (3.5)$$

$$\prod_{(i_1 \ldots i_{k-1}) \in L_{k-1} - L_{k-2}} (g_{n-k+1}(r(i_1 \ldots i_{k-1})))^{q-1}$$

From formula (2.6) we get:

$$\xi_n^{L_{k-1}}(r|k-1|) = Z_n^{-1} \exp(-H_{k-1}(r|k-1|)) \qquad (3.6)$$

$$\prod_{(i_1 \ldots i_k) \in L_k - L_{k-1}} \{\sum_{s \in S} g_{n-k}(s)^{q-1} \exp(<N + Kr(i_1 \ldots i_{k-1}), s>)\}$$

Then the ν_n - conditional probability that $r|k|$ happens if we know that $\sigma_{L_{k-1}} = r|k-1|$ has occurred is:

$$\nu_n\{\sigma_{L_k} = r|k| \ | \ \sigma_{L_{k-1}} = r|k-1|\} =$$

$$(3.7)$$

$$\prod_{(i_1\ldots i_k)\epsilon L_k - L_{k-1}} \{\frac{g_{n-k}(r(i_1..i_k))^{q-1}\exp(<N+Kr(i_1..i_{k-1}),r(i_1..i_k)>)}{\sum_{s\epsilon S}(g_{n-k}(s))^{q-1}\exp(<N+Kr(i_1..i_{k-1}),s>)}\}$$

Now if k = 1 we get the transition probability (depending on n):

$$\nu_n\{\sigma_{L_1} = r|1| \ | \ \sigma(0) = r(0)\} =$$

$$(3.8)$$

$$\prod_{i_1\epsilon Q_1} \{\frac{g_{n-1}(r(i_1))^{q-1}\exp(<N+Kr(0),r(i_1)>)}{\sum_{s\epsilon S}g_{n-1}(s)^{q-1}\exp(<N+Kr(0),s>)}\}$$

Define the following vector $\alpha_n^t = (\alpha_n^t(r): r \in S)$ for $t \in S$; which measures one-step transition probabilities among two neighbor sites:

$$\alpha_n^t(r) = \frac{(g_n(r))^{q-1}\exp(<N+Kt,r>)}{\sum_{s\epsilon S}(g_n(s))^{q-1}\exp(<N+Kt,s>)} \qquad t,r \in S \qquad (3.9)$$

Recall $\sum_{r\epsilon S} \alpha_n^t(r) = 1$. Define:

$$Z_n(t) = \sum_{s\epsilon S} (g_n(s))^{q-1}\exp (< N+Kt,s>) \qquad (3.10)$$

We have $\alpha_n^t(r) = (Z_n(t))^{-1}(g_n(r))^{q-1}\exp(<N+Kt,r>)$.

Now $g_{n-1}(s))^{q-1} = \alpha_{n-1}^t(s)Z_{n-1}(t)\exp(-<N+Kt,s>)$.

By using formula (2.6) we obtain:

$$\{\alpha_n^t(r)Z_n(t)\exp(-<N+Kt,r>)\}^{\frac{1}{q-1}} =$$

$$\sum_{s\epsilon S} \alpha_{n-1}^t(s)Z_{n-1}(t)\exp(<K(r-t),s>).$$

Take the vector $\psi_n^t = (\psi_n^t(r): r \in S)$ defined as:

$$\psi_n^t(r) = \{\alpha_n^t(r)\exp(-<N+qKt,r>)\}^{\frac{1}{q-1}} \tag{3.11}$$

It verifies the relation:

$$\psi_n^t(r) = \tag{3.12}$$
$$c_n(t) \sum_{s \in S} (\psi_{n-1}^t(s))^{q-1}\{expK(<r,s>-<t,r+s>)\}\{\exp(<N+qKt,s>)\}$$

where $c_n(t) = (Z_n(t))^{-\frac{1}{q-1}} Z_{n-1}(t)$.

Consider the following vector $\mu_t = (\mu_t(s): s \in S)$, and the matrices $A_t = (a_t(r,s): r,s \in S)$, $B_t = (b_t(r,s): r,s \in S)$, indexed by $t \in S$ and defined as follows:

$$\mu_t(s) = \exp(<N+qKt,s>) \tag{3.13}$$

$$a_t(r,s) = \exp(K(<r,s> - <t,r+s>)) \tag{3.14}$$

$$b_t(r,s) = a_t(r,s)\mu_t(s) \tag{3.15}$$

Equality (3.12) implies:

$$\psi_n^t = c_n(t)(B_t((\psi_{n-1}^t)^{q-1})) \tag{3.16}$$

From $\sum_{r \in S} \alpha_n^t(r) = 1$, relation (3.11) and definition (3.13) we get:

$$\sum_{r \in S} \mu_t(r)(\psi_n^t(r))^{q-1} = 1 \tag{3.17}$$

Take the following p-norm for functions in S with measure distribution μ^t:

$$\|y\|_p = (\sum_{r \in S} \mu_t(r)(y(r))^p)^{\frac{1}{p}} \tag{3.18}$$

Then (3.17) can be written as $\|\psi_n^t\|_{q-1} = 1$. Apply this equality in (3.16) to get:

$$\psi_n^t = \frac{B_t((\psi_{n-1}^t)^{q-1})}{\| B_t((\psi_{n-1}^t)^{q-1}) \|_{q-1}} \tag{3.19}$$

Let us study an evolution which is similar to the previous one:

$$y_n = \frac{B((y_{n-1})^{q-1})}{\| B((y_{n-1})^{q-1}) \|_q} \quad \text{where } B=B_t, \ A=A_t, \ \mu=\mu_t, \ y_n \in \mathbb{R}^S \tag{3.20}$$

First remark the matrix A is symmetric then B is $<,>_\mu$ self adjoint i.e. B verifies the equality:

$$<By,z>_\mu = <y,Bz>_\mu \quad \text{where } <y,z>_\mu = \sum_{r \in S} y(r)z(r)\mu(r) \tag{3.21}$$

Now the evolution (3.20) can be written as:

$$y_n = h \circ B \circ \eta(y_{n-1}) \quad \text{with } \eta(y) = y^{q-1}, \tag{3.22}$$

$$h(y) = \begin{cases} \dfrac{y}{\|y\|_q} & \text{if } y \neq 0 \\ 0 & \text{if } y = 0 \end{cases}$$

Take:

$$f(y) = \eta \circ h(y) = \frac{y^{q-1}}{\| y \|_q^{q-1}} \quad \text{if } y \neq 0, \ f(0) = 0 \tag{3.23}$$

Then f is a subgradient of the norm $\| y \|_q$ with respect to the inner product $<,>_\mu$, moreover the following equality is verified:

$$<y,f(y)>_\mu = \| y \|_q \quad \text{for any } y \in \mathbb{R}^S \tag{3.24}$$

Note $T = h \circ B \circ \eta$, so (3.22) can be written as $y_n = T(y_{n-1})$. We have (see ref.|4|):

Lemma. For any initial condition $y_o \in \mathbb{R}_+^S - \{0\}$ the points of the limit set of the sequence $(y_n = T(y_{n-1}))$ are of period 2, i.e. any accumulation point $\overline{y}$ of $\{y_n\}_{n \geq 0}$ verifies:

$$T^2 \overline{y} = \overline{y} \tag{3.25}$$

Proof. Define the following quantities:

$$H_n = -\langle B\eta(y_{n-1}), \eta(y_n)\rangle_\mu + \|B\eta(y_{n-1})\|_q + \|B\eta(y_n)\|_q \qquad (3.26)$$

$$\Delta_n = H_n - H_{n-1} \quad \text{for} \quad n \geq 2 \qquad (3.27)$$

If we put $u_n = B\eta(y_n)$ we get $\eta(y_n) = f(u_{n-1})$, with f defined by (3.23), so:

$$H_n = \|u_n\|_q \qquad (3.28)$$

Also, if we use the fact that B is $\langle,\rangle_\mu$ self-adjoint we find:

$$\Delta_n = -\langle f(u_{n-2}), u_n - u_{n-2}\rangle_\mu + \|u_n\|_q - \|u_{n-2}\|_q \qquad (3.29)$$

As f is a subgradient of $\|\ \|_q$ we deduce $\Delta_n \geq 0$. So $H_n = \|u_n\|_q$ increases along the orbit of y_n for $n \geq 1$ to a fixed quantity $\varepsilon > 0$ (because $\|u_n\|$ is bounded) and also $\Delta_n \to 0$ when $n \to \infty$.

Let $\bar{y}$ be a limit point of $\{y_n\}$, consider the sequence $\bar{y}^n = T^n\bar{y}$ for $n \geq 0$. Define $v_n = B\eta(\bar{y}_n)$. Then by developping the same quantities H_n, Δ_n as those employed for y_n we get:

$$\Delta'_n = -\langle f(v_{n-2}), v_n - v_{n-2}\rangle_\mu + \|v_n\|_q - \|v_{n-2}\|_q = 0$$
$$\qquad (3.30)$$
for any $n \geq 2$

As $\|v_n\|_q = \varepsilon = \lim_{n\to\infty} \|u_n\|_q$ we deduce, by using the conditions which make the Hölder inequality becomes an equality, that $v_n = v_{n-2}$ for $n \geq 2$, then $\bar{y} = T^2\bar{y}$. q.e.d.

Now the evolution (3.19) can be written as:

$$w_n = \frac{B\eta(w_{n-1})}{\|B\eta(w_{n-1})\|_{q-1}} = \tilde{h} \circ B \circ \eta(w_{n-1})$$
$$\qquad (3.31)$$
with $\tilde{h}(w) = \dfrac{w}{\|w\|_{q-1}}$ if $w \neq 0$, $h(0) = 0$

It is easily shown that the dynamics $T = h \circ B \circ \eta$ and $\tilde{T} = \tilde{h} \circ B \circ \eta$ acting respectively on $\{y \in \mathbb{R}^S: \|y\|_q = 1\}$ and $\{w \in \mathbb{R}^S: \|w\|_{q-1} = 1\}$

are isomorphic, being $\gamma(y) = \dfrac{y}{\|y\|_{q-1}}$ an isomorphism, that is γ is a bijection such that $\gamma \circ T = \tilde{T} \circ \gamma$. So the evolution (3.19) of ψ_n^t is such that any limit point $\overline{\psi}^t$ of $\{\psi_n^t\}_{n>0}$ satisfies:

$$\overline{\psi}^t = (T_t)^2 \overline{\psi}^t \quad \text{where} \quad T_t = \tilde{h} \circ B_t \circ \eta \qquad (3.32)$$

From equality (3.11) it also follows that any limit point $\overline{\alpha}^t$ of $\{\alpha_n^t\}_{n>0}$ is deduced from a limit point of $\overline{\psi}^t$ by the inverse function of (3.11).

Let us evaluate ν_n. We have:

$$\xi_n^{L_k}(r|k|) = \nu_n\{\sigma_{L_k} = r|k|)\} =$$

$$(\prod_{\ell=1}^{k} \nu_n\{\sigma_{L_\ell} = r|\ell| \mid \sigma_{L_{\ell-1}} = r|\ell-1|\})\nu_n\{\sigma_0 = r(0)\},$$

where $L_0 = \{0\}$. Write $\alpha_n^\phi = \xi_n^0$ (see 2.1) where ϕ means a "void spin". Use (3.7), (3.8) to get:

$$\xi_n^{L_k}(r|k|) = \prod_{\ell=0}^{k} \prod_{(i_1..i_\ell)\in L_\ell - L_{\ell-1}} \alpha_{n-\ell}^{r(i_1..i_{\ell-1})}(r(i_1..i_\ell)) \qquad (3.33)$$

where $\alpha_n^\phi(r(0)) = \alpha_{n-\ell}^{r(i_1..i_{\ell-1})}(r(i_1..i_\ell))$ for $\ell = 0$. With the same techniques as those of the above lemma it was shown in $|1|$ that the limit points $\overline{\psi}^\phi$ of the sequence $(\overline{\psi}^\phi = (\alpha_n^\phi)^{\frac{1}{q}})$ satisfy the equation $\overline{\psi}^\phi = \hat{T}^2 \overline{\psi}$ with $\hat{T} = h \circ \hat{A} \circ \eta$, where h, η are those functions given by (3.22) and $\hat{A} = (\hat{a}(r,s):r,s \in S)$ verifies:

$$\hat{a}(r,s) = \exp(K\langle r,s \rangle + \frac{1}{q}\langle N, r+s \rangle) \qquad (3.34)$$

Any limit point $\overline{\alpha}^\phi$ of the sequence (α_n^ϕ) being such that $\overline{\alpha}^\phi = (\overline{\psi}^\phi)^q$. Define the vector:

$$\alpha_n(r|k|) = (\alpha_{n-\ell}^{r(i_1...i_{\ell-1})} : \ell = 0...k) \qquad (3.35)$$

Take the following transformation δ which depends on $r|k|$:

$$\delta(\alpha_n(r|k|)) = \psi_n(r|k|), \text{ where:}$$

$$\delta(\alpha_{n-\ell}^{r(i_1..i_{\ell-1})})(r) = (\psi_{n-\ell}^{r(i_1..i_{\ell-1})}(r))^{q-1}\mu_{r(i_1..i_{\ell-1})}(r)$$

$$\text{if } \ell = 1...k$$

$$\delta(\alpha_n^{\phi})(r) = (\psi_n^{\phi}(r))^q \quad \text{if } \ell = 0$$

(3.36)

δ is a one-to-one continuous transformation whose inverse δ^{-1} is also continuous. Then the set of limit points of the sequence $(\alpha_n(r|k|))$ is obtained as the inverse, by δ^{-1}, of the set of limit points of the sequence $(\psi_n(r|k|))$. Note by $\bar{\psi}(r|k|)$ any one of this limit points, then:

$$\bar{\psi}(r|k|) = T*^2\bar{\psi}(r|k|) \tag{3.37}$$

where (see (3.32)):

$$T*(\bar{\psi}^{r(i_1...i_{\ell-1})}) = T^{(r(i_1...i_{\ell-1}))}(\bar{\psi}^{r(i_1...i_{\ell-1})})$$

$$\text{if } \ell = 1...k \tag{3.38}$$

$$\text{and} \quad T*(\bar{\psi}^{\phi}) = \hat{T}(\bar{\psi}^{\phi}) \quad \text{if } \ell = 0.$$

Now for any thermodynamic limit τ, its value on the cylinder $r|k|$, $\tau(r|k|) = \tau\{\sigma \in \Omega: \sigma_{L_k} = r|k|\}$, is a limit point of the sequence $\xi_n^{L_k}(r|k|)$, then it can be written as:

$$\tau(r|k|) = \prod_{\ell=0}^{k} \prod_{(i_1...i_\ell)\in L_\ell - L_{\ell-1}} \bar{\alpha}^{r(i_1..i_{\ell-1})}(r(i_1..i_\ell)) \tag{3.39}$$

where $\bar{\alpha}(r|k|) = \delta^{-1}(\bar{\psi}(r|k|))$, with δ given by (3.36) and the vector $\bar{\psi}(r|k|)$ satisfying equation (3.37).

3.   THE CASE L = ℤ

For $q = 2$ the Bethe lattice is $L = \mathbb{Z}$. In this case the functions $\alpha_n^{(t)}(r)$ do not depend on n. In fact from (3.9), (3.10) we deduce that

$$\alpha_n^t(r) = \alpha(r) = (\bar{Z}(t))^{-1} \exp(<N + Kt, r>),$$

(4.1)

where $\bar{Z}(t) = \sum_{s \in S} \exp(<N + Kt, s>)$

Then:

$$\xi_n^{L_k}(r|k|) = \nu_n\{\sigma_o = r(0)\} \cdot$$

$$\prod_{\ell=1}^{k} \prod_{(i_1..i_\ell) \in L_\ell - L_{\ell-1}} \bar{\alpha}^{r(i_1..i_{\ell-1})}(r(i_1..i_\ell))$$

(4.2)

which depends on n only on one-site distribution $\nu_n\{\sigma_o = r(0)\} = \xi_n^o(r(0)) = \alpha_n^\phi(r(0))$.

Now any limit point $\bar{\alpha}^\phi$ of the sequence (α_n^ϕ) verifies the equation:

$$(\bar{\alpha}^\phi)^{\frac{1}{2}} = \hat{T}^2(\bar{\alpha}^\phi)^{\frac{1}{2}} \quad \text{where} \quad \hat{T} = h \circ \hat{A} \circ \eta$$

(4.3)

with η, h given by (3.22) and $\hat{A}$ by (3.34). In the particular case $q = 2$ we have: $\eta(y) = y$, $h(y) = \frac{y}{\|y\|_2}$ if $y \neq 0$. So :

$$(\bar{\alpha}^\phi)^{\frac{1}{2}} = \frac{\hat{A}(\bar{\alpha}^\phi)^{\frac{1}{2}}}{\|\hat{A}(\bar{\alpha}^\phi)^{1/2}\|_2}.$$

(4.4)

Hence the vector $\bar{\psi}^\phi = (\bar{\alpha}^\phi)^{1/2}$ verifies:

$$A\bar{\psi}^\phi = \lambda\bar{\psi}^\phi$$

(4.5)

But $\hat{A} > 0$, then the Perron-Frobenius theorem asserts there is a unique strictly positive vector $\bar{\psi}^\phi > 0$ with strictly positive eigenvalue $\lambda > 0$ such that $\sum_{r \in S} (\bar{\psi}^\phi(r))^2 = 1$. Then there exists a unique probability vector $\bar{\alpha}^\phi$ verifying (4.3). So the set of thermodynamic limits $\{\tau\}$ have cylinder values given by:

$$\tau(r|k|) = \overline{\alpha}^{\phi}(r(0)) \prod_{\ell=1} \prod_{(i_1..i_\ell) \in L_\ell - L_{\ell-1}} \overline{\alpha}^{r(i_1..i_{\ell-1})}(r(i_1..i_\ell))$$

$$(4.6)$$

5. FINAL COMMENTS

We have found the explicit equations that must satisfy the vectors which determine the cylinder distribution of thermodynamic limits on Bethe lattices L for any finite spin set. In general the one-step conditional probabilities and the one-site distribution depend on the size (L_n) (which is exponential on n) and the limit value can be found from a period 2 equation. When $L = \mathbb{Z}$ the one step transition probabilities do not depend on the size, so the thermodynamic limit, which is shown to be unique, depends only on the one-site distribution which is deduced from a Perron-Frobenius equation.

ACKNOWLEDGMENTS

I thank Prof. E. Goles and P. Collet for fruitful discussions. I am also indebted to Prof. A. Verbeure from Leuven University for indicating me references |2,3| and for valuable suggestions. This work was partially supported by DTI, Universidad de Chile and Fondo Nacional de Ciencias 88 grant 0553.

REFERENCES

|1| R.J. Baxter, *Exactly Solved Models in Statistical Mechanics*, Academic Press, 1982.

|2| H.J. Brascamp, 'Equilibrium States for a One Dimensional Lattice Gas', *Comm. Math. Phys.* 21: 56-70 (1971).

|3| M. Fannes and A. Verbeure, 'On Salvable Models in Classical Lattice Systems', *Comm. Math. Phys.* 96: 115-124 (1984).

|4| E. Goles and S. Martínez, 'The One-Site Distribution of Gibbs States on Bethe Lattice are Probability Vectors of Period $\leq$ 2 for a Nonlinear Transformation', *J. Stat. Phys.* 52: 281-299 (1988).

|5| S. Martínez, 'Lyapunov Functionals on Bethe Lattice'. *Proceedings Escuela de Sistemas Desordenados*, Bogotá, World Scientific (1987).

RIGOROUS RESULTS ABOUT DETAILED BALANCE AND CRITICAL SLOWING DOWN

A. Verbeure
Instituut voor Theoretische Fysica
Katholieke Universiteit Leuven
Celestijnenlaan 200 D
B-3030 Leuven
Belgium

ABSTRACT. The material of this lecture contains the following topics.
1. Introduction on the conventional theory of critical slowing down
2. The detailed balance condition
3. Applications for classical lattice systems
4. Applications for quantum systems
The material is mainly based on the references [1-5]

1. INTRODUCTION : THE CONVENTIONAL THEORY OF CRITICAL SLOWING DOWN

The theory of critical phenomena contains typically static properties such as thermodynamic coefficients, single-time correlation functions, etc. which are fully determined by the equilibrium distribution. But this theory contains also dynamic properties of the system such as transport coefficients, relaxation rates, multi-time correlation functions, etc. These are properties which do depend not only on the equilibrium state but also on the equations of motion. The special features of the system which one observes near a critical point are called dynamical critical phenomena. Among the ideas and results on these properties one can distinguish : the conventional theory of critical slowing down, the "mode coupling" theory of transport anomalies, the hypotheses of dynamic scaling and universality, the renormalization group approach to critical slowing down, etc.

In this lecture we limit ourself to the conventional theory, due to Van Hove [6] and, Landau and Khalatnikov [7]. In particular we consider the time auto correlation function

$$t \in \mathbb{R} \rightarrow \omega_\beta(A^* A_t) - |\omega_\beta(A)|^2$$

where t is the time parameter, A is an observable of the system, A_t is the time evolved observable under a time evolution which we will specify later and ω_β is the equilibrium state of a system at inverse temperature $\beta = 1/kT$. From an experimental point of view this is an interesting object because it can be measured rather directly. For many systems one observes an exponential relaxation to equilibrium i.e. for

131

E. Tirapegui and D. Villarroel (eds.), Instabilities and Nonequilibrium Structures II, 131–143.
© *1989 by Kluwer Academic Publishers.*

large t :

$$\omega_\beta(A^*A_t) - |\omega_\beta(A)|^2 \underset{\sim}{} \exp - \frac{t}{\tau(\beta,A)} \tag{1}$$

It is clear that the time evolution $A \to A_t$ will depend on the equilibrium state ω_β in order that $\omega_\beta(A^*A_t) - |\omega_\beta(A)|^2 \to 0$ as $t \to \infty$. This is the subject of the next section. The quantity $\tau(\beta,A)$ in (1) is called the relaxation time of the observable A at β.

If the system shows a phase transition at some critical value β_c, then there might exist an observable A such that

$$\lim_{\beta \to \beta_c} \tau(\beta,A) \underset{\sim}{} \lim_{\beta \to \beta_c} \chi(\beta,A) \to \infty \tag{2}$$

i.e. the relaxation time $\tau(\beta,A)$ as well as the fluctuation of the observable A behave in a divergent manner at $\beta = \beta_c$. The content of the conventional theory of the phenomenon of critical slowing down consists in the fact that the dynamical quantity $\tau(\beta,A)$ behaves proportional to the static equilibrium quantity $\chi(\beta,A)$. One has developed many models in order to compute numerically or to obtain by heuristic arguments the behaviour of these quantities around T_c. A typical form of the relaxation time is

$$\tau(\beta,A) \underset{\sim}{} (T - T_c)^{-\gamma}$$

then γ is called a dynamical critical exponent; for the fluctuation one proposes also the form

$$\chi(\beta,A) \underset{\sim}{} (T - T_c)^{-\gamma'}$$

where γ' is called the static critical exponent. The conventional theory of critical slowing down suggests a relation between the static quantity γ' and the dynamic quantity γ as illustrated by formula (2).

Although the conventional theory is insufficient in many cases, we tried to prove some facts about this conventional theory. In particular we tried to understand in a rigorous mathematical frame the basic mechanism behind this theory. We are discussing classical lattice systems as well as quantum systems. For classical lattice systems, we have a Hamiltonian describing the interactions but we have no natural time evolution. Therefore we have to invent one. This is the content of the next section. The type of evolution which we will study is one satisfying the condition of detailed balance with respect to a particular state of the system. It will turn out that the property of detailed balance and locality is sufficient to yield the phenomena of critical slowing down. We derive also a relation between the relaxation time of an observable and its fluctuations. This is our main result for classical systems.

For quantum systems a Hamiltonian H does describe a natural time evolution by the unitary operator exp itH. But except for maybe very special cases this evolution does not yield the behaviour as described in (1). Again one should invent the right model to investigate some dynamical aspects of the equilibrium states near the critical point. It will turn out from our analysis that also here the property of detailed balance is the right condition on the evolution. For quantum systems we have not such general theorem as for classical lattice systems, because we have difficulties in implementing the property of locality which is expressed by the local commutativity of the observables. We are able to find an explicit form of a detailed balance evolution and to apply it to soluble models.

2. NOTION OF DETAILED BALANCE

The notion of detailed balance is fairly well studied for classical systems. Here we give a formulation of it which is extendable to quantum systems.

Consider the extreme simple model with a phase space $K = \{1,\ldots,n\}$ consisting of n events. A state is described by the probability measure $\omega = (p_1,\ldots,p_n)$, $0 \leq p_i \leq 1$, $\sum_i p_i = 1$. The observables are the complex functions f on K i.e. $f : i \in K \to f_i \in C$, and the expectation value of f in ω is given by

$$\omega(f) = \sum_{i=1}^{n} p_i f_i$$

There is no natural time evolution, but one can define a Markov process on the state space (Schrödinger picture) :

$$\frac{d\, p_i(t)}{dt} = \sum_{k=1}^{n} D_{ki} p_k(t) - D_{ik} p_i(t) \tag{3}$$

where the D_{ik} are suitably normalized constants which are non-negative, describing the transition probability from event i to event k. In Heisenberg picture i.e. as an evolution on the observables (3) becomes :

$$\frac{d\, f_i(t)}{dt} = \sum_{k=1}^{n} D_{ik}(f_k(t) - f_i(t)) \tag{4}$$

This Markov process satisfies the <u>Detailed Balance condition</u> with respect to the state ω if for all i,k :

$$p_i D_{ik} = p_k D_{ki} \tag{5}$$

Remark that this condition implies that the state is time invariant :

$$\frac{dp_i}{dt} = 0 \qquad \text{or} \qquad \frac{d}{dt}\, \omega(f) = 0$$

but that stationarity does not imply detailed balance. Now we reformulate this condition. Denote by L the generator of the Markov process :

$$(L f)_i = \sum_{k=1}^{n} D_{ik}(f_k - f_i)$$

then one checks :
  (i) $L(1) = 0$ (the evolution is unity preserving)
  (ii) $L(f^2) - 2f L(f) \geqslant 0$ for all f ; (the evolution is dissipative)
  (iii) the detailed balance condition, by multiplying on both sides of
        (5) by g_i and f_k :

$$g_i P_i D_{ik} f_k = f_k P_k D_{ki} g_i$$

and

$$\sum_{i,k} g_i P_i D_{ik}(f_k - f_i) = \sum_{i,k} f_k P_k D_{ki}(g_i - g_k)$$

or equivalently : for all f and g

$$\omega(g L(f)) = \omega(L(g)f) \tag{6}$$

Conversely, it is clear that (6) yields (5).
     We have rewritten the detailed balance condition into the condi-
tion on L to be a symmetric (self-adjoint) linear operator on the Hilbert space of observables with a scalar product $<f,g> = \omega(\bar{f} g)$.
     In the form (6) the notion is extendable to arbitrary classical
systems, but also to quantum systems. Take e.g. a quantum state ω_ρ
determined by a density matrix ρ on a Hilbert space $\mathcal{H}$:

$$\omega_\rho(A) = tr_{\mathcal{H}} \rho A \quad ; \quad A \in B(\mathcal{H})$$

and a (semi-)group evolution

$$\frac{d A_t}{dt} = L A_t$$

with solution

$$A_t = e^{tL} A_0$$

such that
  (i) $L(1) = 0$ (unity preserving)
  (ii) $L(A^*) = L(A)^*$
        $L(A^*A) - L(A)^*A - A^*L(A) \geqslant 0$ (dissipativity)
then the evolution satisfies
  (iii) the detailed balance with respect to ω_ρ if for all A and B holds :

$$\omega_\rho (L(A)B) = \omega_\rho (A \; L(B))$$

Remark that also in the quantum case, the D.B. condition with respect to ω_ρ, implies the stationarity of the state : $\omega_\rho(L(B)) = 0$ for all $B \in \mathcal{B}(\mathcal{H})$.

So far we discussed the D.B. condition with respect to any expectation value or state ω. If one wants to analyze properties of equilibrium states, e.g. near a phase transition, and study the phenomenon of critical slowing down, then one has to take state ω to be an equilibrium state at some inverse temperature β. In the next section we give examples of these evolutions for equilibrium states.

3. CLASSICAL LATTICE SYSTEMS

In this section we formulate our main result. In order to clarify the result we develop all the necessary notions for the ν-dimensional Ising model and describe the general situation at the end.

First we consider equilibrium states. Normally the Ising model is described by the following local Hamiltonians; let Λ be any finite subset of the lattice, then

$$H_\Lambda = -J \sum_{\substack{<i,j> \\ i,j \in \Lambda}} \sigma_i \sigma_j$$

where $\sum_{<i,j>}$ stands for the sum over all nearest neighbours (i,j); σ_i is the spin variable, taking the values ± 1.
In fact in order to define equilibrium states the basic notion is the relative Hamiltonian h. Denote by α_k the spin-flip at site k :

$$\alpha_k(\sigma_i) = \sigma_i \qquad \text{if} \qquad i \neq k$$

$$= -\sigma_i \qquad \text{if} \qquad i = k$$

then

$$h(\alpha_k) = \lim_\Lambda (\alpha_k H_\Lambda - H_\Lambda)$$

$$= 2 J \sum_{j, <j,k>} \sigma_k \sigma_j$$

describing the energy difference of any configuration with its transformed configuration under the flip α_k. An equilibrium state at $\beta = 1/kT$ is a state ω_β satisfying the conditions (equivalent to the DLR-conditions [8,9]);

$$\omega_\beta(\alpha_k f) = \omega_\beta(f \; e^{-\beta h(\alpha_k)}) \tag{7}$$

for all f and α_k.

Before we give the solutions of the D.B. equations for this ω_β, we formulate the equilibrium condition for general classical lattice systems.

Consider any lattice S; for each finite subset X of S, denote $K_X = \prod\limits_{j \in X} K_j$; $K_j = K$ with K any compact Hausdorff space; denote $\Omega = \prod\limits_{j \in S} K_j$; C(X) the set of continuous real functions on K_X and $C(\Omega)$ the continuous real function on Ω.

Denote by Q the group of transformations of $C(\Omega)$, the transposed of the invertible transformations of the configuration space Ω, changing the configurations in some finite volume only.

Now a relative Hamiltonian is a map $h : Q \to C(\Omega)$; $\tau \to h(\tau)$ satisfying the cocycle relation

$$h(\tau_1 \tau_2) = \tau_1 h(\tau_2) + h(\tau_1)$$

for τ_1, $\tau_2 \in Q$.

Definition III.1

A measure ω_β of $C(\Omega)$ satisfies the DLR-conditions at inverse temperature $\beta > 0$ with respect to a relative Hamiltonian h iff for all $\tau \in Q$ and all $f \in C(\Omega)$ holds

$$\omega_\beta(\tau^{-1} f) = \omega_\beta(f \ e^{-\beta h(\tau)}) \tag{8}$$

or equivalently iff, the Radon-Nykodim derivative of $\omega_\beta \cdot \tau^{-2}$ with respect to ω_β is given by

$$\frac{\partial \omega_\beta \cdot \tau^{-1}}{\partial \omega_\beta} = e^{-\beta h(\tau)} \qquad \text{for all} \quad \tau \in Q$$

For the Ising model we have the following solutions of the D.B. condition.

Proposition III.2

If ω_β is a measure satisfying (7) then for all sites k the following maps

$$L_k = e^{-\beta h(\alpha_k)/2} (\alpha_k - 1) \tag{9}$$

satisfy the D.B. condition (6) with respect to ω_β.

Proof : This is readily seen from a substitution of (9) into (6)   ∎

One checks also the converse :

Proposition III.3

If ω is any measure such that $\omega \cdot L_k = 0$ for all k, then ω satisfies (7).

∎

This means that the D.B. solutions (9) do characterize dynamically the equilibrium state. This is also the mathematical expression of what is sometimes said that D.B. expresses local equilibrium. This way the D.B. solutions are canonical.

Remark that, because of the linearity of the D.B. condition also the following generator satisfies the D.B. condition with respect to ω_β :

$$L = \sum_k L_k = \sum_k e^{-\beta h(\alpha_k)/2} (\alpha_k - 1)$$

This is the generator of the Glauber dynamics [10].

Even more, consider any finite volume Λ of the lattice, and

$$\alpha_\Lambda = \prod_{k \in \Lambda} \alpha_k$$

one checks that,

$$L_\Lambda = e^{-\beta h(\alpha_\Lambda)/2} (\alpha_\Lambda - 1)$$

because of the cocycle relation of the relative Hamiltonian, also satisfies the D.B. condition for ω_β and hence also

$$L = \sum_\Lambda L_\Lambda$$

So far for the Ising model. Now we are in a position to formulate our main result which holds for arbitrary classical lattice systems, with a compact phase space per lattice site as explained above.

Theorem III.4

Suppose that $\Gamma = \{\Lambda\}$ is a family of finite subsets of the lattice, such that for each Λ, there exists a $L_\Lambda : C(\Omega) \rightarrow C(\Omega)$

$$L_\Lambda : f \rightarrow L_\Lambda f$$

satisfying :
   (i) $L_\Lambda f = 0$ if supp $f \subset \Lambda^c$ (complement of)
           (property of locality)
  (ii) L_Λ is unity preserving and dissipative
 (iii) L_Λ is detailed balance with respect to a measure ω_β

(iv)

$$\sup_{\substack{j \\ k}} \sum_{\substack{k \\ j,k \in \Lambda}} \sum_{\Lambda \in \Gamma} \|L_\Lambda\| < \infty$$

then if the spectrum of $L = \sum_\Lambda L_\Lambda$ has an energy gap ℓ_0, there exists

an observable A and a constant C, such that

$$\ell_0 < \frac{C}{\chi(\beta,A)}$$

or equivalently

$$\tau(\beta) = \frac{1}{\ell_0} > C' \ \chi(\beta,A) \ ; \quad C' \in \mathbb{R}^+$$

Proof : can be found in [3] and uses essentially the locality property and the detailed balance property.   ∎

This theorem proves partially the validity of the conventional theory
of critical slowing down. We got an inequality relation between the
static quantity namely the fluctuation and the relaxation time, the
dynamic quantity. It remains an open problem whether an inverse inequality exists also.

4. QUANTUM SYSTEMS

Now we turn to the question whether an analogous theorem as III.4 also
exists for quantum systems.
     We start from an evolution

$$A_t = e^{tL}A_0$$

with A_0 an element of the bounded operators $B(\mathcal{H})$ on a Hilbert space $\mathcal{H}$,
such that (see section 2.) :
  (i) $L(1) = 0$
  (ii) $L(A^\star) = L(A)^\star$
       $L(A^\star A) - L(A)^\star A - A^\star L(A) \geqslant 0$
  (iii) $\omega_\beta(L(A)B) = \omega_\beta(A \ L(B))$; $A,B \in M$, a von Neumann algebra of ob-
       servables and ω_β a KMS-state for a time-evolution determined by
       a Hamiltonian H (for the definition of a KMS-state of a $W^\star$-dyna-
       mical system, see [11].
To fix our ideas we develop again a special case of a finite system i.e.
for the algebra of observables we take the n × n matrix algebra M_n, a
Hamiltonian H wit spectral resolution

$$H = \sum_k \varepsilon_k \, E_{kk}$$

$$H \, \phi_k = \varepsilon_k \, \phi_k \quad ;$$

$\{\phi_k | k\}$ an orthonormal base, diagonalizing the Hamiltonian.

Denote

$$E_{k\ell} = |\phi_k\rangle\langle\phi_\ell|$$

and

$$\omega_\beta(A) = \frac{\mathrm{tr}\ e^{-\beta H}\, A}{\mathrm{tr}\ e^{-\beta H}} \quad ; \quad A \in M_n \tag{10}$$

as equilibrium state.
Then solutions of the D.B. condition with respect to ω_β are the maps

$$L_{k\ell} = e^{\beta(\varepsilon_k - \varepsilon_\ell)/2} \{E_{k\ell}[\,\cdot\,, E_{\ell k}] + [E_{k\ell}, \cdot\,] E_{\ell k}\}$$

$$+ \text{ symmetric term } (k \leftrightarrow \ell). \tag{11}$$

for all $k, \ell = 1, \ldots n$.
One calls generator of this type of the Lindblad type, which can already be extracted from [12].

Here again one has to ask the question whether the solutions (11) do determine the equilibrium. We have indeed the following property :

Proposition IV.1

Let ω be any state of M_n, then

$$\omega \cdot L_{k\ell} = 0 \quad \text{for all } k, \ell$$

implies $\omega = \omega_\beta$ = equilibrium state (10).
Proof : Consider first

$$\omega(L_{kk}(E_{ks})) = 0 \quad \text{with } k \neq s$$

As

$$L_{kk}(E_{ks}) = -2 \, E_{ks}$$

one gets

$$\omega(E_{ks}) = 0 \tag{α}$$

Take then for $k \neq \ell$

$$L_{k\ell}(E_{kk}) = -2 \, e^{\frac{\beta}{2}(\epsilon_k - \epsilon_\ell)} E_{kk} + 2 \, e^{\frac{\beta}{2}(\epsilon_\ell - \epsilon_k)} E_{\ell\ell}$$

hence

$$\omega(E_{kk}) e^{\beta\epsilon_k} = \omega(E_{\ell\ell}) e^{\beta\epsilon_\ell} = \lambda$$

and

$$1 = \omega(1) = \sum_k \omega(E_{kk}) = \lambda \, \mathrm{tr} \, e^{-\beta H}$$

Therefore

$$\omega(E_{kk}) = \frac{e^{-\beta\epsilon_k}}{\mathrm{tr} \, e^{-\beta H}} \qquad\qquad (\beta)$$

The result follows from (α) and (β). ∎

For general W^*-dynamical systems, one finds also the D.B. solutions [1], corresponding to the ones of (11) for finite systems. One can also prove a theorem of the type of proposition IV.1.

In this way one arrives at the main question. What can be said about the spectrum of L ? Is the energy gap related to fluctuations ? We have no general theorem as for classical lattice systems, the problem being the lack of locality in the quantum mechanical generator. The invariance of the generator under the natural time evolution destroys the locality. However we have rigorous results about soluble models.

a) The free Bose gas

The result for the free Bose gas is the following.
If $T < T_c$ (critical temperature) one derives

$$\lim_V \{\omega_\beta(\alpha_V^+ \, e^{tL} \, \alpha_V) - |\omega_\beta(\alpha_V)|^2\} \simeq \frac{1}{t}$$

for large t,

where $\alpha_V = \frac{1}{V^{1/2}} \int_V dx \, a(x)$

i.e. we have polynomial decay.

If $T > T_c$, we find

$$\lim_V \{\omega_\beta(a_V^+ \, e^{tL} \, a_V) - |\omega_\beta(a_V)|^2\} \underset{\sim}{\sim} e^{-\frac{tC}{(T - T_c)^{-2}}}$$

for large t, i.e. we have dynamical critical slowing down, with a critical exponent equal to two. One checks that also the fluctuation

$$\lim_V \chi(\beta, a_V) \underset{\sim}{\sim} (T - T_c)^{-2}$$

has a critical exponent equal to two. These results prove completely the conventional theory of critical slowing down.

b) The Dicke-Maser model

The Dicke-Maser model is described by the following Hamiltonian

$$H_N = \sum_{k=-N}^{N} (a_k^+ a_k + \varepsilon \, \sigma_k^+ \sigma_k^-)$$

$$+ \frac{\lambda}{2N + 1} \sum_{k,\ell=-N}^{N} (a_k^+ \sigma_\ell^- + a_k \sigma_\ell^+)$$

where a_k, a_k^+ are Boson creation and annihilation operators; σ_k^+, σ_k^- are Pauli-spin matrices. The model is a mean field model showing a phase transition if it is satisfied to the condition

$$0 < \varepsilon < \lambda^2$$

In this case the critical temperature is given by

$$\beta_c = \frac{1}{kT_c} = \left(\frac{2}{\varepsilon}\right) \, th^{-1} \, \frac{\varepsilon}{\lambda^2}$$

The results about critical slowing down are the following.
If $T > T_c$ one finds exponential decay with a relaxation time independent of the temperature.
If $T < T_c$, consider a semigroup generator

$$L = \int_{-\infty}^{\infty} dt \, f(t) \, M_s\{X_t[\cdot, X_{s+t}] + [X_t, \cdot] X_{t+s}\}$$

with

$$X = \alpha \, a_k^+ + \alpha^* \, a_k \; ; \quad k \in \mathbb{N}$$

$$\alpha = \lim_{N \to \infty} \frac{1}{2N} \sum_{i=-N}^{N} a_i \ .$$

$$M_s \ \psi(s) = \lim_{R \to \infty} \frac{1}{2R} \int_{-R}^{R} ds \ \psi(s)$$

$$X_t = \lim_{N} e^{it \ H_N} X \ e^{-it \ H_N}$$

From the special choice of X one remarks that the semigroup generator describes a detailed balance transition between the condensate mode and an arbitrary excited mode.
One can explicitly compute the spectrum of the generator. It is given by :

$$Spectr \ L = \{-n(\beta)p \mid p \in \mathbf{N}\}$$

and hence the energy gap is

$$\ell_0 = n(\beta)$$

where

$$n(\beta) \ \underset{\sim}{} \ (T - T_c)$$

when $T \to T_c$. The dynamical critical exponent equals one and together with the static results one obtains

$$\tau(\beta) \ \underset{\sim}{} \ \chi(\beta,\alpha) \ \underset{\sim}{} \ (T - T_c)^{-1}$$

proving the conventional theory of critical slowing down for this quantum model.

Furthermore one can also show that the dynamical critical exponent is the same for all mean field models.

REFERENCES

[1] J. Quaegebeur, G. Stragier, A. Verbeure; Ann. Inst. H. Poincaré 41, 25 (1984)
[2] A. Verbeure; Commun. Math. Phys. 95, 301 (1984)
[3] R. Alicki, M. Fannes, A. Verbeure; J. Stat. Phys. 41, 263 (1985)
[4] J. Quaegebeur, A. Verbeure; Lett. Math. Phys. 9, 93 (1985)
[5] D. Goderis, A. Verbeure, P. Vets; J. Math. Phys. 28, 2250 (1987)
[6] L. Van Hove; Phys. Rev. 93, 1374 (1954)
[7] L.D. Landau, I.M. Khalatnikov; Dokl. Akad. Nauk. SSSR 96, 469 (1954)
[8] R.L. Dobrushin; Theory Prob. Appl. 13, 197 (1968)
[9] O.E. Lanford, D. Ruelle; Commun. Math. Phys. 13, 194 (1969)
[10] R.J. Glauber; J. Math. Phys. 4, 294 (1963)

[11] O. Bratteli, D.W. Robinson; Operator Algebras and Quantum Statisti-
       cal Mechanics II; Springer-Verlag 1979
[12] R. Alicki; Rep. Math. Phys. 10, 249 (1976)

QUANTUM CHAOS

T. Dittrich and R. Graham
Fachbereich Physik
Universität Essen
D-4300 Essen, W. Germany

ABSTRACT. This review is concerned with dynamical consequences of the
quantization of systems with classically chaotic dynamics. It is argued
that such systems display an instability in the classical limit. The
standard map is used as a prototype. Besides a discussion of the phenomenon of localization in action space due to quantum interferences, we
consider in particular the effects of repeated measurements and of dissipation which tend both to reduce or destroy localization phenomena
and trigger the instability otherwise only encountered when taking the
classical limit.

1. INTRODUCTION

The present lecture is concerned with the topic of 'quantum chaos', and,
like the other lectures of this conference, it deals with an instability, the instability of quantum mechanics in the classical limit $\hbar$ /(relevant scale of action) $\rightarrow$ 0 under circumstances which render the classical dynamics chaotic. To see that an instability occurs in this limit
consider a finite bounded nonlinear autonomous Hamiltonian system with
two or more freedoms. Classically, for sufficient strength of the nonlinearity the dynamics in some parts of phase space is chaotic in general, i.e. there is sensitive dependence on initial conditions and the
classical trajectories are exponentially unstable with a positive average exponentiation rate (Lyapunov-exponent) of small perturbations.
Quantum mechanically, however, i.e. before the classical limit is taken,
bounded systems have a discrete energy spectrum. Therefore, all expectation values, correlation functions, and response functions are quasiperiodic functions of time. Such functions are not sensitively dependent on initial conditions, there is no exponential instability, and
the Lyapunov exponents vanish. The classical limit therefore produces
the instability which is present in classical chaotic motion. The asymptotics of quantum systems in that limit is commonly referred to as
quantum chaos. It is the aim of the present lecture to discuss some
aspects of this instability. The emphasis of the work to be reviewed
is on dynamical properties. For static properties, in particular for

145

E. Tirapegui and D. Villarroel (eds.), Instabilities and Nonequilibrium Structures II, 145–162.
© 1989 by Kluwer Academic Publishers.

the very interesting results on the universal properties of the statistics of energy levels which appear asymptotically as the classical limit is approached, the reader is referred to the literature [1].

   The lecture is organized around a rather simple specific example
of a dynamical system, the 'standard map' [2]. In section 2 we shall
explain why this is an interesting and typical dynamical system to
study [2,3]. This map is then considered in quantized form in section 3
[4]. The origin of the higher stability of the quantum dynamics as compared to the chaotic dynamics is due to a phenomenon called localization
[4-7], which is explained in section 4. The disappearance of localization is responsible for the instability in the classical limit. The
remaining sections of this lecture are then concerned with a review of
our own work in this area [8-10]. We discuss the effects of perturbing
environments (reservoirs) on quantum systems close to the instability
leading to classical chaos. The coupling of quantum systems to environments is necessary if quantum measurements are to be performed. As we
shall discuss in section 5 the quantum back-action of measurements of
the localized (action) variables tends to destroy their localization
via randomizing the conjugate (angle) variables, thereby triggering the
instability present in the classical system. A second physical reason
for introducing reservoir couplings is the necessity to account for
dissipative effects, which are discussed in section 6. We shall find
there that due to the proximity of the classical instability, dissipation in quantum systems may act in a destabilizing way even if classically it tends to stabilize the motion in a steady state. In a final
section we summarize our conclusions.

2. CLASSICAL STANDARD MAP

In the study of chaos in dynamical systems with one or two freedoms the
device of discrete maps has proven an invaluable tool (cf. e.g. [3]).
We recall that one may define basically two kinds of maps (i) the Poincaré return maps for Hamiltonian systems with two freedoms in which we
take a two-dimensional cross section of the three-dimensional energy
hypersurface and (ii) stroboscopic maps for periodically driven systems
with one freedom in which we take isochronous snap-shots of the two-
dimensional phase space. Only the stroboscopic maps permit to follow in
time the fate of a state described by a phase-space distribution pre-
pared initially, e.g. at time 0. Also for the purpose of quantization
stroboscopic maps are particularly useful because they allow us to
follow a quantum state in time, at least for an infinite sequence of
discrete points on the time axis. In the following we shall therefore
restrict our attention to stroboscopic maps. It is useful to represent
the map in action and angle variables of the undriven system with vanishing dissipation. The standard map [2,3] then arises naturally and
rather generally by linearizing the actual map with respect to deviations p of the action variable from a picked value and neglecting all
but the first term of a Fourier expansion with respect to the angle

variable q . In this manner one obtains in dimensionless variables

$$P_{n+1} = P_n - \frac{K}{2\pi} \sin(2\pi q_n),$$

$$q_{n+1} = (q_n + p_{n+1})(\mod 1),$$

(2.1)

where p_n and q_n are canonically conjugate and the single parameter K is a measure of the strength of nonlinearity. The classical map (2.1) has been studied in great detail [3]. The phase space is the cylinder $0 \leqslant q < 1, -\infty < p < +\infty$. For K=0 the motion is regular and lies on Kolmogorov Arnold Moser (KAM) curves which close around the cylinder. For K>0 but small the KAM theorem [3] ensures that nearly all KAM curves survive, even though in a slightly deformed way; however, all KAM curves containing periodic points and all those sufficiently nearby are destroyed and break up into chains of alternating elliptic and hyperbolic fixed points with domains of chaos appearing in the vicinity of the hyperbolic fixed points. For K< 0.9716... there exists at least one KAM curve which 'localizes' all orbits within a unit interval $\Delta p = 1$ of the action variable. For K>0.9716... the last KAM curve has disappeared by developping holes and turning into a Cantor set [11] (called a Cantorus).

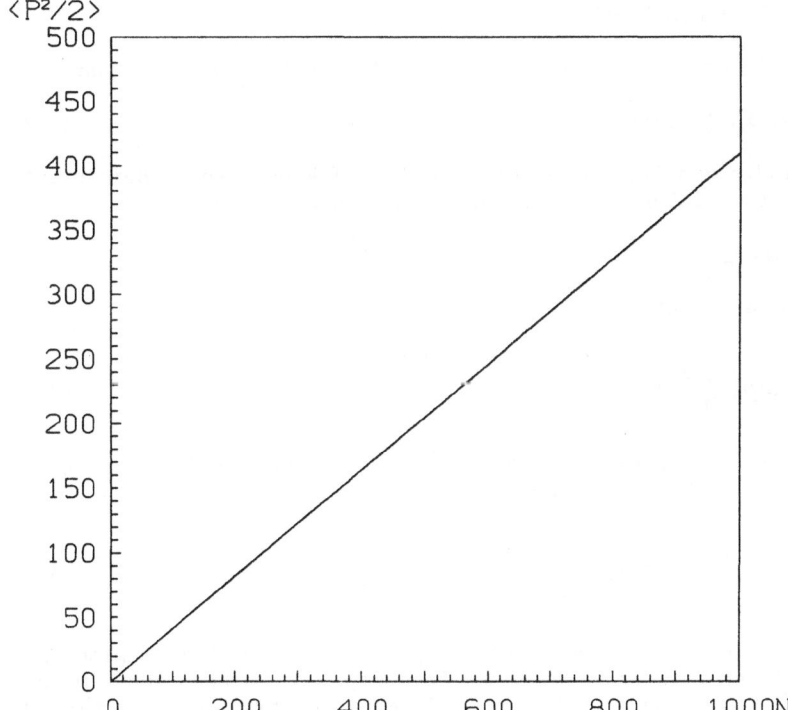

Figure 1. Diffusion of the action variable in the classical standard map for K=10, initial state p=0, $0 \leqslant q < 1$ random.

The chaotic orbits are now free to diffuse to unbounded values of the action variable p. For sufficiently strong nonlinearity K the diffusion constant is easily estimated from eq. (2.1) in the random phase approximation, which assumes that the phase q is quickly randomized during the motion. One finds [2,3]

$$\langle (p_{n+1} - p_n)^2 \rangle = \frac{K^2}{4\pi^2} \langle (\sin(2\pi q_n))^2 \rangle \simeq \frac{K^2}{8\pi^2}. \tag{2.2}$$

Corrections to this estimate taking into account residual correlations in the motion of the phase variable have also been calculated [12]. A numerical example showing the classical diffusion of the action variable is given in fig. 1. An initial ensemble of 2.5×10^5 points with $p = 0$ and equally distributed phases q was propagated over 10^3 iterations with $K = 10$ and the mean value of $p^2/2$ was plotted after each time-step. For this example the estimate (2.2) is too high. However, the observed lower value of the diffusion constant is well understood in terms of the corrections to (2.2) mentioned before [12].

3. QUANTIZED STANDARD MAP

A periodically driven quantum system satisfies a Schrödinger equation

$$i\hbar \frac{d}{dt} |\psi(t)\rangle = H(t) |\psi(t)\rangle \tag{3.1}$$

with a periodic Hamiltonian $H(t+T) = H(t)$. Taking stroboscopic cross-sections at time intervals T one obtains a quantum map

$$|\psi_{n+1}\rangle = U |\psi_n\rangle \tag{3.2}$$

with the unitary operator

$$U = T \exp\left(\frac{-i}{\hbar} \int_{t_n}^{t_n+T} dt\, H(t)\right) \tag{3.3}$$

where T is the chronological ordering operator. The standard map is generated by the Hamiltonian [4]

$$H(t) = \frac{p^2}{2} - \frac{K}{4\pi^2} \cos(2\pi q) \sum_{n=-\infty}^{+\infty} \delta(t - n), \tag{3.4}$$

where time is measured in units of the period T. The Hamiltonian describes a 'kicked rotator'. It is not the only, but perhaps the simplest, Hamiltonian which generates the standard map. Quantization is achieved by demanding the canonical commutation relations

$$[q,p] = i\hbar. \tag{3.5}$$

It should be noted that $\hbar$ in eq. (3.5) is an effective dimensionless version of Planck's constant, because p, q and the time-scale are all dimensionless in eq. (3.4). Hence $\hbar$ in eq. (3.5) depends on system parameters and can be used as a control parameter in the study of instabilities.

The quantum map (3.2)-(3.5) was first defined and iterated on a computer by Casati et.al. [4], starting with an initial state $|\psi_0\rangle = |0\rangle$, where the notation $|l\rangle$ is used to denote the eigenstates of the operator p for the action variable (angular momentum) of the rotator

$$p|l\rangle = 2\pi\hbar l|l\rangle, \quad l = 0, \pm 1, \pm 2, \ldots \tag{3.6}$$

A result of such a computation is shown in fig. 2 where the expectation values $\langle p_n^2/2\rangle$ for 10^3 iterations of the map are plotted for $K = 10$ and $2\pi\hbar = 0.15g^{-1}$, $g = (1-\sqrt{5})/2$. In fact, the result one obtains is highly

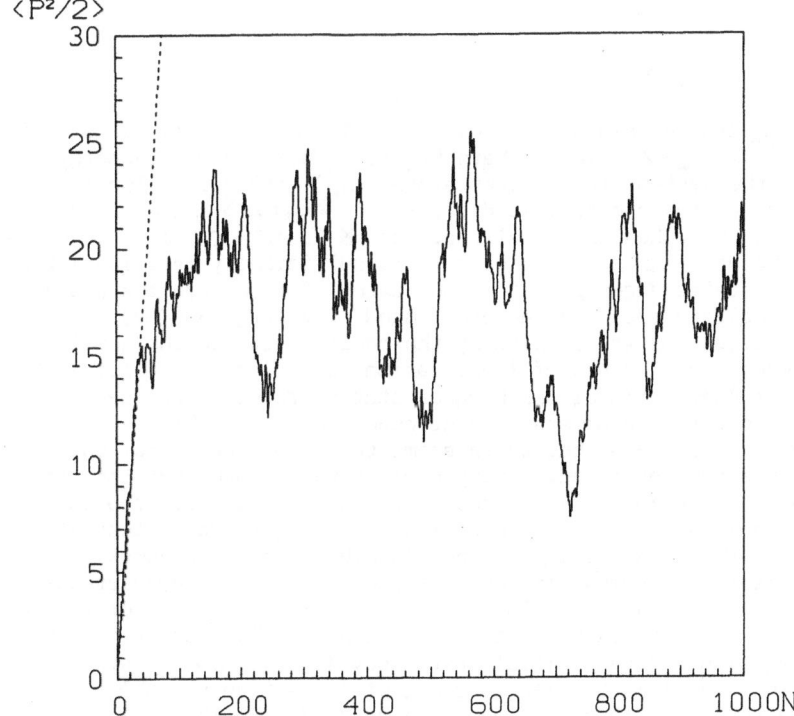

Figure 2. Expectation value of $p^2/2$ as a function of n, the number of iterations of the map, for $K = 10$, $2\pi\hbar = 0.15g^{-1}$, $g = (1-\sqrt{5})/2$, initial state $|\psi_0\rangle = |0\rangle$. Classical result shown by dashed line.

sensitive to the degree of irrationality of the parameter $2\pi\hbar$, as was demonstrated in [13], because rationality of $2\pi\hbar$ gives rise to quantum resonances with a quadratical increase of $\langle p^2/2\rangle$ for long times [14]. The dashed line in fig. 2 indicates the classical diffusion shown in fig. 1. Obviously, there exists a rather well defined time n^* after which the quantum result differs qualitatively from the classical result [4]. While classically the action variable diffuses in an unconstrained way the quantum result, for sufficiently irrational $2\pi\hbar$, displays localization of the action variable in a certain neighbourhood $2\pi\hbar L$ of its initial value $p=0$.

4. LOCALIZATION

The Floquet theorem [15] ensures that the Schrödinger equation (3.1) has solutions of the form

$$|\psi_{\varkappa}(t)\rangle = e^{-i\omega_{\varkappa}\frac{t}{T}} |u_{\varkappa}(t)\rangle \tag{4.1}$$

with

$$|u_{\varkappa}(t+T)\rangle = |u_{\varkappa}(t)\rangle. \tag{4.2}$$

The phases $\omega_{\varkappa}$ are defined by eq. (4.1) $mod\,2\pi$ and can all be taken in the interval $0 \le \omega_{\varkappa} < 2\pi$. The numbers $\hbar\omega_{\varkappa}/T$ form the quasi-energy spectrum of the system. The Floquet states $|u_{\varkappa}\rangle$ are called localized and the quasi-energy spectrum is discrete if $\langle u_{\varkappa}|u_{\varkappa}\rangle < \infty$, and the $|u_{\varkappa}\rangle$ can then be taken as normalized. For $K/(2\pi\hbar)^2 \to 0$ but independent of the value of $2\pi\hbar$ as long as it is sufficiently irrational the quasi-energy states degenerate to eigenstates of p and are therefore trivially localized [16]. On the other hand, for large $K/(2\pi\hbar)^2$ many eigenstates of p are coupled by the nonlinearity in the Hamiltonian and the presence or absence of localization is no longer a trivial question. For $2\pi\hbar$ rational the Floquet states are not localized and the quasi-energy spectrum consists of continuous bands. For $2\pi\hbar$ sufficiently irrational it appears (but there seems to be no proof) that the Floquet states are localized. A very suggestive argument for localization has been given in [6,7] by mapping the kicked rotator problem on a 1-dimensional solid state problem with a pseudo-random potential, in which Anderson localization can be plausibly argued to appear. A simple qualitative argument in favor of localization has been proposed by Chirikov et.al. [5] which also leads to a simple estimate of the localization length. It starts by supposing that the Floquet states which have overlap with a given initial state, say $|\psi(0)\rangle = |0\rangle$, are localized within a localization length $2\pi\hbar L$ of the action variable p and it proceeds to show, via a self-consistency condition that L has, in fact, a finite value. The self-consistency condition follows from the observation that about $2L$ localized states can have overlap with the initial state and therefore also overlap with each other. As a conse-

quence their eigenphases $\omega_{\varkappa}$ repel each other and spread more or less
uniformly over the interval $0 \leq \omega_{\varkappa} < 2\pi$, defining a mean level distance
$\delta\omega_{\varkappa} = 2\pi/2L$. The inverse of this distance defines the break time
$n^* \simeq 2\pi/\delta\omega_{\varkappa}$ by the uncertainty principle, i.e. the time after which
the classical diffusion of the action variable stops. The localization
length $2\pi\hbar L$ on the other hand, can also be defined as the scale of
angular momentum reached via the initial diffusion with the classical
rate (2.2) within the time n^* , which closes the self-consistency
argument:

$$\langle p_{n^*}^2 \rangle = (2\pi\hbar L)^2 \simeq \frac{K^2}{8\pi^2} n^* = \frac{K^2}{8\pi^2} 2L. \tag{4.3}$$

Eq. (4.3) has a solution for finite L , $L \simeq (K/4\pi^2\hbar)^2$, which is qualitative evidence for localization. More evidence comes from numerical
studies. In summary the case for localization in the quantized standard
map for sufficiently irrational $2\pi\hbar$ looks very convincing.

5. REPEATED MEASUREMENTS

The analysis and even the definition of chaos in classical systems pre-
supposes that a large number of measurements is performed on the same
system during its evolution in time. E.g. the largest Lyapunov exponent
[3] or the Kolmogorov Sinai entropy, whose positive signs are the con-
stituent characteristics of chaos, are operationally defined by and can
be extracted from time series of data taken by repeated measurements
on the same system. In classical dynamics the back-action of the measurement can be assumed to be negligibly small, at least in principle.
In quantum mechanics this is, of course, not the case. If repeated
measurements are performed on a system their effect alters the dynamics
and has to be taken into account in a complete calculation of the time-
evolution of the observed system. One may therefore argue that a discussion of chaos in quantized dynamical systems must deal with the
complete dynamics of the system which is generated by the combined
action of the Hamiltonian and the back-action of the measuring device.
The back-action of the measuring device may, at least in principle, be
understood as the result of the (Hamiltonian) interaction of the observed microscopic system with a macroscopic system - the measuring
apparatus. It is the macroscopic nature of the measuring device (forma-
lized by taking an appropriate limit) which gives rise to the seeming-
ly non-Hamiltonian back-action on the observed microscopic system like
the collapse of the wave function.
     In view of the remarks made above we now wish to consider the
effect of taking a time-series of precise measurements of the action
variable p in the quantized standard map. We suppose that the measurement is made periodically, always after m iterations of the unperturbed
map have been performed, where m is a positive integer $m \geq 1$. The precise measurement of p at time n leads to the collapse of the wave

function $|\psi_n\rangle$ to a mixture g_n

$$|\psi_n\rangle \rightarrow g_n = \sum_{l=-\infty}^{+\infty} W_n(l) \, |l\rangle\langle l|, \tag{5.1}$$

where $W_n(l)$ is the probability that the measurement turns out the eigen-value $p_l = 2\pi\hbar l$. The statistical back-action of the macroscopic measuring device thus makes it necessary to abandon a deterministic description of the time-evolution as given by eq. (3.2) and to use a statistical description instead. In fact, it follows immediately from eqs. (3.2), (5.1) that the statistical map describing the combined evolution due to the Hamiltonian and the indicated measurements is

$$W_{n+m}(l) = \sum_{l'=-\infty}^{+\infty} |\langle l | U^m | l' \rangle|^2 \, W_n(l'). \tag{5.2}$$

We are now interested, in particular, how the results for the expectation value $\langle p^2/2 \rangle$ change due to the back-action of the measurements. If $m = 1$, i.e. if a measurement is performed after each iteration of the map, the desired expectation value can easily be evaluated in closed form for arbitrary nonlinearity K. We find in this case

$$\langle \frac{p_n^2}{2} \rangle = \frac{K^2}{8\pi^2} \, n, \tag{5.3}$$

which is the same diffusion law as derived for the classical map under the (only approximately valid) assumption of completely random and un-correlated phases. Under the present circumstances this basic assumption is exactly satisfied due to exact phase randomization by the measurement after each step. We also note that, unlike in the classical case (2.2) the diffusion law (5.3) now holds even for arbitrarily small values of K. In this sense the dynamics of the periodically measured quantum system is more random than the corresponding classical system for which back-actions from the measuring device are neglected. In fig. 3 we present numerical results for $\langle p_n^2/2 \rangle$ over 10^3 iterations for $m = 1, 2, 5, 10, 20$. The values of K and $\hbar$ are the same as in fig. 2. As one should expect the effective diffusion constant decreases monotonically with increasing m due to the tendency of the time-evolution, generated by U alone, to localize with respect to the action variable. On the other hand it can also be seen that the differential change of the diffusion constant with m is not monotonous. This is due to the strong interference effects present in the coherent time-evolution U^m which are clearly displayed in fig. 2. In summary, repeated measurements of the action variable destroy localization and lead to a diffusion of the action variable similar to the one found for the classical chaotic map. The mechanism behind this diffusion in both cases is an effective randomization of the angle variable. In the classical map the randomization is due to chaos, i.e. the positivity of a Lyapunov exponent, and

appears therefore only for sufficiently strong nonlinearity. In the
quantum map it is caused by the measuring device and not tied to the
strength of nonlinearity.

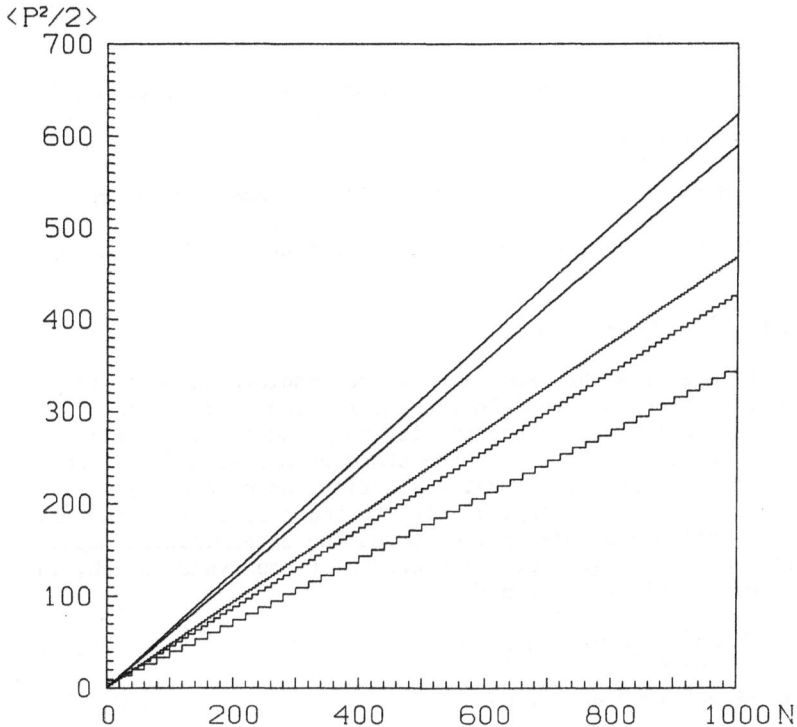

Figure 3. Expectation value of $p^2/2$ as a function of n for the quantized
standard map measured repeatedly after m iterations for $K=10$,
$2\pi\hbar = 0.15g^{-1}, g=(1-\sqrt{5})/2$, $m = 1,2,5,10,20$ (from top to bottom).

6. DISSIPATION [8-10]

Dissipation has a profound influence on dynamical systems both classically and quantum mechanically. In classical systems dissipation can
give rise to the appearance of low dimensional attractors, which are
called 'strange attractors' in the case of chaotic dynamics [3]. In
quantum systems dissipation tends to reduce or destroy quantum mechanical coherence effects. E.g. quantum tunnelling between two degenerate
classical attractors tends to be reduced by dissipation, i.e. dissipation in this case tends to localize the system near one of the two
attractors [17]. On the other hand, as discussed in section 4, a classical chaotic system may be delocalized by diffusion of the action vari-

able while the corresponding quantum system is localized. In this case
one may expect that dissipation tends to delocalize the quantum system.
The details of this effect are considered in the present section. A
quantum version of the dissipative standard map

$$p_{n+1} = \lambda p_n - \frac{K}{2\pi} \sin(2\pi q_n),$$

$$q_{n+1} = (q_n + p_{n+1})(mod\,1),$$

(6.1)

is constructed by coupling the kicked rotator (3.4) to a reservoir [8]

$$H = H_o + \sum_i \hbar\omega_i\, b_i^\dagger b_i + \sum_i \hbar g_i(\Gamma b_i^\dagger + \Gamma^\dagger b_i)$$

(6.2)

Here H_o is the Hamiltonian given in eq. (3.4), the Bose operators $b_i, b_i^\dagger$
describe harmonic modes of the reservoir with frequencies ω_i , the g_i
are coupling constants. The operator Γ (and its adjoint $\Gamma^\dagger$) is defined by

$$\Gamma = \sum_{l \geq 0} \sqrt{l l} \, (|l-1\rangle\langle l| + |-l+1\rangle\langle-l|)$$

(6.3)

and designed to describe the reduction of the absolute value of the
angular momentum by one quantum. In the interaction part of the Hamiltonian (6.2) a rotating wave approximation has been made. The elimination of the reservoir variables in the standard weak-coupling, and
Markov approximation leads to a master equation, which one may formulate again as a stroboscopic map. Similar to the case of repeated measurements, the presence of dissipation turns the deterministic map into
a stochastic map for the statistical operator of the system which, in
the $|\iota\rangle$ -representation, has the form

$$\langle\iota'|g_{n+1}|m'\rangle = \sum_{\iota,m=-\infty}^{+\infty} G(\iota',m'|\iota,m)\langle\iota|g_n|m\rangle.$$

(6.4)

The kernel G is obtained as [8,10]

$$G(\iota',m'|\iota,m) = \lambda^{\frac{1}{2}(|\iota|+|m|)} \left[U(\iota',m'|\iota,m) + \right.$$

$$\left. + \frac{1-sgn(\iota m)}{2} \sum_{j=1}^{min(|\iota|,|m|)} (\frac{1-\lambda}{\lambda})^j \sqrt{\binom{|\iota|}{j}\binom{|m|}{j}} U(\iota',m'|\iota-sgn(\iota)j, m-sgn(m)j) \right]$$

with

$$U(\iota',m'|\iota,m) = \langle\iota'|U|\iota\rangle\langle m|U^\dagger|m'\rangle.$$

(6.6)

U is defined by eqs. (3.3), (3.4) and λ is the constant already appearing in eq. (6.1). It is connected to the coupling constant and the den-

sity of modes $D(\omega_i)$ in the reservoir by

$$|\ln(\lambda)| = \pi g_i^2 D(\omega_i) . \tag{6.7}$$

To reproduce the form of the dissipative term assumed in eq. (6.1) we have to assume that the right hand side of eq. (6.7) is independent of i within the frequency range of interest. We have iterated the quantum map (6.4), (6.5) numerically using a finite basis of $|\iota\rangle$ states for $-256 \leq \iota \leq 256$. In the following we discuss our results [9,10]. First we consider the case of comparatively strong dissipation $\lambda = 0.3$ where a steady state is easily reached. The corresponding relaxation time following from the classical map is

$$n_o = \frac{1}{1-\lambda} \tag{6.8}$$

In fig. 4a the expectation value $\langle p_n^2/2 \rangle$ is plotted for $\lambda = 0.3$, $K = 5$, $2\pi\hbar = 0.01$ (full line) and compared with the classical (dotted line) and the semi-classical result (dashed line). The latter is generated by a c-number Langevin equation [18]

$$p_{n+1} = \lambda p_n - \frac{K}{2\pi} \sin\left(2\pi(q_n + \xi_{n+1})\right) + \eta_{n+1} ,$$

$$q_{n+1} = (q_n + p_{n+1} + \xi_{n+1})(\bmod 1), \tag{6.9}$$

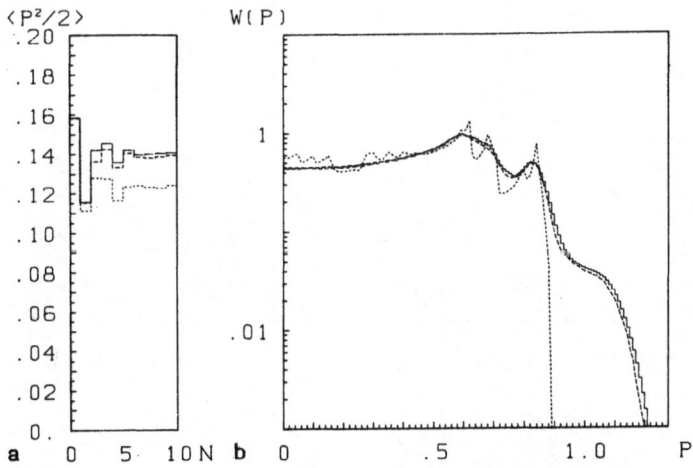

Figure 4.   (a) Expectation value of $p^2/2$ as a function of n for the quantized dissipative standard map for $\lambda = 0.3$, $K = 5$, $2\pi\hbar = 0.01$. Full line quantum mechanical result, dashed line semi-classical result, dotted line classical result (from [9])

(b) Probability distribution of the action variable p in the steady state (for $n = 10$) for the same cases as shown in fig. (a) (from [9]).

with Gaussian white noise sources η, ξ , whose strengths

$$\langle \eta_{n+1}^2 \rangle = \hbar \frac{\lambda(1-\lambda)|p_n|}{2\pi} ,$$

$$\langle \xi_{n+1}^2 \rangle = \hbar \frac{1-\lambda}{8\pi\lambda |p_n|} , \qquad\qquad (6.10)$$

follow from the asymptotics of the master equation (6.5) for $\hbar \to 0$.
As can be seen from fig. 4a a steady state is reached rapidly, in agreement with (6.8), and the semi-classical description accounts for most of
the quantum noise. In fig. 4b the probability distribution of the action
variable in the steady state is given for the three cases of fig. 4a.
The pronounced cusps of the classical distribution which reflect the
different branches of the classical attractor (cf. fig. 5) are smoothened by the quantum noise. Again the semi-classical approximation is
quantitatively a good representation of the quantum results. In fig. 5
we give a representation of the stationary phase-space density $W(p,q)$
of the classical map, again for $\lambda = 0.3$, $K = 5$, by plotting $W(p,q)$ for
fixed values of p against q on an abscissa displaced proportional to
the chosen p above the base-line. By symmetry it is not necessary to
plot also values $p < 0$. Different arms of the classical strange attractor are resolved in fig. 5a and an apparent high non-uniformity of the
distribution along the attractor is visible, caused by an insufficient
resolution of the attractor's fine structure. In fig. 5b we plot a
stationary Wigner phase space density of the quantum map for $2\pi\hbar = 0.01$

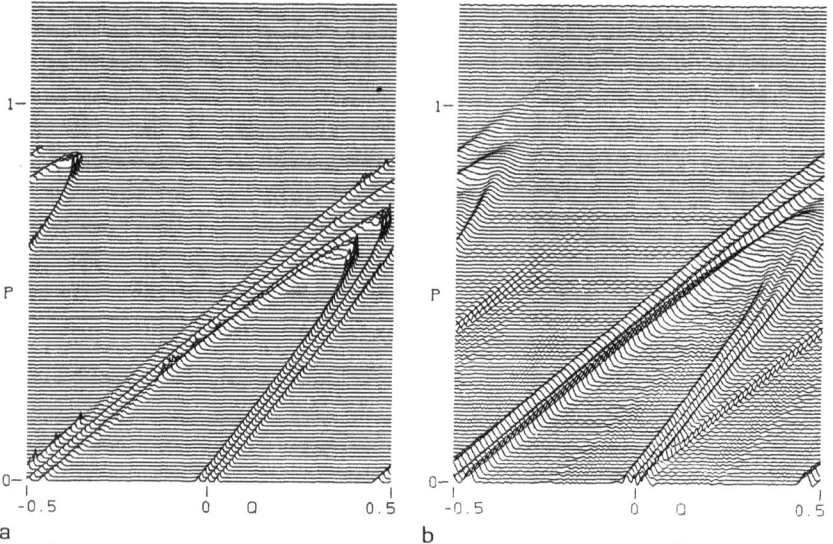

Fig. 5.   (a) Phase space distribution in the steady state (for $n = 10$)
of the classical dissipative standard map for $\lambda = 0.3$, $K = 5$ (from [9]).
    (b) Wigner distribution (6.11) with $2\pi\hbar = 0.01, \lambda = 0.3$, $K = 5$ for the
quantum case corresponding to fig. 5a.

and $\lambda = 0.3$, $K = 5$. The Wigner distribution plotted in fig. 5b is defined
by [*]

$$W_n(p,q) = Tr\left[g_n \sum_{m=-\infty}^{+\infty} \int_{-\frac{1}{2}}^{\frac{1}{2}} d\xi \exp(2\pi i(m(q-\hat{q})+\xi(p-\hat{p})))\right],$$
(6.11)

where we have now to distinguish the operators $\hat{p}$, $\hat{q}$ and the c-numbers
p , q by notation. The quantum mechanical distribution in fig. 5b is
smoothened compared to the classical distribution and extends to higher
values of p . Some quantum interferences (turning the Wigner distribution to negative values) are also visible in fig. 5b, but they occur
in regions with small statistical weight and they are not able to influence expectation values appreciably, as was already shown by the
success of the semi-classical approximation in fig. 4. Next we turn to
a regime of weaker dissipation, where a steady state is not reached as
easily. From the master equation (6.4), (6.5) one derives the time scale

$$n_{\varkappa} = \frac{2\pi\hbar}{(1-\lambda)\langle u_{\varkappa}||p||u_{\varkappa}\rangle}$$
(6.12)

in which a Floquet state $|u_{\varkappa}\rangle$ decays due to dissipation. We are in-
terested in the case where the initial state is $g_0 = |0\rangle\langle 0|$. This
state has overlap only with Floquet states within a localization length
$2\pi\hbar L$ of $p=0$, i.e. in (6.12) we may put $\langle u_{\varkappa}||p||u_{\varkappa}\rangle = 2\pi\hbar L$. The decay time of the Floquet states of interest is therefore

$$n_a \simeq \frac{1}{(1-\lambda)L} .$$
(6.13)

A regime of weak dissipation may reasonably be defined by the condition

$$\delta\omega_{\varkappa} > \frac{2\pi}{n_a}$$
(6.14)

where $\delta\omega_{\varkappa} \simeq 2\pi/2L$. Eqs. (6.13), (6.14) imply the condition

$$\langle 1-\lambda \rangle < \frac{1}{2L^2} = \frac{1}{2}\left(\frac{4\pi^2\hbar}{K}\right)^4$$
(6.15)

---

[*]   We remark that this definition circumvents some inconveniences of an
alternative version of the Wigner distribution on cylindrical phase
spaces, which we used in our earlier work [9,10].

for weak dissipation. If eq. (6.15) is satisfied the quantum decay time
(6.13) is longer than the break-time n^* defined in section 4. Therefore,
one expects classical diffusion to occur initially like in the absence
of dissipation, which stops at the time n^* , then one expects localization to occur for $n^* < n < n_\varphi$, whereas for $n > n_\varphi$ one might expect a return to classical diffusion. In fig. 6 we present the numerical results
for the expectation $\langle p_n^2/2 \rangle$ as a function of time n for $2\pi\hbar = 0.3/(1-\sqrt{5})$,
$K = 10$, and various values of $(1-\lambda)$ also covering the regime (6.15).
First of all it can be seen that switching on dissipation tends to
<u>increase</u> the action variable in the quantum map, quite different from
the classical map, where dissipation tends to decrease the action variable. The reason is, of course, the breaking of localization by dissipation in the quantum map. The behavior for $n < n^*$ and for $n^* < n < n_\varphi$ (in
the case of $(1-\lambda) = 5 \cdot 10^{-6}$ and $1 \cdot 10^{-4}$) is as expected. However, for
$n > n_\varphi$, the system does not return to classical diffusion, which is also
shown in fig. 6 for reference. Instead the system diffuses at a strongly
reduced rate. We have analyzed the reason for this reduction and found
it in the $\langle |p| \rangle$ -dependence of the quantum decay rate (6.12). It follows

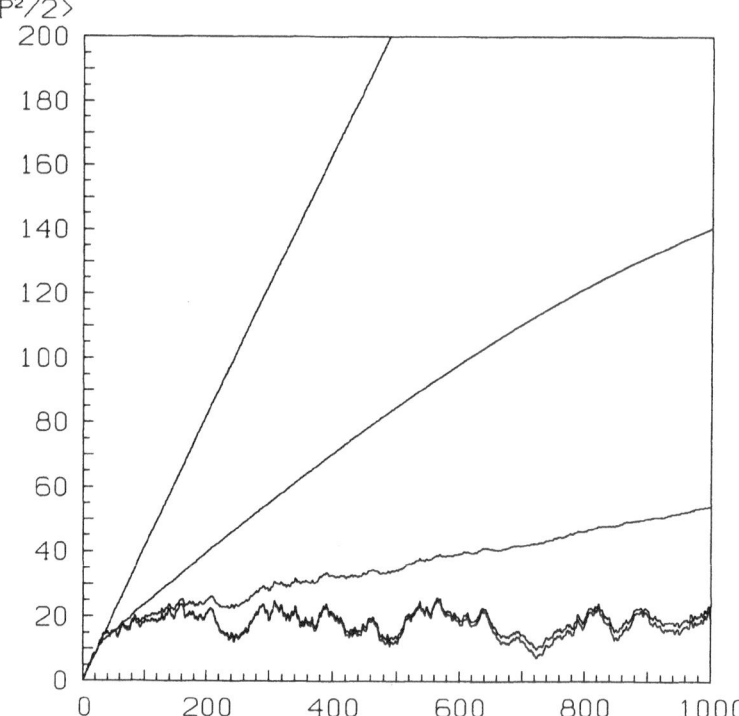

Fig. 6. Expectation value of $p^2/2$ as a function of n for the quantized
dissipative standard map with $K = 10$, $2\pi\hbar = 0.3/(1-\sqrt{5})$, and $(1-\lambda)$
increasing from lower curve to next to upper curve through $(1-\lambda) = 0$,
$5 \cdot 10^{-6}$, $1 \cdot 10^{-4}$, $1 \cdot 10^{-3}$. Upper curve for $\hbar = 0$, $(1-\lambda) = 0$; initial
state $|\psi_0\rangle = |0\rangle$.

from (6.12) that states with very small $\langle |p| \rangle$ are not destroyed by
dissipation for a very long time. This severely inhibits the transport
of the probability weight from very small values of $|p|$ to larger $|p|$.
In fig. 7 we plot the probability distribution $W(p)$ reached after 10^3
iterations for $2\pi\hbar = 0.3/(1-\sqrt{5})$, $K = 10$, $(1-\lambda) = 1 \cdot 10^{-4}$, and compare
with the corresponding classical distribution. It can clearly be seen
that even after 10^3 iterations a large fraction of the probability
weight (the logarithmic scale in fig. 7 should be noted) remains frozen
at or near $p = 0$. The time-scale on which the probability weight near
$p = 0$ can decay can be obtained by taking the supremum of $n_{\varkappa}$, eq. (6.12),
over all quasi-energy states $|u_{\varkappa}\rangle$ which have appreciable overlap with
the initial state. This time-scale may become very long, much longer
than n_{a} , but it is finite due to the discreteness of the quasi-energy
spectrum. A typical value for this time-scale may be obtained by esti-
mating $\inf(\langle u_{\varkappa}||p||u_{\varkappa}\rangle) \simeq 2\pi\hbar$ and we then find that $\sup(n_{\varkappa}) \simeq (1-\lambda)^{-1} = n_{0}$.
    We have checked numerically that the spreading of the probability
weight at large p in fig. 7 is, in fact, satisfactorily described by a
diffusion with the classical diffusion constant. Therefore, it is

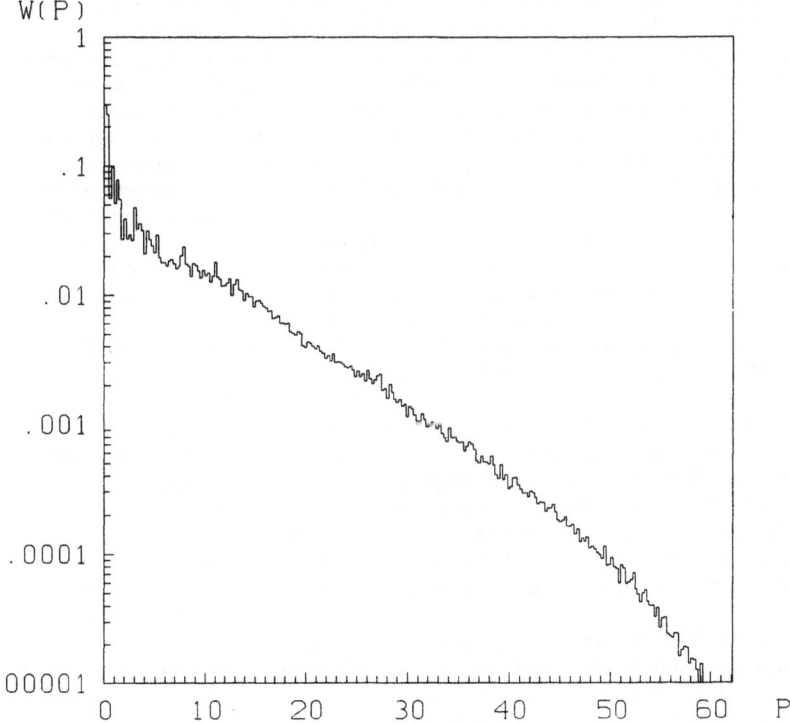

Fig. 7. Probability distribution of p for the quantized dissipative
standard map for $(1-\lambda) = 1 \cdot 10^{-4}$, $K = 10$, $2\pi\hbar = 0.3/(1-\sqrt{5})$, after 10^3
iterations; initial state $\varrho_{0} = |0\rangle\langle0|$.

clearly the reduced flux of probability current from small $|p|$ to larger $|p|$ for $n < (1-\lambda)^{-1}$ that leads to reduced overall diffusion in the dissipative quantum map for times $n_a < n < n_b$. If n approaches $n_b \approx (1-\lambda)^{-1}$ a steady state is again reached. Indications for the approach of this regime can be seen in fig. 6 for the curve with $(1-\lambda)^{-1} = 1 \cdot 10^{-3}$.

7. CONCLUSION

Classical chaotic Hamiltonian systems and their quantum mechanical counterparts show qualitatively different dynamical behavior. The quasi-periodic dynamics of bounded Hamiltonian quantum systems gives way to chaotic dynamics if the classical limit $\hbar \to 0$ is taken, i.e. the quantum system develops an instability in that limit. Unbounded Hamiltonian quantum systems, like the kicked rotator, may or may not display quasi-periodic dynamics. If they do, like the kicked rotator seems to do for sufficiently irrational $2\pi\hbar$, the quasi-energy spectrum is discrete and the quasi-energy states are localized. There is then a qualitative change, i.e. an instability, of the dynamics in the classical limit also in this case. If quantum systems are periodically measured or if the dissipation effects are not negligible the pure deterministic quantum dynamics $|\psi(0)\rangle \to |\psi(t)\rangle$ must be modified by coupling the system to macroscopic reservoirs describing the measuring device or the environment which absorbs the dissipated quantities (energy, angular momentum etc.). The perturbations of the original Hamiltonian system due to these reservoir couplings are essential perturbations in the sense that they change the nature of the instability of the pure quantum dynamics at $\hbar \to 0$. In the quantized kicked rotator with dissipation, which we have studied in this lecture, the parameters which control the instability are $2\pi\hbar$, the dissipation $(1-\lambda)$ and the time-interval n over which the dynamics is considered. In this example dissipation is an essential perturbation of the instability at $\hbar \to 0$ because it tends to destroy localization, which is the qualitative difference between the quantum dynamics and the classical dynamics. It is well known that in the conservative system $(1-\lambda) = 0$ the limits $\lim\limits_{\hbar \to 0} \lim\limits_{n \to \infty} \neq \lim\limits_{n \to \infty} \lim\limits_{\hbar \to 0}$ do not commute, in general. The reason is that the limit on the left is approached for $n > n^*$ i.e. through localized states, while on the right it is approached for $n^* > n$ i.e. through a regime in parameter space where diffusion prevails. This is no longer true in the system with finite dissipation $(1-\lambda) > 0$, where a steady state exists both in the classical and in the quantum system, and the classical steady state is a smooth limit of the quantized steady state. Hence $\lim\limits_{\hbar \to 0} \lim\limits_{n \to \infty} = \lim\limits_{n \to \infty} \lim\limits_{\hbar \to 0}$ if $(1-\lambda) \neq 0$. As $(1-\lambda) \neq 0$ destroys localization the limit is approached on both sides in such a way that diffusion prevails. On the other hand, if $(1-\lambda)$ becomes very small, it may take a very long time, $n > n_a$, before diffusion can be seen in the quantum system, and, since n_x given by (6.12) depends on the inverse of the scale of the action variable there exist quantum corrections which decay much more slowly than on the time-scale n_a. Depending on the duration n of observation and the dissipation parameter $(1-\lambda)$ the asymptotics in the following different limits is relevant, provided the quantum system is weakly damped and close to its classical limit:

$n < n^*$, $n^* \gtrsim n_a$, $\hbar \to 0$ - the classical limit of the dissipative standard map, semi-classical Langevin equation (6.9); cf. fig. 4,5.

$n^* < n < n_a < n_b$, $(1-\lambda) \to 0$ - the quantized conservative standard map (3.3), (3.4); cf. fig. 6 for $(1-\lambda) = 5 \cdot 10^{-6}$.

n^*, $n_a < n < n_b$ - no limit may be taken because the map has not yet reached a steady state and behaves classically for large actions, but quantum effects are still important for small actions; cf. fig. 6 for $(1-\lambda) = 1 \cdot 10^{-4}$, $1 \cdot 10^{-3}$ and fig. 7.

n^*, $n_a < n_b < n$, $n \to \infty$ - the steady state of the quantized standard map with dissipation. A numerical example for the last case is difficult to generate because of the necessity of a large number of iterations. However, the approach towards a steady state for this case after 10^3 iterations is already clearly visible in fig. 6 for $(1-\lambda) = 1 \cdot 10^{-3}$.

ACKNOWLEDGEMENTS

Financial support of this work by the Deutsche Forschungsgemeinschaft through the Sonderforschungsbereich 237 "Unordnung und große Fluktuationen" is gratefully acknowledged. One of us (T.D.) acknowledges additional support by the Studienstiftung des Deutschen Volkes in 1985-1987. Thanks are due to Prof. G. Eilenberger, KFA Jülich, for his interest and special cooperation.

REFERENCES

[1]   O. Bohigas, M.J. Giannoni, C. Schmit, Phys. Rev. Lett. 52, 1
       (1984); E. Haller, H. Köppel, L.S. Cederbaum, Phys. Rev. Lett.
       52,1165 (1984).
[2]   B.V. Chirikov, Preprint no. 267, Inst. Nucl. Phys., Novosibirsk
       (1969); Phys. Rep. 52, 263 (1979).
[3]   A.J. Lichtenberg, M.A. Lieberman, Regular and Stochastic Motion,
       Springer, Berlin 1983.
[4]   G. Casati, B.V. Chirikov, F.M. Izrailev, J. Ford, in Lecture
       Notes in Physics vol. 93, p. 334, Springer, Berlin 1979.
[5]   B.V. Chirikov, F.M. Izrailev, D.L. Shepelyansky, Sov. Sci. Rev.
       C2, 209 (1981).
[6]   S. Fishman, D.R. Grempel, R.E. Prange, Phys. Rev. Lett. 49, 509
       (1982); Phys. Rev. A29, 1639 (1984).
[7]   D.L. Shepelyansky, Phys. Rev. Lett. 56, 677 (1986).
[8]   T. Dittrich, R. Graham, Z. Physik B62, 515 (1986).
[9]   T. Dittrich, R. Graham, Europhys. Lett. 4, 263 (1987).
[10]   T. Dittrich, R. Graham, in Chaos and Related Nonlinear Phenomena;

     Where Do We Go From Here?, ed. I. Procaccia, Plenum, New York
     1988.
[11] R.S. MacKay, J.D. Meiss, I.C. Percival, Physica D13, 55 (1984).
[12] A.B. Rechester, R.B. White, Phys. Rev. Lett. 44, 1586 (1980).
[13] G. Casati, J. Ford, I. Guarneri, F. Vivaldi, Phys. Rev. A34, 1413
     (1986).
[14] F.M. Izrailev, D.L. Shepelyansky, Teor. Mat. Fiz. 43, 417 (1980).
[15] G. Floquet, Ann. de l'Ecole Norm. Sup. 12, 47 (1983).
[16] E.V. Shuryak, Zh. Eksp. Teor. Fiz. 71, 2039 (1976). (Sov. Phys.
     JETP 44, 1070 (1976)).
[17] A.O. Caldeira, A.J. Leggett, Phys. Rev. Lett. 46, 211 (1981);
     Ann. Phys. (N.Y.) 149, 374 (1983); 153, 445 (1984).
[18] R. Graham, Europhys. Lett. 3, 259 (1987).

REDUCTION OF QUANTUM NOISE IN ATOMIC SYSTEMS

M. Orszag
Catholic University of Chile
Casilla 6177, Santiago 22
Chile.

ABSTRACT. The quantum noise of an atomic system can be reduced by the idea of correlated emission laser (CEL). Here, we analize various CEL systems and how the phase diffusion coefficient is reduced in each case.

1. CORRELATED EMISSION LASERS

In the optical detection of small changes of a given physical quantity, the change is converted into a phase shift (passive scheme) or frequency shift (active scheme) of a laser field. This is accomplished by sending a laser light through an optical cavity whose optical path length is sensitive to the physical effect to be measured. This shift is then detected by beating the output light with that from a reference laser. The typical examples are gravitational wave detectors and laser gyroscope.

In the active detection scheme, the limiting noise source is the fluctuation, caused by independent spontaneous emission events in the relative phase between the two lasers. It has been shown recently [1] that the linewidth and the associated uncertainty in the relative phase can be eliminated by preparing the laser medium in a coherent superposition of the two upper states in a three level atoms, via a resonant microwave field as in the quantum beat experiments or by coherent pumping as in the Hanle Laser or correlated emission free electron lasers.

In general, in all these examples of correlated emission laser, phase locking between the two modes is achieved in such a way that a complete quenching of the relative phase a noise of the two modes is achieved.

Finally, we notice that the quantitative treatment of the Correlated Emission Laser is based on the following decomposition of the phase noise [2]:

$$< (\delta\theta)^2 > = \frac{1}{4\bar{n}} + <:(\delta\theta)^2 >,\tag{1}$$

163

E. Tirapegui and D. Villarroel (eds.), Instabilities and Nonequilibrium Structures II, 163–167.
© 1989 by Kluwer Academic Publishers.

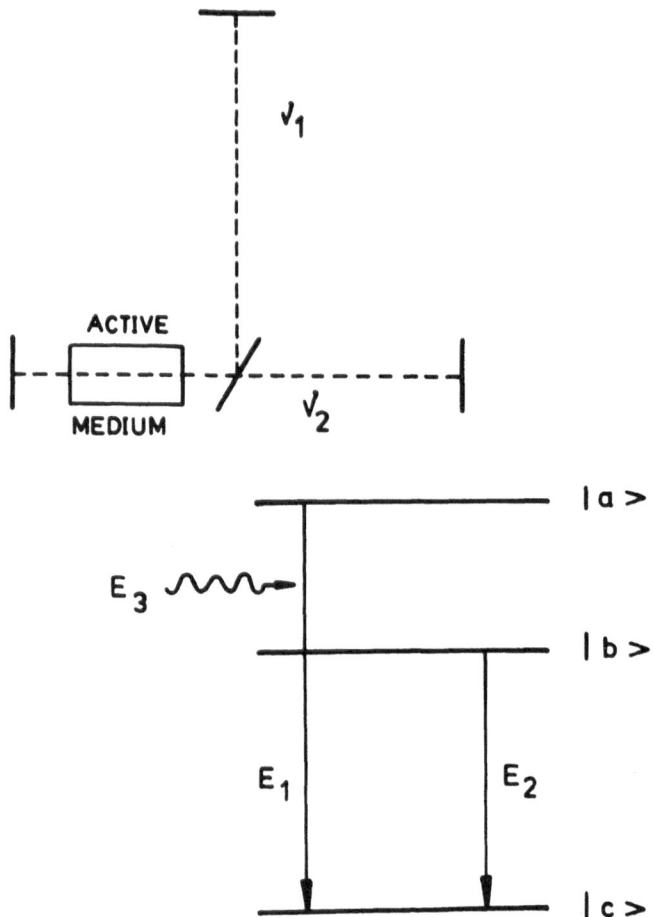

Figure 1. The quantum beat laser. A collection of three level atoms inside a double cavity. A coherent superposition of the upper levels is achieved by pumping resonantly with a microwave field E_3.

detuning conditions. Defining $\Delta_1 = \omega_{ac} - \nu_1$ and $\Delta_2 = \omega_{bc} - \nu_2$ as the detunings of the two atomic transitions with respect to the cavity modes, then if $\Delta_1 = \Delta_2 = \Omega/2$, Ω being the Rabi frequency of the driving microwave field at frequency ν_3, then the diffusion coefficient of the relative phase $D(\theta)$ is given by [6]:

$$D(\theta) = \frac{g^2 r_a}{4\gamma^2 \bar{n}} (1 - \cos\psi), \tag{2}$$

where $1/4\bar{n}$ represents the shot noise, whose origin is in the vacuum fluctuations of the reservoir, :: stands for normal ordering, so that the second term in the equation (1), represents the normal ordered added noise in an active system and it can be obtained from a Fokker Planck treatment of the system. This last term can be uniquely related to the relative phase diffusion constant obtained from the relevant Fokker Planck equation in each correlated emission laser system (CEL system).

Recently, there have been reports of experimental evidence supporting the CEL effect. P.E. Toschek and J. Hall [3,4], reported the first observation of the CEL effect. In a He-Ne Zeeman Laser, operating at 633 nm, they had a two mode laser system. The beat signal, in the free running case, between the σ_+ and σ_- polarizations, gave a linewidth of $1/4$ H_z for a measurement time of a few seconds. When the r.f. magnetic field H_1 (perpendicular to the axial Zeeman field H_0) was supplied to the gain medium, the fluctuations of the beat signal were reduced by a large factor, to less than $1/10$ of the Schawlow-Townes limit. A more recent second experiment [5] was performed in a semiconductor laser with an extended cavity. As a result, the heterodyned spectral width between the two lasing modes in a grating extended cavity laser was reduced to below the spontaneous emission noise level.

A key point in the theory of the CEL operation is that the relative phase diffusion constant depends or is an explicit function of this relative phase between the two laser signals θ. In all cases to be discussed in the next section, $D(\psi) \backsim (1-\cos\psi)$, with $\psi = \theta-(\varphi_a - \varphi_b)$, where φ_a and φ_b are the phases associated with the upper two atomic levels a and b.

2. VARIOUS CEL SYSTEMS

2.1. The quantum beat laser

In the quantum beat laser, an active medium, consisting of a collection of three level atoms, where the upper two levels are coherently excited, drives a doubly resonant cavity, as shown in the figure 1.

The coherence of the upper two levels is achieved by shining a strong microwave signal E onto the system, resonant with the $|a>$ -$|b>$ transition.

A recent publication [6] shows that in a linear theory, one can write a Fokker Plank equation for a Glauber's P distribution in polar

coordinates $P(\rho_1,\rho_2,\theta,\mu)$, where $\alpha_1 = \rho_1 e^{i\theta_1}$

$\alpha_2 = \rho_2 e^{i\theta_2}$, $\theta = \theta_1 - \theta_2$, $\mu = (\theta_1 + \theta_2)/2$

The main result of this analysis shows that the injected atomic coherence generates a strong correlation between the two signals at frequencies ν_1 and ν_2 in such a way that the diffusion constant of the relative phase of the two signals will vanish under certain where in Eq. (20) we have assumed that the two coupling constants of the two transitions are equal ($g_1 = g_2 = g$), γ i the inverse atomic relaxation time of the two upper levels (taken to be the equal) to the lower level. The constant r is the atomic injection rate and $\psi = -\theta + \Phi$, Φ being a characteristic phase of the input microwave field.

A fully non-linear theory of the quantum beat laser has been also develope [7] thus generating a Fokker Planck equation with non-linear term up to g^4 included. The relative phase diffusion constant has, now, a non-linear contribution, which also vanishes when $\psi = 0$. Therefore, the CEL effect persists in higher orders.

2.2. Other CEL systems

There are various other schemes, besides the quantum beat laser, that exhibit the CEL effect. A first example, is the Hanle laser, where the active medium is prepared initially in a coherent excitation of the upper two levels $|a>$ and $|b>$, which will decay to the lower state $|c>$, via emission of the different polarization states. This is the Hanle effect and it can be achieved with a system similar to that shown in the figure 1, except that the center mirror is polarization sensitive and that the two modes have the same frequency but orthogonal polarizations.

One can show [8] that a linear theory gives a vanishing relative phase diffusion constant $D(\theta)$ for equal detunings. A fully non-linear theory gives similar results [9].

Another example is the holographic laser, where one has a ring laser with two counterpropagating modes coupled by a spatial modulation of the gain medium. In a holographic laser [10], each beam is reflected in part by the thin atomic layers of the gain medium. When the reflected light interferes constructively with the counterpropagating beam, noise quenching is achieved.

A third example is the correlated emission free electron laser. In the two previous examples, the strong microwave signal of the quantum beat laser was not necessary. One rather prepared initially the atom in a coherent superposition state. In a similar way, one can think of preparing an electron in a coherent superposition of the two momentum states in a compton regime free electron laser configuration. Once more, such an initial preparation of the electrons can lead to noise quenching, provided the difference of the two momenta does not exceed twice the momentum of the emitted photon [11]. This result is somewhat restrictive, but shows the possibility of CEL effect in the different modes of a free electron laser. Finally, a two photon correlated emission laser is also possible, where now the atomic coherence is established between the $|a>$ and $|c>$ levels, assuming and intermediate b level near the center of the $|a> - |c>$ transition. The $|a> - |c>$ transition is not dipole allowed to emit a single photon. However a two photon laser, each photon with half of the $|a> - |c>$ energy gap, can be achieved. One is again interested in the role of the atomic coherence between the most distant levels $|a>$ and $|c>$, in quenching the quantum noise. Referring back to the equation (1) and under certain conditions, the phase noise is reduced below the shot noise level, that is one can generate squeezed light [12].

3. DISCUSSION

The Correlated Emission Laser effect has been demonstrated both experimentally and theoretically, thus opening a new avenue for measurements of ultrasmall effects such as detection of gravitational waves. It also shed light in a fundamental problem. That is, when looking at the equation (1), a relevant question is how far can one reduce the phase fluctuations of a given optical system. Thus far, it is obvious that the normally ordered phase fluctuations, coming from a Fokker Planck equation, can be reduced to zero in a phase sensitive system. However, in the last example, we mentioned that one can even go beyond that point, with the phase fluctuations below the shot noise level, thus generating squeezed light. A more detailed and general discussion on this topic will be published [2].

REFERENCES

1. M.O.Scully, Phys. Rev. Lett. **55**, 2802 (1985).
2. J. Bergou, M. Orszag, M. O. Scully, K. Wodkiewicz (to be published).
3. P.E. Tosckek and J. Hall, Abstract in XV International Conference on Quantum Electronics J.O.S.A., B, **4**, 124 (1987).
4. For a theoretical account of the experiment, see: J. Bergou, M. Orszag, J.O.S.A. B, **5**, 249 (1988).
5. M. Ohtsu and K. Y. Liou (preprint).
6. M. O. Scully and M.S. Zubairy, Phys. Rev. A. **35**, 752 (1987).
7   J. Bergou, M. Orszag, M.O. Scully (to be published).
8. J. Bergou and M.O. Scully (to be published).
9. J. Bergou, M. Orszag, M.O. Scully (to be published).
10. J. Krause, M.O. Scully, Phys. Rev. A36, 1771 (1987).
11. M. Orszag, W. Becker and M.O. Scully, Phys. Rev. A, **36**, 1310 (1987).
12. M.O. Scully, K. Wodkiewicz, M.S. Zubairy, J. Bergou, Ning Lu and J. Meyer ter Vehn (to be published).

PART II

INSTABILITIES IN NONEQUILIBRIUM SYSTEMS

DIRECT SIMULATION OF TWO-DIMENSIONAL TURBULENCE

M. E. Brachet
CNRS, Laboratoire de Physique Statistique
Ecole Normale Supérieure
24 rue Lhomond 75231 Paris Cedex 05
FRANCE

ABSTRACT. Direct numerical simulations of decaying high Reynolds number turbulence are presented at a resolution of 2048^2 for a periodic flow with large scale symmetries. We characterize a transition of the inertial energy spectrum exponent from $n \simeq -4$ at early times to $n \simeq -3$ at latter times. In physical space, the first regime is associated with isolated vorticity gradient sheets, as predicted by Saffman (1971). The second regime corresponds to an enstrophy cascade (Kraichnan 1967, Batchelor 1969) and reflects the formation of layers resulting from the packing of vorticity gradient sheets.

We review the linear description of two-dimensional turbulence of Weiss (1981), which predicts that coherent vortices will survive in regions where vorticity dominates strain, while vorticity gradient sheets will be formed in regions where strain dominates and show that this analysis remains valid after vorticity gradient sheets have been formed.

1.Introduction

Two-dimensional incompressible turbulence is governed by the Navier-Stokes equation

$$\partial_t \mathbf{v} + (\mathbf{v} \cdot \nabla)\mathbf{v} = -\nabla p + \nu \nabla^2 \mathbf{v}$$

$$\nabla \cdot \mathbf{v} = 0 \tag{1.1}$$

initial and boundary value.

At large Reynolds numbers, two-dimensional turbulence displays interesting specific properties: the vorticity $\omega = \nabla \times \mathbf{v}$ is perpendicular to the plane of the flow and satisfies

$$\partial_t \omega + (\mathbf{v} \cdot \nabla)\omega = \nu \nabla^2 \omega. \tag{1.2}$$

It is thus a (pseudo) scalar $\omega = \partial_x v - \partial_y u$, conserved along the fluid trajectories in the inviscid limit. This prevents vorticity stretching and thus the development of an energy cascade to the small scales, a central feature of three-dimensional turbulence. In two-dimensions, the mechanism for small scale generation is the stretching of vorticity gradients. In the inviscid limit, the latter satisfies

$$\partial_t \nabla\omega + (\mathbf{v} \cdot \nabla)\nabla\omega = -(\nabla\mathbf{v}) \cdot \nabla\omega. \tag{1.3}$$

It is also convenient to consider

$$\eta = \nabla \times \omega = \begin{bmatrix} \partial_y \omega \\ -\partial_x \omega \end{bmatrix}, \tag{1.4}$$

171

E. Tirapegui and D. Villarroel (eds.), Instabilities and Nonequilibrium Structures II, 171–179.
© *1989 by Kluwer Academic Publishers.*

which is deduced from $\nabla\omega$ by a rotation of $\pi/2$ and satisfies

$$\partial_t \eta + (\mathbf{v} \cdot \nabla)\eta = \eta \cdot (\nabla\mathbf{v}). \tag{1.5}$$

Vorticity conservation and vorticity gradient production are the basic elements of small scale dynamics in two-dimensional turbulence. As noticed by Saffman (1971), the advection of vorticity along the fluid trajectories may bring close together different values of ω, producing thin layers between macro-eddies across which vorticity jumps. Such quasi-discontinuities of vorticity along rectilinear structures would lead to an inertial range with a k^{-4} energy spectrum. A different point of view was presented by Kraichnan (1967) and Batchelor (1969) who used a statistical approach. By analogy with the direct energy cascade in three-dimensions, they conjectured the existence in two-dimensions, of an enstrophy (mean square vorticity) cascade to the small scales. Phenomenology and dimensional considerations then lead to a k^{-3} inertial energy spectrum (with a possible logarithmic correction suggested by Kraichnan (1971) to take into account the effect of non-local interactions).

2. Numerical Method

We assume that the flow is 2π - periodic, and use Fourier spectral methods for the space variables because they are both precise and easy to implement (Gottlieb and Orszag 1977). We further assume that the stream function can be written

$$\psi(x,y,t) = \sum_{k_x=1}^{N/2} \sum_{k_y=1}^{N/2} a(\mathbf{k},t) \sin(k_x x) \sin(k_y y), \tag{2.1}$$

where the $a(\mathbf{k},t)$ coefficients are non zero only when k_x and k_y are both even or both odd. This representation, which is compatible with the Navier Stokes equations, corresponds in physical space to the following symmetries: (i) invariance by rotation of π around the point $x = \pi/2$, $y = \pi/2$, (ii) reflectional symmetry on the sides of an impermeable box $x = 0$ and π, $y = 0$ and π. This so-called "sparse mode technique" was first implemented in (Brachet, Meiron, Orszag, Nickel, Morf and Frisch 1983) for the three dimensional Taylor-Green vortex. For a given ratio between the larger and the smaller scales retained in the system, this method leads to a significant reduction of both storage and operation number.

The equation for the vorticity is integrated in the form

$$\partial_t \omega + \partial_x(u\omega) + \partial_y(v\omega) = \nu\nabla^2\omega, \tag{2.2}$$

where u and v are the two velocity components.

The time stepping is done in Fourier space with a second order leap-frog-Crank-Nicolson scheme of the form:

$$\frac{\omega_{n+1}(\mathbf{k}) - \omega_{n-1}(\mathbf{k})}{2\delta t} = -\nu k^2 \frac{\omega_{n+1}(\mathbf{k}) + \omega_{n-1}(\mathbf{k})}{2} + F_n(\mathbf{k}), \tag{2.3}$$

where $\omega_n(\mathbf{k})$ denotes the k-vorticity mode at time $n\delta t$ and $F_n(\mathbf{k})$ is the Fourier transform of the non linear term at time $n\delta t$. $F_n(\mathbf{k})$ is computed from $\omega_n(\mathbf{k})$ by the following procedure. In Fourier space, the stream function is related to the vorticity by

$$\psi_n(\mathbf{k}) = \frac{\omega_n(\mathbf{k})}{k^2}, \tag{2.4}$$

and the Fourier transform of the velocity

$$u = \partial_y \psi$$
$$v = -\partial_x \psi$$

reads

$$u_n(\mathbf{k}) = ik_y \psi_n(\mathbf{k})$$
$$v_n(\mathbf{k}) = -ik_x \psi_n(\mathbf{k}). \tag{2.5}$$

After transforming the velocity and the vorticity to physical space, one computes $\alpha_n(\mathbf{x}) = u_n(\mathbf{x})\omega_n(\mathbf{x})$ and $\beta_n(\mathbf{x}) = v_n(\mathbf{x})\omega_n(\mathbf{x})$ at the collocation points. One finally transforms back to Fourier space to calculate $ik_x\alpha_n(\mathbf{k})$, $ik_y\beta_n(\mathbf{k})$ and finally $\omega_{n+1}(\mathbf{k})$.

The initial data is a realisation of the gaussian (pseudo)-random field such that the energy spectrum

$$E(k) = \frac{1}{2} \sum_{k-1/2 < |\mathbf{k}'| < k+1/2} |\mathbf{v}(\mathbf{k}')|^2, \tag{2.6}$$

is initially given by

$$E_0(k, t = 0) = Ck \exp(-(k/k_o)^2), \tag{2.7}$$

with $k_0 = 5$ and total energy $E = 6 \ 10^{-2}$. The viscosity was $\nu = 2.33 \ 10^{-5}$.

3. Results

Among the significant integral quantities that describe the development of a two-dimensional turbulence initially concentrated in the large scales, one can consider the (two-dimensional) skewness and the enstrophy dissipation. The 2D-skewness S which is a non-dimensional measure of the rate of production of mean square vorticity gradients by non-linear effects, is defined as follows (Herring et al. 1974). Let

$$T_\omega(k) = \sum_{k-1/2 < |\mathbf{k}'| < k+1/2} \sum_{\mathbf{k}''} |\omega(-\mathbf{k}')\mathbf{v}(\mathbf{k}' - \mathbf{k}'')i\mathbf{k}''\omega(\mathbf{k}'')|, \tag{3.1}$$

denote the enstrophy transfer through the k-shell at time t. Then

$$S = \frac{\sum_k k^2 T_\omega(k)}{\sum_k k^4 E(k)(\sum_k k^2 E(k))^{1/2}}. \tag{3.2}$$

Fig. 1a shows the evolution of the skewness. We see that after an early rapid growth corresponding to the period of small scale generation, S reaches a maximum around $t = 1$ and then decays slowly. Fig. 1b shows the evolution of the enstrophy dissipation

$$\Sigma = \nu \int (\nabla \times \omega)^2 d\mathbf{x}. \tag{3.3}$$

It reaches a maximum at a time $t \simeq 6$ (significantly later than the skewness maximum), and then decays.

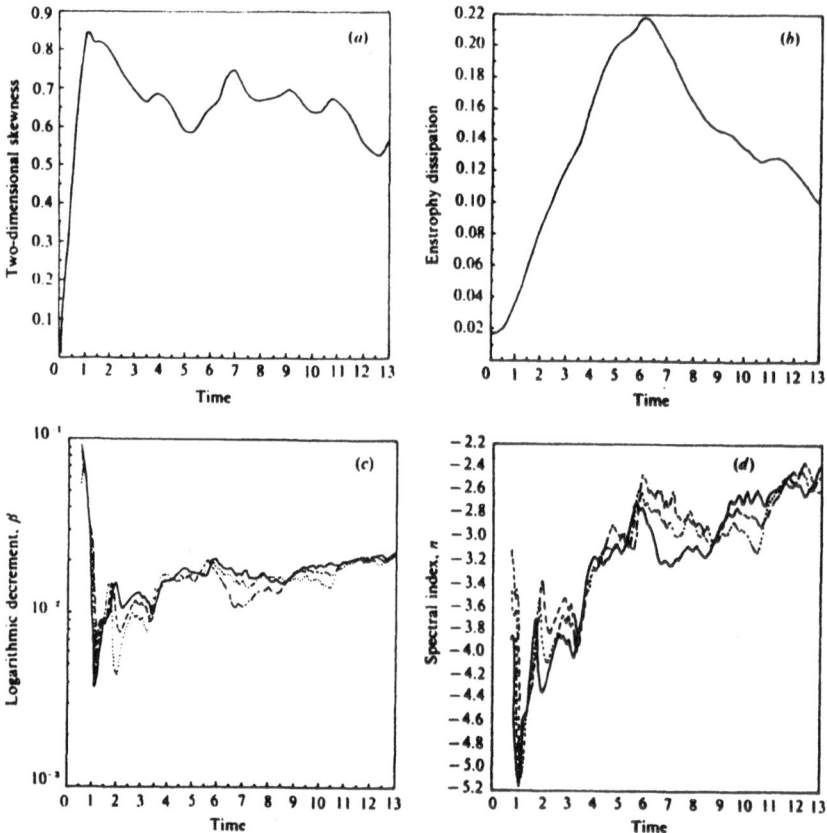

Figure 1. Spectral evolution of the run. Fig. 1.a skewness, Fig. 1.b total enstrophy dissipation, Fig. 1.c logarithmic decrement and Fig. 1.d spectral exponent of the energy spectrum versus time.

In order to extract quantitative information about the power law exponent and the high wavenumber exponential decay, we have resorted to analyzing the energy spectrum in terms of an assumed functional form. We fit the numerically computed energy spectrum with a function $A(t)k^{-n(t)} \exp(-\beta(t)k)$. In this way, we estimate both the smallest significantly excited scale by the logarithmic decrement β (Sulem, Sulem and Frisch 1983, Brachet et al. 1983), and the spectral exponent n.

Fig. 1c shows that the early time dynamics is characterized by an exponential decay of the logarithmic decrement. This process stops when skewness reaches its maximum. At this time, scales small enough to be dissipated have been excited. Vorticity gradient production is then inhibited and the width of the analyticity strip $\delta(t) \simeq \beta(t)/2$ of the Navier Stokes solution stabilizes. Just after the stabilization, the spectral exponent is close to the value $n = -4$ (Fig. 1d), predicted by the Saffman (1971) theory. Later, around the time of maximum enstrophy dissipation ($t \simeq 6$), the spectral exponent displays a sharp transition to a value close to $n = -3$, a value consistent with the enstrophy cascade (Kraichnan 1967, Leith 1968, Batchelor 1969.

One of the main interests of direct numerical simulations is that, when visualized, they give a complete picture of the turbulent flow. At $t = 3$ (Fig. 2), a time for which the energy spectrum displays a k^{-4} range, we observe thin and isolated layers across which vorticity changes drastically. As suggested by Saffman (1971), these sheets may be viewed as boundary layers between large scale eddies. We also see that these layers correspond to vortex-tails produced by stretching of vorticity gradients.

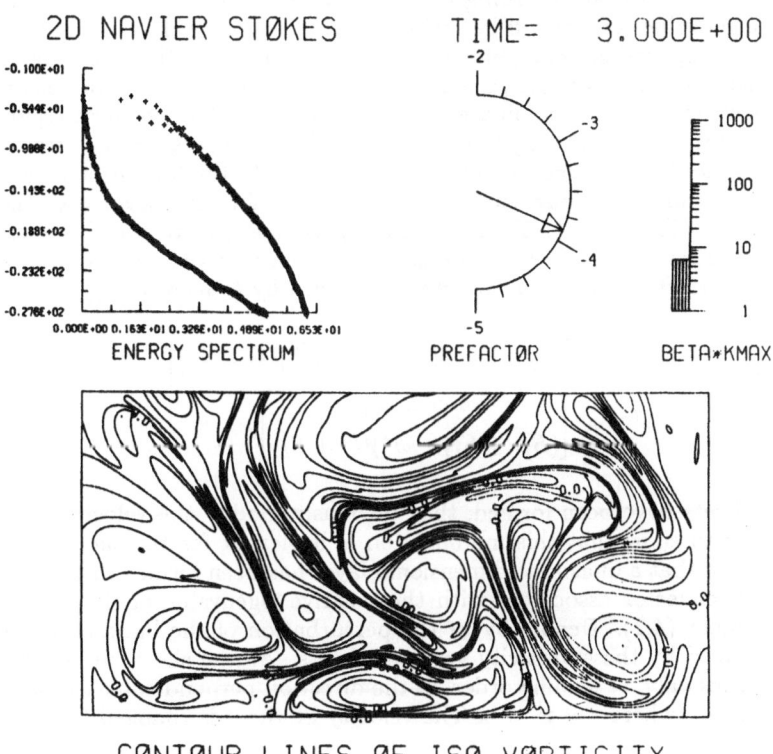

Figure 2. Physical space visualizations at $t = 3$: contour lines of vorticity ω.

It is possible to characterize the regions where vorticity gradient sheets will be formed: following Weiss (1981) we assume a (temporal and spatial) scale separation between velocity and vorticity gradients. In other words, $\nabla \mathbf{v}$ is assumed to be frozen as far as the dynamics of $\nabla \times \omega$ is concerned. Eq.(1.5) then becomes linear in Lagrangian coordinates and the evolution of $\nabla \times \omega$ is prescribed by the eigenvalues of the velocity gradient. It is convenient to rewrite $\nabla \mathbf{v}$ in terms of the strain, $S = \nabla \mathbf{v} + \nabla \mathbf{v}^{tr}$ and the vorticity:

$$\nabla \mathbf{v} = \frac{1}{2} \begin{pmatrix} S_{11} & S_{12} + \omega \\ S_{12} - \omega & -S_{11} \end{pmatrix}, \tag{3.4}$$

where $S_{11} = 2\partial_x u = -2\partial_y v$, $S_{12} = \partial_x v + \partial_y u$, and $\omega = \partial_x v - \partial_y u$. The eigenvalues of $\nabla \mathbf{v}$ then read

$$\lambda_{\pm} = \pm\sqrt{-\det(\nabla \mathbf{v})} = \pm\frac{1}{2}\sqrt{S_{11}^2 + S_{12}^2 - \omega^2}. \tag{3.5}$$

This equation shows that according to the relative importance of strain and vorticity, the eigenvalues of $\nabla \mathbf{v}$ will be real or purely imaginary. In the regions where strain dominates, vorticity gradients are stretched, leading to the formation of vorticity gradient sheets. In these regions, the motion is hyperbolic. In contrast, in the regions where vorticity dominates, vortices will be stable. In these regions the motion is elliptic. It is thus expected that vorticity gradients will be stretched in the region where the $\nabla \mathbf{v}$-eigenvalues are real, leading to the formation of vorticity gradient sheets directed along the left eigenvector of $\nabla \mathbf{v}$ associated with the positive eigenvalue. From eq. (1.3) the direction perpendicular to the sheets is given by the right eigenvector of $\nabla \mathbf{v}$ associated with the negative eigenvalue. These (unnormalized) left and right eigenvectors are given by (see eq. (3.4))

$$\begin{aligned} \mathbf{l}^{(+)} &= \left[S_{12} - \omega, (S_{11}^2 + S_{12}^2 - \omega^2)^{1/2} - S_{11} \right] \\ \mathbf{r}^{(-)} &= \begin{bmatrix} -\omega - S_{12} \\ (S_{11}^2 + S_{12}^2 - \omega^2)^{1/2} + S_{11} \end{bmatrix}. \end{aligned} \tag{3.6}$$

When the sheets have been formed, they obviously modify the velocity gradients. We are going to show that this modification is irrelevant for the sheet dynamics. Indeed, assume that a quasi-one dimensional layer has been formed in the direction of the left eigenvector associated with the positive eigenvalue of the background velocity gradient (4.1). Denote by ψ_I the perturbation of the background stream function induced by these vorticity gradient sheets. As the sheets are quasi recti-linear, ψ_I varies only, at leading order, in the direction perpendicular to the sheets. Thus $\psi_I = \psi_I(s)$, where $s = \mathbf{r}^{(-)} \cdot \mathbf{x}$ The velocity gradient is changed to

$$\nabla(\mathbf{v} + \mathbf{v}_I) = \frac{1}{2} \begin{pmatrix} S_{11} & S_{12} + \omega \\ S_{12} - \omega & -S_{11} \end{pmatrix} + \begin{pmatrix} \mathbf{r}_x^{(-)}\mathbf{r}_y^{(-)} & \mathbf{r}_x^{(-)\,2} \\ \mathbf{r}_y^{(-)\,2} & \mathbf{r}_x^{(-)}\mathbf{r}_y^{(-)} \end{pmatrix} \frac{d^2\psi_I}{ds^2}(s). \tag{3.7}$$

A little algebra shows that the eigenvalues and eigenvectors of $\nabla(\mathbf{v}+\mathbf{v}_I)$ are identical to those of $\nabla\mathbf{v}$. This result is best understood by going into the orthonormal frame with x-axis along $\mathbf{r}^{(-)}$ and y-axis along $\mathbf{l}^{(+)}$. In this frame the sheet is directed along the y-axis, and the values of the strain components are such that (see eq. (3.5) and (3.6)) $S_{11} = -\sqrt{S_{11}^2 + S_{12}^2 - \omega^2}$, $S_{12} - \omega = 0$. At leading order ψ_I depends on x only and eq. (3.7) reads

$$\nabla(\mathbf{v} + \mathbf{v}_I) = \frac{1}{2}\begin{pmatrix} S_{11} & 2\omega - \nabla^2(\psi_I) \\ 0 & -S_{11} \end{pmatrix}, \qquad (3.8)$$

whose eigenvectors and eigenvalues are the same as that of $\nabla\mathbf{v}$. This explains why scale separation holds even after vorticity gradient sheets have been formed (see Brachet et al. (1988) for Raster visualizations of $\det(\nabla\mathbf{v})$).

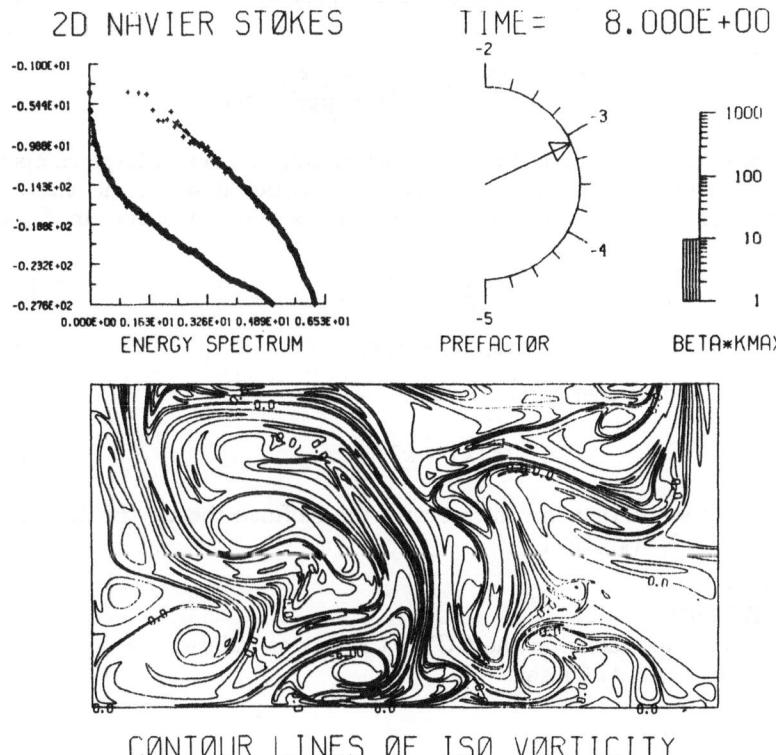

Figure 3. Physical space visualizations at $t = 8$: contour lines of vorticity ω.

Figures 3 correspond to a later time $t = 8$ within the k^{-3} regime. We see that the vorticity gradient sheets are no longer isolated but have been packed together. In the regions where strong (quasi one dimensional) layers are formed, we can locally approximate the velocity gradient by a constant strain, corresponding to a velocity field

$$u = -x$$
$$v = y.$$

(3.9)

Vorticity is then viewed as a passive scalar and we can apply the Batchelor (1959) analysis (see also Leslie 1973), whose relevance in the description of the enstrophy cascade was pointed out by Kraichnan (1975). Such a velocity field will pack together the vortex sheets in the direction of the y-axis. Furthermore, it is easy to write the equation satisfied by the correlation function of the vorticity field. Assuming homogeneity and taking into account the one-dimensionality of the problem (to leading order), one has

$$\Omega(\xi, t) = < \omega(x, t)\omega(x + \xi, t) >,$$

(3.10)

which satisfies

$$\frac{\partial \Omega(\xi, t)}{\partial t} - \xi \frac{\partial \Omega(\xi, t)}{\partial \xi} + 2\nu \frac{\partial^2 \Omega(\xi, t)}{\partial \xi^2} = 0,$$

(3.11)

where ξ denotes the x-component of point separation. At scales large compared to the dissipative scales but small compared to the scales at which the sheets are generated, an equilibrium is established for which the vorticity spectrum $E_\omega(k, t)$ satisfies

$$\frac{\partial(kE_\omega(k))}{\partial k} = 0$$

(3.12)

$$E_\omega(k) \sim 1/k,$$

(3.13)

which corresponds to an energy spectrum

$$E(k) \sim k^{-3}.$$

(3.14)

It is important to stress that in contrast with the Saffman theory, this Batchelor-Kraichnan analysis is statistical.

Acknowledgments
The computations were done on the CCVR Cray-1s at Palaiseau, using the Fast Fourier Transforms of S.A. Orszag.

References
Batchelor G.K. (1959) J. Fluid Mech. **5** , 113.

Batchelor G.K. (1969) Phys. Fluids **12** , 233.

Brachet M.E., Meiron D.I., Orszag S.A., Nickel B.G., Morf R.H. and Frisch U. (1983) J. Fluid Mech. **130** , 411.

Brachet M.E, Meneguzzi M., Politano H. and Sulem P.L. (1988) J. Fluid Mech. **194** , 333.

Gottlieb D., Orszag S.A. (1977) *Numerical Analysis of Spectral Methods: Theory and Applications*, SIAM, Philadelphia.

Herring J.R., Orszag S.A., Kraichnan R.H. and Fox D.G. (1974) J. Fluid Mech. **66**, 417.

Kraichnan R.H. (1967) Phys. Fluids **10** , 1417.

Kraichnan R.H. (1971) J. Fluid Mech. **47** , 525 (1971).

Kraichnan R.H. (1975) J. Fluid Mech. **67** , 155 .

Leith C. (1968) Phys. Fluids **11** , 671.

Leslie D.C.(1973) *Development in the Theory of Turbulence*, Clarendon, Oxford.

Saffman R.G. (1971) Stud. Appl. Math. **50** , 377.

Sulem C., Sulem P.L.and Frisch H. (1983) J. Comp. Phys. **50**, 138.

Weiss J.(1981) *The dynamics of enstrophy transfer in two-dimensional hydrodynamics*, La Jolla Inst. La Jolla, California , LJI-TN-81-121.

TRANSITION TO TURBULENCE IN OPEN FLOWS
A 'metaphoric' approach

Yves Pomeau
Department of Physics
Ecole Normale Superieure, 24 rue Lhomond
Paris 75230
France.

ABSTRACT. As well known, the equations of hydrodynamics cannot be solved in many situations of interest. Here I advocate that has been called a "metaphoric" approach, that is one that relies upon the qualitative properties of some PDE's, with the hope that they are still relevant for flows, although the solutions of the true equations are hopeless. The example presented here concerns the formation of localized patches of turbulence in transition flows, as observed already by Osborne Reynolds.

1. INTRODUCTION

One of the themes of this Conference is the study of the chaotic behavior of solutions of PDE's as the equations of Fluid Mechanics. As is well known, the general solutions of such complicated equations is completely hopeless so that one has to reduce them to more simple ones that may eventually be seen as giving "metaphors" for studying the true fluid behavior even outside of the strict domain of validity of the limits where the reduction to an Amplitude approach is valid. To make more concrete that I have in mind, let me consider a classical phenomenon in Fluid Mechanics, that is the formation of localized patches of turbulence in an otherwise laminar flow. As I argued [1] some time ago this may be interpreted by analogy with the bahavior of solutions of reaction-diffusion equations. Let us assume that the local state of the fluid can be represented via an 'order parameter', a scalar quantity that changes continuously from a definitive value in the turbulent state to another one in the laminar state. The most simple example of such a quantity would be the mean amplitude of the turbulent fluctuations. The dynamics of this amplitude may be derived through a systematic small amplitude expansion "a la Landau". Suppose now that both the turbulent state and the laminar one are stable against small fluctuations, thus it is natural to represent the dynamics of the order parameter as a gradient flow in a two well potential, each well corresponding to a locally stable state of the global system. Then it is still natural to represent

181

E. Tirapegui and D. Villarroel (eds.), Instabilities and Nonequilibrium Structures II, 181–187.
© 1989 by Kluwer Academic Publishers.

the tendency of the system to become spatially uniform, say because of molecular diffusion effects by adding a diffusion term to the equation of motion for the order parameter. This yields a reaction-diffusion equation of a kind that has been much studied [2]. In good cases it is even possible to derive this sort of equation from the first principles in a convenient limit, this giving [3] the so called Amplitude theory. The main property that interests us here is that those reaction-diffusion(like) equations have as an asymptotic solution to a rather large class of initial conditions a front separating the two possible uniform stable states of the system and moving at constant speed. This is of course what one would expect from a metaphor for the phenomenology of the transition flows. Recalling below some simple features of this theory, I consider the additional complications arising from a nongeneric property of the reaction-diffusion model, namely the existence of a Lyapunov functional. In a realistic (or at least less schematic) picture of the development of turbulent patches one should also include their feedback on the basic properties of the flow field. I claim that this feedback explains [4] the barberpole turbulence in the Taylor-Couette geometry with counterrotating cylinders.

This paper is dedicated to Willem Malkus who warned its author against his unrefrained enthousiasm for the metaphoric approach in fluid mechanics.

2. LOCALISED STRUCTURES IN TRANSITION FLOWS

As explained in reference [1] one can understand the dynamics of the localized patches in turbulence through the solutions of reaction diffusion (RD) equations with the general structure:

$$dA/dt = D^{ij}A_{,ij} - dV/dA, \qquad (1)$$

where everything is real-for the moment-, where D^{ij} is symmetric and positive "diffusion tensor" and where $A_{,ij}$ is the second derivative of the amplitude A with respect to the Cartesian coordinates of index i and j. In equation [1] one has applied the Einstein convention for the index summation and V(A) is the two well potential, finally dA/dt is the time derivative of the amplitude.

As said before this equation has solutions attracting a large set of initial datas and representing moving fronts separating two linearly stable regions. Those fronts are sometimes called the Zeldovich-Frank Kamenetskii [5.a] fronts to help to distinguish them from the KPP [5.b] fronts separating stable from linearly unstable regions. Fronts moving at speed u are represented by solutions of (1) with a dependence on one space coordinate, say x, and on time through the combination z = (x - ut), and with the condition that A(z) reaches each of the two stable equilibria at z equals plus and minus infinity. There are many important differences between this picture and the real front separating two different flow regimes as for instance observed in pipes. Before to come to this let us emphasize however some striking similarities: first of all, in the RD picture as for real flows those fronts are rather independent on the details of the initial conditions because of the metastability of the system, a consequence itself of the subcritical character of the bifurcation yielding one of the two states from the other one. Then the front velocity is a smooth function of any parameter controlling deformations of the potential V(.). Now there are also some qualitative differences between real fronts as observed in pipes for instance and what follows from the RD picture. Below I shall consider them one after the other.

(i) Indeed one expects that in most parallel flows, but for the real plane Couette, such fronts are convected at some constant speed by the mean flow. Those advection terms appear through first order derivative in space added to equation (1). They have been already introduced [6] in the amplitude theory for tilted Taylor vortices by Tabeling. This is not enough, as even with those advection terms the trailing and leading fronts are deduced from each other through a reflection, although all experimental datas point to the absence of such a symmetry [7]. This symmetry may be broken, for instance by adding third order derivatives in space to the amplitude equations, as allowed from the basic symmetries of parallel flows, since no Galilean frame can get rid of all possible effects of the advection.

(ii) The amplitude equations should be written for complex amplitudes and, when some of their coefficients are complex they do not have a Lyapunov functional, as equation (1) has. However even in those nonvariational cases the notion of linearly stable state persists and one may reasonnably assume that the moving front solutions still attract a large class of initial conditions with the two different states at + and - infinity. As one cannot compare the energies of the two possible states to determine which one is metastable or stable this is decide now (i.e. for those nonvariational systems without any energy function) through the sign of the velocity of the front. It was predicted in reference [1] and verified [8] on a concrete PDE by Manneville and Chaté that this velocity is a critical quantity (in the sense of critical phenomena in statistical physics) when it almost vanishes and for fronts separating regular from turbulent domains. Other subtelties have been also predicted in [1] and observed recently (Boris Shraiman, David Bensimon; private communication) when fronts propagate on an underlying periodic pattern.

(iii) As well known the question of linear stability of parallel flows is complicated [9] by the distinction between absolute and convective instability. Without going into too many details, one can say that the relevant stability here is the absolute stability of the two states on each side of the front in the frame of reference moving with the front itself. One possible explanation for the observed breakup of the Emmons's spots as they travel downstream is the transition from convective to absolute instability in the frame of reference of the spot boundary. This could also explain why such a moving spot acts as a source for unstable Tollmien-Schlichting waves, but growing away from the spot, because they are convectively unstable in its frame of reference, as often observed in experiments.

(iv) Another problem which has not yet been looked at is the extension of this theory to the development of "turbulent" patches of turbulence in anisotropic 2d (as plane Poiseuille or Couette) or 3d flows (as a Blasius boundary layer). Indeed a theory in the form of equation (1) can model at least in part the anisotropy of such flows through a nonisotropic tensor D. However this is not enough. As said before, in this gradient picture the sign of the front velocity is completely determined by the potential difference between the two equilibria. This implies in particular that this velocity vanishes in all possible directions for the same value of the control parameter. This is presumably a nongeneric situation for nonpotential systems.

Thus let us describe what we hope to be the generic situation, for instance in 2 spatial dimensions that could apply to Poiseuille or real Couette. For a given value of the control parameter one may reasonably assume that a front velocity is defined for every direction of the normal to the front. This yields a velocity indicatrix by plotting a vector having this velocity as length in the corresponding direction. Let us assume furthermore that the flow has an axial symmetry so that this indicatrix has this axis of symmetry too. One may decide to draw it in a Galilean frame of reference such that the 2 possible velocities along the axis of symmetry are exactly opposed (see figure 1 for such an example of indicatrix). A turbulent spot will grow if the velocity of the front edge is larger than the velocity of the trailing edge in this frame of reference. In a potential system the crossover (as a relevant control parameter varies) between an expanding and a recessing spot will occur when the two potential minima are equal. The whole indicatrix reduces then to a single point for this steady coexistence between the two possible homogeneous states of the system. But for a nonpotential system one does not expect such a thing: the first crossover will occur for instance between two opposite velocities, but not for all directions at a time. Suppose that this happens first along the axis of symmetry, as shown qualitatively on figures (1.a-c), and let us describe what will happen after this has occured (i.e. when the indicatrix is as in figure 1.c).

First one has to describe in general the expansion of such a turbulent spot, once the velocity indicatrix is known. We will consider the simple case where all velocities point outward. Long after the initiation of the turbulent spot, its size will be some constant depending on the initial data plus something proportional to time multiplied by a constant velocity depending on the direction. This part proportional to time is dominant at those later times and it may be drawn by a Wulff's like construction [10]. The distance run along a given orientation is proportional to the velocity along this direction, which implies that, up to a multiplication by the time, the expanding spot must be everywhere tangent to the normal to the velocity indicatrix drawn at its end, and this is identical to the Wulff's construction. Indeed this construction has problems when the velocity indicatrix is not homeomorphic to a circle (i.e.

has self crossings in a plane). Following the reasoning made before this indicatrix would show two self crossings at velocities almost parallel to the axis of symmetry when the onset of expansion of the turbulent spot is reached along this axis (see figure 1.b). In that case the Wulff-like construction given before cannot apply directly. It would generate a spot having itself a self crossing boundary. Physically this would lead to a crescentshaped spot with two singular points at the tips of the crescent. At those two singular points, the merging of the front and rear boundaries should be on length scales of the order of the front thickness, outside of the domain of application of the picture provided by the velocity indicatrix, since this one assumes that the radius of curvature of the spot boundary is much larger than the front thickness. Indeed this crescentshape - or arrowhead as often said - is quite reminescent of the Emmons spots, although the external parameters here change, because the Reynolds number increases downstream in a Blasius boundary layer. Note however that this fits well with the observation that turbulent spots in open flows grow by dilation and proportionnaly to time.

Let us now look at the barberpole turbulence observed [4] in Taylor-Couette flows, as I want to show that this can be understood by this RD "metaphor" with consideration of the peculiarities of this geometry. This barberpole turbulence is localised, as are the turbulent spots in pipes for instance, and it is experimentally hysteretic, as a subcritical instability should be. But there is an important difference with what we discussed before: in the RD metaphor the turbulent spots either expand or shrink, they stay in equilibrium for well defined values of the control parameter only, whereas the barberpole stays fixed in a well chosen rotating frame of reference. It is only the relative size of the spiral (and not the front velocity) that changes as one varies the control parameters (= rotation speeds of the outer and inner cylinders). This can be explained by a feedback between the mean flow and the growth of the turbulent domain: as this turbulent domain grows, it diminishes the Reynolds number, because of pressure effects discussed below and this reacts upon the growth of the turbulent spot until it stops growing at some well defined Reynolds number depending both on the external constraint and on the relative size of the turbulent domain. The feedback through the pressure may be understood as follows: in the fully laminar regime the pressure is independent on the azimuthal angle, whereas there is a mean azimuthal flow pulled by the rotating cylinders. When the turbulent spot is small, this mean flow is to be considered as a constraint put on the hydrodynamical conditions in this spot, and this yields a mean pressure drop through the turbulent spot, but because of the geometry this drop acts also upon the laminar part of the flow, adding a Poiseuille component to it. This is the feedback effect alluded to before. There the turbulent domains have a barberpole structure because the basic instability is for perturbations at an angle with the axis of the cylinder, being in between the Taylor Couette and purely Couette situation. Indeed this feedback effects is always present in the development of

subcritical instabilities in pressure driven flows. However it may
be neglected if the size of the turbulent domain is negligible compared
to the one of the global flow. In general one would expect that
this turbulent spot perturbs the outer laminar flow field as a
localised force.

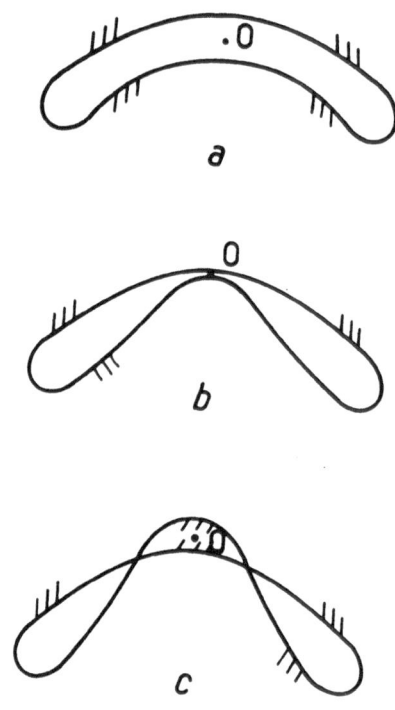

Figure 1. Guessed velocity indicatrix (see text for definition) in
a two dimentional geometry. The velocity origin is at 0 and the
shaded area indicates on which side of the front the turbulent domain
is, this to leave the sign degeneracy. The indicatrix changes its
shape as a control parameter changes, and is drawn with an assumed
axisymmetry, as expected for a Poiseuille or a Couette flow. The
frame of reference of these drawings is such that the 2 velocities
along the symmetry axis are opposite.

    In Fig. 1.a, the spot is always recessing.
    In Fig. 1.b, the spot is marginally recessing, because the
velocities of the leading and trailing edges along the axis of symmetry
are equal. Note that for a potential system this marginal situation
would correspond to a single point or perhaps to a single arc.
    In Fig. 1.c, the spot may expand at least in some directions
and is bounded by two arcs cutting at an angle where an inner
description must be used.

Finally let us mention the possibility that the Great Red Spot on the planet Jupiter has the same explanation as the spiral turbulence in Taylor Couette: the flow inside the GRS could be another state of the flow in the atmosphere of this planet and yield a globally stable spot because of the above discussed feedback effect. This would explain in particular why it is has been there over a huge period of time compared to any other observed time scale for the fluctuations of the fluid motion on this planet.

3. CONCLUSION

It is often useful to apply the general ideas of thermodynamics to rather general systems, following a line of thinking proposed by Robert Graham (this conference) and Enrique Tirapegui (this conference too). Here I have tried to get some results from this that could be applicable to the real situation of open flows. It turns out that some qualitative differences may appear between the thermodynamic situation and the one of fluid dynamics. Perhaps the most striking one could be in the predicted angular dependence on the onset of expansion of localised spots of the "different" state. This could perhaps be tested on experiments.

REFERENCES

[1]   Y. Pomeau; Physica 23D, 3 (1986).
[2]   J. Smoller. Shock Waves and reaction diffusion equations. Springer, Berlin (1983).
[3]   A.C. Newell, J. Whitehead. J. of Fluid Mech. 38, 79 (1969). L.A. Segel, ibid, p. 203.
[4]   C. van Atta. J. of Fluid Mech. 25, 495 (1966). C.P. Andereck, S.S., Liu and H.L. Swinney, J. of Fluid Mech. 164, 155 (1986).
[5.a]   Ya. B. Zel'dovich. Theory of combustion and detonation of gases. Moscow (1944).
[5.b]   A.N. Kolmogorov, A.N. Petrovskii and N.S. Piskunov. Bulletin de l'Universite d'Etat a Moscou, Sec. A, Vol. 1, Math. et Mec., 1 (1937).
[6]   Tabeling. J. de Phys. Lettres 44, 665 (1983). P. Hall, Phys. Rev. A29, 2921 (1984).
[7]   I.J. Wygnansky and F.H. Champagne. J. of Fluid Mech. 59, 281 (1973); see also van Atta in ref. 5.
[8]   H. Chaté and P. Manneville. Phys. Rev. Lett. 58, 112 (1987).
[9]   P. Huerre and P.A. Monkewitz. J. of Fluid Mech. 159, 15 (1985).
[10]   L. Landau and E.M. Lifshitz. Statistical Physics. Pergamon, New York (1973).

THE ROLE OF TOPOLOGICAL DEFECTS IN SUBCRITICAL BIFURCATIONS.

P. Coullet*, L. Gil and D. Repaux
Laboratoire de Physique Théorique
Parc Valrose. 06034 Nice Cedex, France.

ABSTRACT. A deterministic mechanism of transition from a metastable phase to a stable one and associated with topological defects is described. The distinction between kink and vortex-type defects is emphasized.

1. INTRODUCTION

The aim of this paper is to describe a deterministic mechanism of transition from a metastable phase to a stable one. One of the main aspects of the classical nucleation theory [1] is the study of the dynamics of a seed of stable phase with spherical shape plunged into the bulk of the metastable phase. Generally a full range of parameters exists for which the system possesses a stationary but unstable seed solution with a critical radius. Above this critical size the stable solution undergrows the whole system and disappears in the reversed situation when the initial seeds is too small. One is then led to a statistical description [2] since the initial seed of stable phase is assumed to be generated by thermal fluctuations or impurities. On the contrary, the mechanism we describe in this paper is essentially deterministic because the initial seed of stable phase is induced by the core of a defect [3]. Defects are inhomogeneous solutions of a system related to the breaking of a symmetry enclosed in a small localized region of the space and whose stability is mainly due to topological considerations [4]. The relationship between defects and subcritical transition is due to the existence in the defect solutions of regions where stable and metastable phases are respectively reached. In the following we study the details of this mechanism by using some typical examples on one or two space dimensions.

---

* also Observatoire de Nice. Mont Gros.

E. Tirapegui and D. Villarroel (eds.), Instabilities and Nonequilibrium Structures II, 189–205.

2. ONE SPACE DIMENSION

The simultaneous existence of several linearly stable homogeneous
solutions can be realized in the neighbourhood of a subcritical
bifurcation. The most simple model of a subcritical broken symmetry
transition in one space dimension reads :

$$\frac{\partial A}{\partial t} = \mu A + A^3 - A^5 + \frac{\partial^2 A}{\partial x^2} \tag{1}$$

where $A(x,t)$ is a real function. This previous equation possesses
three kinds of homogeneous stationary solutions :

$$A_0 = 0$$

$$A_{1\pm} = \pm \sqrt{\frac{1}{2} - \sqrt{\mu + \frac{1}{4}}}$$

$$A_{2\pm} = \pm \sqrt{\frac{1}{2} + \sqrt{\mu + \frac{1}{4}}}$$

A_0 is linearly stable for μ negative and unstable in the other case.
A_1 and A_2 exist only for $\mu \geq \mu_\ell = -1/4$. $A_{1\pm}$ are unstable and $A_{2\pm}$ stable
solutions for $\mu_\ell < \mu$ (see Fig. 1).

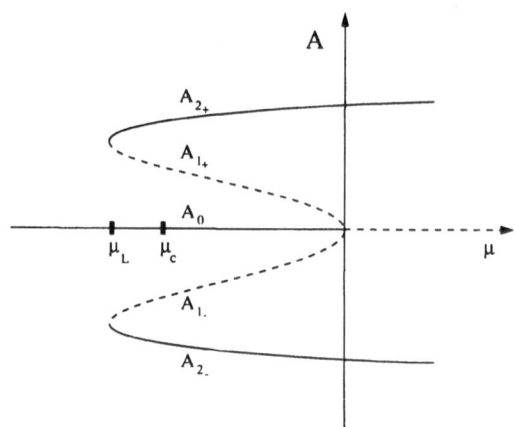

Figure 1. Bifurcation diagram associated with the equation
$\partial A/\partial t = \mu A + A^3 - A^5 + \partial^2 A/\partial x^2$. Stable and unstable solutions
respectively correspond to solid and dashed lines.

Equation (1) can be expressed in a variational form:

$$\frac{\partial A}{\partial t} = - \frac{\partial F}{\partial A} \tag{2}$$

with

$$F = \int_{-\infty}^{+\infty} \rho (A, \frac{\partial A}{\partial x}) dx$$

and where ρ, the analog of a density of free energy is defined as follows:

$$\rho (A, \frac{\partial A}{\partial x}) = (\mu \frac{A^2}{2} + \frac{A^4}{4} - \frac{A^6}{6}) + \frac{1}{2} (\frac{\partial A}{\partial x})^2 \tag{3}$$

F is minimized by the time evolution. In the range of parameters where both A_0 and A_{2+} are linearly stable ($\mu_\ell < \mu < 0$) a good tool of comparison is provided by the computation of the free energies of the two solutions. When $F(A_0) > F(A_2)$, A_0 correspond to the metastable phase and A_2 to the stable one, and roles are obviously exchanged when $F(A_0) < F(A_2)$. The critical value μ_c is defined as:

$$F(A_0) = F(A_{2\pm}) \quad \rightleftharpoons \quad \mu = \mu_c \tag{4}$$

Equation (1) possesses also interesting solutions which involve strongly localized space variations. Let us assume the existence of initial conditions which connect asymptoticaly A_{2-} at $-\infty$ to A_{2+} at $+\infty$ by means of a small region of strong gradients. Then time evolution is expected to conserve these asymptotic behaviours because of the linear stability of the homogeneous stationary connected solutions and of their infinite "weight" in the initial space distribution. Solutions connecting A_{2-} to A_{2+} or the opposite are termed as kink-like defects and are stationary because of the symmetry of the asymptotic behaviours. On the contrary non-symmetrical solutions connecting A_{2-} or A_{2+} to A_0 are moving with a finity velocity and correspond to fronts. The direction of propagation of the front depends on the value of $F(A_0)$ - $F(A_2)$ and always corresponds to the decreasing of the region of metastable phase.

The equation describing the stationary solutions:

$$0 = \mu A + A^3 - A^5 + \frac{\partial^2 A}{\partial x^2}$$

contains and summarizes all the previous informations (see Fig. 2).

Phase space analysis is simplified by taking into account the existence of a space independent physical quantity $\&$

$$0 = \frac{\partial A}{\partial x} (\mu A + A^3 - A^5 + \frac{\partial^2 A}{\partial x^2})$$

$$\Rightarrow \frac{\partial}{\partial x} [\frac{\mu}{2} A^2 + \frac{1}{4} A^4 - \frac{1}{6} A^6 + \frac{1}{2} (\frac{\partial A}{\partial x})^2] = 0$$

$$\Rightarrow \quad \frac{\mu}{2} A^2 + \frac{1}{4} A^4 - \frac{1}{6} A^6 + \frac{1}{2} \left(\frac{\partial A}{\partial x} \right)^2 = \mathcal{E} \tag{5}$$

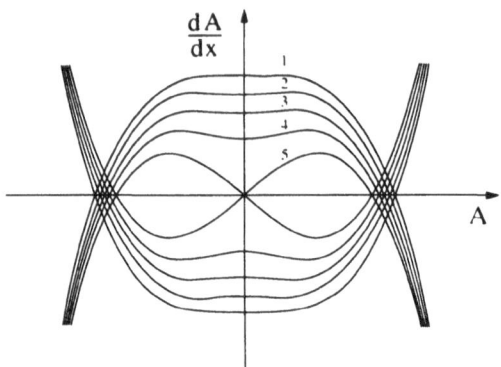

Figure 2. Phase space of the stationary problem: $0 = \mu A + A^3 - A^5 + \partial^2 A/\partial x^2$. Only heteroclinic curves corresponding to defect solutions have been drawn. The numbers $(1,2\ldots5)$ refer to various values of μ: $1 \rightarrow \mu = 0$; $2 \rightarrow \mu = \mu_c/4$; $3 \rightarrow \mu = 2\mu_c/4$; $4 \rightarrow \mu = 3\mu_c/4$; $5 \rightarrow \mu = \mu_c$.

Few remarks are in order. First, heteroclinic curves connecting A_{2-} to A_{2+} (respectively A_{2+} to A_{2-}) which are associated with defects exist only for $\mu > \mu_c$. The value of $|\partial A/\partial x|$ corresponding to $A = 0$ is a decreasing function of $\mu - \mu_c$. At the critical value $\mu = \mu_c$ four heteroclinic curves are present and $\partial A/\partial x = 0$ when $A = 0$. Thus the defect core (i.e. the region where A is close to 0) increases as the homogeneous corresponding phase becomes more and more stable. When the critical value μ_c is reached, the disappearance of stationary solutions connecting A_{2+} to A_{2-} (respectively A_{2-} to A_{2+}) is explained by a splitting of defect solutions into two opposite propagative fronts (see Fig. 3) [5].

The logarithmic divergence of the size of the defect's core is related to the fact that the fixed point ($A = 0$, $\partial A/\partial x = 0$) is a real saddle. Actually the "defect curves" become closer and closer to the stable and unstable manifolds as $\mu - \mu_c$ goes to zero [6]. The saddle character of the trivial fixed point is generic since a stationary solution of an equation, linearly stable with respect to time evolution is associated with opposite eigenvalues for the stationary problem [7].

Analytical results can be obtain by using standard perturbation techniques. For $\mu = \mu_c$, there exists a stationary solution $A^*(x)$ (see Fig. 4) connecting A_{2-} at $-\infty$ to A_0 at $+\infty$ and solution of:

$$0 = \mu_c A + A^3 - A^5 + \frac{\partial^2 A}{\partial x^2} \tag{6}$$

$A_0 = 0$ is also a solution of the previous equation with the associated eigenvalues:

$$\lambda_{\pm} = \pm \sqrt{-\mu_c} \qquad (7)$$

Let us call $A_{\ell}(x) = A*(x)$ and $A_r(x) = -A*(-x)$. We look for a solution of equation (1) of the form (see Fig. 5) [8]:

$$A(x,t) = A_{\ell}[\ x + \frac{d_0}{2} - r_f(t)] + A_r[\ x - \frac{d_0}{2} + r_f(t)] + w(x,t) \qquad (8)$$

where $\partial r_f(t)/\partial t$ and $w(x,t)$ are assumed to be small corrections of the same order of magnitude and $\partial w/\partial t$ to be negligible with respect to w.

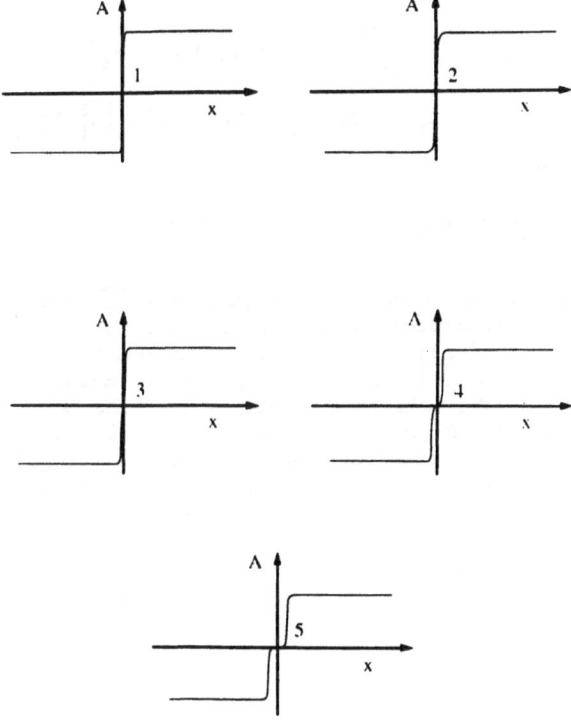

Figure 3. In this figure, the curves $A(x)$ associated with the different values of μ have been displayed. The last plot correspond in fact to a value of μ slighty smaller than μ_c [$(\mu - \mu_c)/\mu_c = 5 \cdot 10^{-4}$] because of the logarithmic divergence of the size of the defect's core.

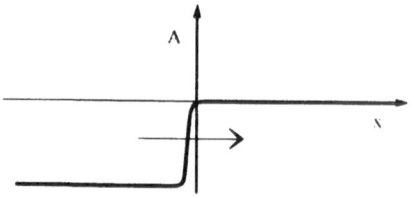

Figure 4. $A^*(x)$ is a stationary solution connecting A_{2-} to A_0 for $\mu = \mu_c$.

The substitution of (8) in (1) leads to:

$$L_{A_\ell,A_r}w = - \frac{\partial r_f}{\partial t}(\frac{\partial A_\ell}{\partial x} - \frac{\partial A_r}{\partial x}) - (\mu - \mu_c)(A_r + A_\ell)$$
$$-3(A_r^2 A_\ell + A_r A_\ell^2) + 5(A_r^4 A_\ell + A_r A_\ell^4) + 10(A_r^3 A_\ell^2 + A_r^2 A_\ell^3) \tag{9}$$

where L_{A_ℓ,A_r} is defined by:

$$L_{A_\ell,A_r}w = \mu_c w + 3(A_\ell + A_r)^2 w - 5(A_\ell + A_r)^4 + \frac{\partial^2 w}{\partial x^2} \tag{10}$$

With the usual scalar product $< A(x),B(x) > = \int_{-\infty}^{+\infty} A(x)B(x)dx$ defined in a space of functions with vanishing derivatives at infinity, the operator $L_{A,B}$ is self-adjoint. It is an important property since the kernel of L^* can then be easily computed by means of symmetry considerations: the Nambu-Goldstone modes (i.e. the kernel of L^*) are found by taking the derivative of equation (6) with respect to the continuous variable (x) associated with the translational symmetry. The fact that A_r (respectively A_ℓ) is a solution of equation (1), leads to:

$$\frac{\partial}{\partial x}(\mu_c A_r + A_r^3 - A_r^5 + \frac{\partial^2 A_r}{\partial x^2}) = 0 \tag{11}$$

or in a more condensed way

$$L_{A_r,0}(\frac{\partial A_r}{\partial x}) = 0 \tag{12}$$

The assumption of weak interaction between the two solutions A_r and A_ℓ (in the other term w small) is realized when the distance $(d_0 - 2r_f)$ between the two solutions is large enough. In order to be able to exhibit a solvability condition, we just have now to divide

the whole space in three parts as in Fig. 5:
in the first region $] - \infty, - d_0/2 + r_f [$, A_r is completely negligible
because the asymptotic behaviour $(A = 0)$ is reached exponentially.
$\partial A_r/\partial x$ is then very close to zero and therefore:

$$L_{A_r,A_\ell} \left(\frac{\partial A_r}{\partial x} \right) \approx 0 \tag{13}$$

in the second region $] - d_0/2 + r_f, \ d_0/2 - r_f [$ and in the third
one $] d_0/2 - r_f, + \infty [$ A_ℓ is negligible and

$$L_{A_r,A_\ell} \approx L_{A_r,0} \tag{14}$$

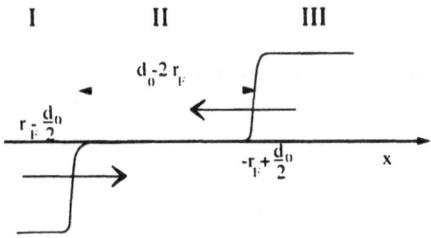

Figure 5. Graph of the solution defined by equation (8) :

$$A(x,t) = A_\ell [x + \frac{d_0}{2} - r_f(t)] + A_r [x - \frac{d_0}{2} + r_f(t)] + w(x,t)$$

Equations (12) and (14) imply that:

$$L_{A_r,A_\ell} \left(\frac{\partial A_r}{\partial x} \right) \approx 0 \tag{15}$$

Under these approximations [8], $\partial A_r /\partial x$ belongs to the kernel of
$L^*_{A_r,A_\ell}$ in the whole space and consequently the solvability condition
is expressed as:

$$\nu \frac{\partial r_f}{\partial t} = \kappa_1 (\mu - \mu_c) + \kappa_2 \tag{16}$$

where

$$\nu = - \int_{-\infty}^{+\infty} \left(\frac{\partial A_\ell}{\partial x} - \frac{\partial A_r}{\partial x} \right) \frac{\partial A_r}{\partial x} \, dx \tag{17}$$

and

$$\kappa_1 = \int_{-\infty}^{+\infty} (A_\ell + A_r) \frac{\partial A_r}{\partial x} \, dx \tag{18}$$

$$\kappa_2 = \int_{-\infty}^{+\infty} [3(A_r^2 A_\ell + A_r A^2) - 5(A_r^4 A_\ell + A_r A_\ell^4) - 10(A_r^3 A_\ell^2 + A_r^2 A_\ell^3)] \frac{\partial A_r}{\partial x} \, dx \tag{19}$$

Because of the exponentially falling down of the asymptotic behaviours of A_r and A_ℓ, the following approximations can be made:
1) all the integrals can be taken between $] - d_0/2 + r_f, \; d_0/2 - r_f \; [$ instead of $] - \infty, +\infty \; [$
2) in this interval we can write:

$$A_r(x,t) \simeq \sigma e^{\lambda_- (-x + d_0/2 - r_f)} \qquad A_\ell(x,t) \simeq -\sigma e^{\lambda_- (x + d_0/2 - r_f)} \tag{20}$$

where σ is a constant, verifying $\sigma > 0$, and λ_- defined as before is negative. With these assumptions, it is found that :

$$\kappa_1 = -\frac{\nu}{\lambda_-}$$

and the following asymptotic relations:

$$\frac{1}{\nu} \to (\frac{-2}{\sigma^2 \lambda_-}) \; (1 - 2\lambda_- (d_0 - 2r_f) \, e^{\lambda_- (d_0 - 2r_f)})$$

$$\kappa_2 \to \sigma^4 \, (-\frac{3}{2} + \frac{5}{4}\sigma^2) \, e^{\lambda_- (d_0 - 2r_f)} \tag{21}$$

as $(d_0 - 2r_f) \to \infty$. The asymptotic behaviour at large $d_0 - 2r_f$ of the front's velocity is then given by:

$$\frac{\partial r_f}{\partial t} = \frac{(\mu - \mu_c)}{(-\lambda_-)} + \zeta e^{\lambda_- (d_0 - 2r_f)} \tag{22}$$

ζ depends on σ and is assumed to be negative for physical meaning. Since we are interested in stationary solutions of the form (8), we have to solve:

$$-\frac{1}{\zeta} \frac{(\mu - \mu_c)}{(-\lambda_-)} = e^{\lambda_- x_c} \tag{23}$$

where x_c stands for the size of the core of the defect $(d_0 - 2r_f)$. Then, under these assumptions, the logarithmic divergence of x_c with $\mu - \mu_c$ already noticed in the phase space considerations as well as the disappearance of stationary solutions for $\mu - \mu_c < 0$ is confirmed by this perturbation analysis.

In order to study the genericity of these results, a four order derivative is added to the equation (1):

$$\frac{\partial A}{\partial t} = \mu A + A^3 - A^5 + \alpha \frac{\partial^2 A}{\partial x^2} + \beta \frac{\partial^4 A}{\partial x^4} \qquad \alpha > 0 \;,\; \beta < 0 \qquad (24)$$

For $\mu < 0$, A_o and A_2 are always stationary homogeneous stable solutions. As before, there exist kink-type defects connecting A_{2-} to and A_{2+} and for $\mu = \mu_c$ a stationary solution $A^*(x)$ connecting A_o to $A_{2\pm}$ and governed by:

$$0 = \mu_c A + A^3 - A^5 + \alpha \frac{\partial^2 A}{\partial x^2} + \beta \frac{\partial^4 A}{\partial x^4} \qquad (25)$$

$(A = 0)$ is always solution of equation (25) but now the associated eigenvalues are given by:

$$\lambda = \pm \sqrt{\frac{-\alpha}{2\beta} \pm i \sqrt{\frac{\mu}{\beta} - \frac{\alpha^2}{4\beta^2}}} \qquad \mu < \frac{\alpha^2}{4\beta} \qquad (26)$$

Hence, there exists a full range of parameter where the associated eigenvalues are generically complex [9]. The equation giving the critical sizes x_c of the stationary solutions of the form (8) is then modified and in the limit of large $(d_o - 2r_f)$ is expected to be of the form:

$$-\frac{(\mu - \mu_c)}{\zeta} = e^{(-x_c \lambda_1)} \cos(\lambda_2 x_c) \qquad \lambda_1 > 0 \;,\; \lambda_2 > 0 \qquad (27)$$

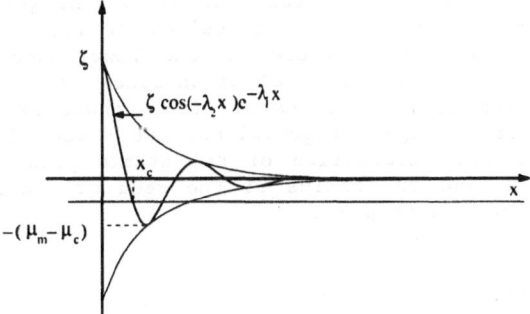

Figure 6. Graphical construction of the solutions of equation (27):

$$-\frac{(\mu - \mu_c)}{\zeta} = e^{(-x_c \lambda_1)} \cos(\lambda_2 x_c)$$

As consequences (see Fig. 6): first the critical value μ_n associated with the transition of the stable phase A_0 is slightly smaller than μ_c, secondly the size of the nucleus of stable phase does not diverge to infinity when μ tends towards μ_n, and finally there exist several critical values of x_c which could be observed in an experiment with a large radius of the initial nucleus of stable phase and a decreasing μ. Numerical experiments confirm the existence of $\mu_n < \mu_c$ (see Fig. 7) as well as the oscillation of $\partial r_f / \partial t$ with the size of the core of the defect (see Fig. 8).

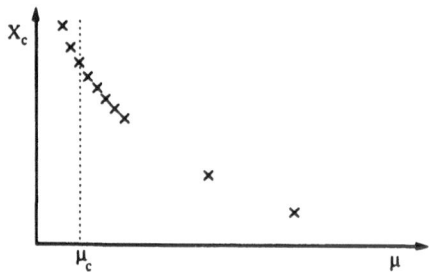

Figure 7.   Numerical computation of the radius of nucleus of stable phase in the case of complex eigenvalues in one dimension. The non-divergence of the core at μ_c as well as the finite value of x_c at μ_n are observed.

In conclusion, in one space dimension, two situations generically occur depending on the nature of the eigenvalues of the stationary problem linearized in the neighbourhood of the homogeneous stable solution A_0. In the case of purely real eigenvalues the divergence to infinity of the radius of the nucleus of stable phase is observed for $\mu = \mu_c$. For complex conjugate eigenvalues, the value of μ which corresponds to a spontaneous transition of the stable phase (μ_n) is slightly smaller than μ_c and the radius of the seed of stable phase does not diverge to infinity with $\mu - \mu_n$.

3.   TWO SPACE DIMENSIONS.

The simplest model corresponds to the subcritical two dimensional Ginzburg-Landau equation:

$$\frac{\partial A}{\partial t} = \mu A + \left(\frac{\partial^2}{\partial x^2} + \frac{\partial^2}{\partial y^2} \right) A + |A|^2 A - |A|^4 A \tag{28}$$

where A is now a complex order parameter. Homogeneous solutions A_0, $A_{1\pm}$ and $A_{2\pm}$ with arbitrary phase, obvious generalization of the

real case, are still present. The topological defects brought into
play are here the vortex solutions. They are associated with a
circulation of the phase gradient (∇_φ, $A = Re^{i\varphi}$) along a closed line
surrounding a core of a vortex, equal to $\pm 2\pi$. This geometrical
property is responsible for the topological stability of the solutions.
At the core of the defect where the real and imaginary part of A
vanish, the homogeneous phase A_0 is approached, while the other phase
A_{2+} corresponds to the asymptotic behaviour ($r \rightarrow +\infty$) of the defect
(see Fig. 9).

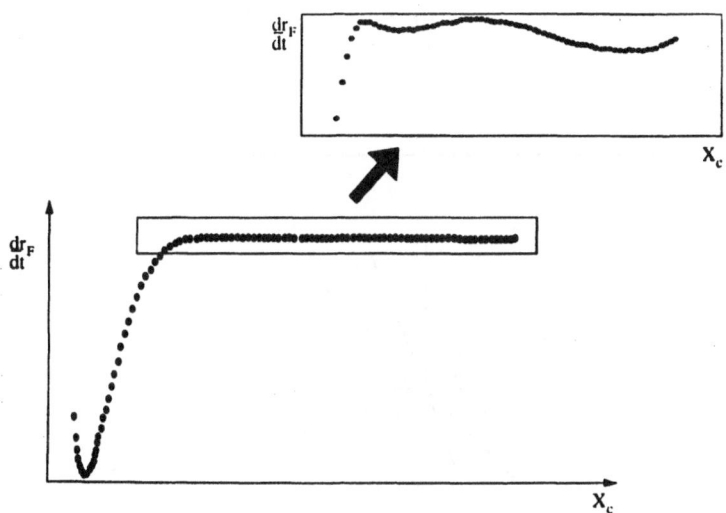

Figure 8.   The numerical mesure of the front's velocity for $\mu < \mu_n$ in
the case of complex eigenvalues in one dimension confirms the
oscillation of $\partial r_f/\partial t$ as well as its exponential   falling down.   (b)
is an enlargement of the asymptotic behaviour of the velocity of (a).

In two dimensional space, the dynamical analysis seems to be
quite different since the coefficients of the linearized problem
are no more constant. When the initial seed of phase A_0 is induced
by a thermal fluctuation or by a impurity localized at $r = 0$, the
order parameter $A = R(r)\exp(i\varphi)$ where φ is an arbitrary phase, is
governed by:

$$\frac{\partial^2 R}{\partial r^2} + \frac{1}{r}\frac{\partial R}{\partial r} + \mu R + R^3 - R^5 = 0 \tag{29}$$

while for the topological solution $A = R(r)\exp(\pm i\theta)$, where θ is the
azimuthal angle, the equation for the modulus is expressed as:

$$\frac{\partial^2 R}{\partial r^2} + \frac{1}{r} \frac{\partial R}{\partial r} - \frac{1}{r^2} R + \mu R + R^3 - R^5 = 0 \qquad (30)$$

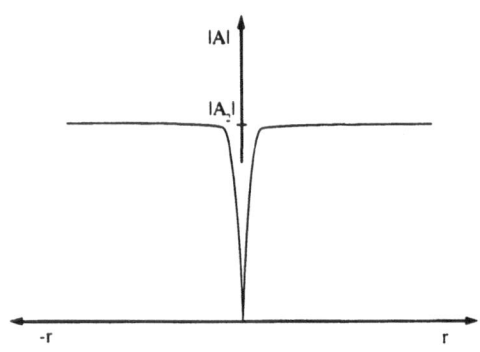

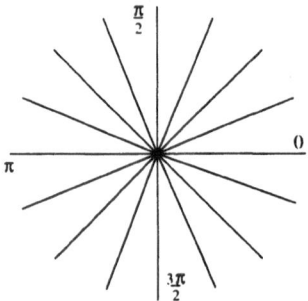

Figure 9.   Modulus R versus r and the   radial   distribution   of   the
isophase lines in the (x,y) plane.

As before equation (28) can be expressed in a variational way:

$$\frac{\partial A}{\partial t} = \frac{\delta F}{\delta \overline{A}}$$

where F is defined by:

$$F(A,\overline{A}) = \int_{-\infty}^{+\infty} \int_{-\infty}^{+\infty} \rho \left(A, \overline{A}, \frac{\partial}{\partial x}, \frac{\partial}{\partial y} \right) dxdy$$

and the density of free energy ρ, by:

$$\rho = -(\mu|A|^2 + \frac{|A|}{2} - \frac{|A|^6}{3}) + \nabla A \cdot \nabla \overline{A} \tag{31}$$

In order to be able to roughly describe the dynamics of the transition of the stable phase into the metastable one from a core of a topological defect, the space dependence of R is assumed to look like those of Fig. 10.

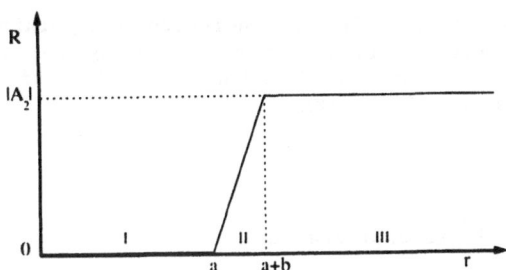

Figure 10. Form of function R(r) which is assumed for the dynamical analysis in two space dimensions.

As A is searched under the form $A = R(r)e^{(im\theta)}$ where m = 0 corresponds to the classical situation of a thermal drop, and m = 1 to the topological one, the density of free energy is then expressed:

$$\rho(r) = -(\mu R + \frac{R^4}{2} - \frac{R^6}{3}) + (\frac{\partial R}{\partial r})^2 + (\frac{mR}{r})^2 \tag{32}$$

Let us call S the surface of the physical system. The free energy is a function of the parameters a and b and is given by: in the first region:

$$F_1 = 0 \tag{33}$$

in the second region:

$$F_2 = -\mu \frac{R_\infty}{6}[3(b^2+ab)-b^2] - \frac{R_\infty^4}{60}[6(b^2+ab)-b^2] + \frac{R_\infty^6}{168}[8(b^2+ab)-b^2]$$

$$+ \frac{R_\infty^2}{2}(\frac{(a+b)^2 - a^2}{b^2})$$

$$+ (\frac{mR_\infty}{b})^2 [\frac{b^2 - 2ab}{2} + a^2\ln(\frac{a+b}{a})] \tag{34}$$

with $R_\infty = \sqrt{\dfrac{1}{2} + \sqrt{\mu + \dfrac{1}{4}}}$

in the third region

$$F_3 = - \left(\frac{S - \pi(a+b)^2}{2\pi}\right)\left(\mu R_\infty + \frac{R_\infty^4}{2} - \frac{R_\infty^6}{3}\right) + (mR_\infty)^2 \ln\left(\frac{\sqrt{\frac{S}{\pi}}}{a+b}\right) \qquad (35)$$

We would have now to take into consideration the minimisation of the free energy with respect to a and b. Assuming the form of the front is chosen really more quickly than the value of the radius of the stable phase, we are left with:

$$\frac{\partial F}{\partial a} = 0 \qquad (36)$$

In the limit $a \gg b$, $\dfrac{\partial F}{\partial a}$ is expressed as:

$$\frac{\partial F_1}{\partial a} = \rho(A_0)a$$

$$\frac{\partial F_2}{\partial a} = \sigma = \left(-\frac{\mu}{2}R_\infty - \frac{R_\infty^4}{10} + \frac{R_\infty^6}{21}\right)b + \frac{(R_\infty)^2}{b} \qquad (37)$$

$$\frac{\partial F_3}{\partial a} = -a\rho(A_2) - (mR_\infty)^2\frac{1}{a}$$

With these assumptions F_2, which represents the free energy of surface tension, is proportional to the length of the interface ($\sigma > 0$). Finally we are left with:

$$[\rho(A_0) - \rho(A_2)]a + \sigma = \frac{(mR_\infty)^2}{a} \qquad (38)$$

The difference between the classical situation ($m = 0$) and the topological one ($m = 1$) is now brought to the fore (see Fig. 11). Actually for $\mu \gg \mu_c$ the homogeneous solution A is more stable than A_0, $\rho(A_0) - \rho(A_2)$ is positive and the previous equation possesses one stable solution r_{t_1} for $m = 1$ and no solution for $m = 0$. r_{t_1} which is found to be stable, corresponds to the defect solution whose stability is assumed by geometrical considerations [10]. The occurrence of a solution r_c for $m = 0$ is obtained for $\mu \leq \mu_c$ and is associated with the appearance of a new but unstable solution r_{t_2} for $m = 1$. As $\mu - \mu_c$ is more and more negative, the classical solution r_c becomes smaller and smaller. On the contrary, the existence of a critical value μ_n associated with the disappearance of the topological stationary solutions and defined by $r_{t_1} = r_{t_2}$ is observed. For $\mu < \mu_n$ the stable phase A_0 undergrows the whole system. The existence of such solutions as well as their classification with respect to the size of the nucleus or the existence of various critical values of μ have been confirmed numerically.

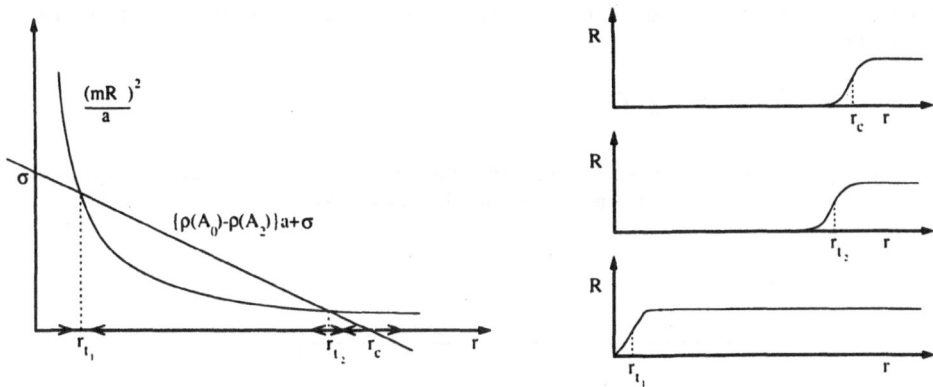

Figure 11. (a) shows the graphic computation of solutions of equation (38), and (b) are the corresponding functions $R(r)$.

In conclusion, the presence of a vortex-type defect implying the existence of a nucleus of stable phase A_0 modifies significantly the value of the critical thermal fluctuation required for spontaneous transition. The corresponding new critical value of fluctuations defined as $\Delta F_{top} = | F_{r_{t_2}} - F_{r_{t_1}} |$ where $F_{r_{t_2}}$ stands for the free energy of the unstable topological solution and $F_{r_{t_1}}$ for the free energy of the stable one, is always smaller than $\Delta F_{cla} = | F(r_c) - F(A_2) |$ which corresponds to the fluctuation of energy needed for transition in the classical theory. A good confirmation of this mechanism is provided by the observation of a saddle node bifurcation for the disappearance of the vortex solution (see Fig. 12), where for $\mu = \mu_n$ stable and unstable topological stationary solutions coincide ($\Delta F_{top} = 0$) while r_c and A_0 solutions are still really distinct ($\Delta F_{cla} > 0$).

4. CONCLUSION

In three dimensions, defect solutions are closed vortex lines, and the topological mechanism expected should correspond to a transition of the stable phase from the line of core both in tangential and perpendicular directions. For illustration, let us assume a system with a volume ν governed by a three dimensions subcritical generalized Ginzburg-Landau equation, and the presence of a defect line solution with a circle shape. The free energy is then expressed as:

$$F = [\nu - (\frac{a^2\ell}{2} - \frac{4a^3}{3})] \rho(A_2) + (\frac{a^2\ell}{2} - \frac{4a^3}{3}) \rho(A_0) + \ell a\sigma$$

Then the minimization of F with respect to a and ℓ defined in Fig. 13 leads to:

$$a = \frac{2\sigma}{\rho(A_0) - \rho(A_2)} \qquad , \qquad \ell = \frac{(\frac{16}{3}\sigma)}{\rho(A_0) - \rho(A_2)}$$

When μ approaches μ_c the radius of the torus of the phase A_0 as well as its length diverge to infinity in the same way.

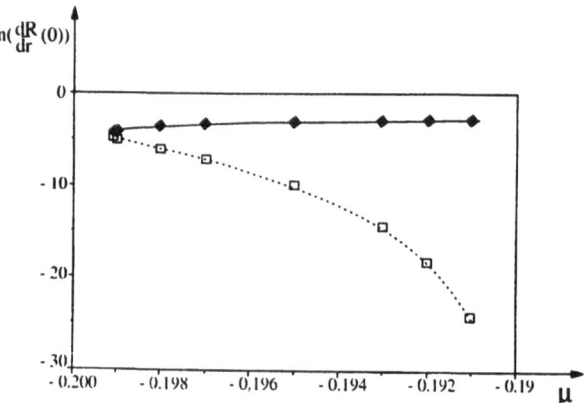

Figure 12. The stable (solid line) and unstable (dashed line) stationary solutions of equation (30) are simply characterized by the value of $\partial R/\partial r$ at $r = 0$ which is plotted as a function of μ.

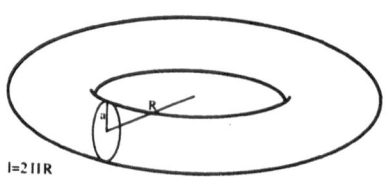

Figure 13. This picture represents a vortex line solution. $\ell = 2\pi R$ is the length of the line and a the radius of the defect's core.

The new mechanism of transition described in this paper and associated with topological defects brings into play purely deterministic solutions of the system. However the mechanism of creation, which is associated with thermal fluctuations, as well as the initial positions of the cores of the defects (randomly distributed) may play an important role. Actually a region containing two cores of defects of opposite topological charges may be considered as a classical (i.e. neutral) seed of phase A_0 generated by a thermal fluctuation. The increase or decrease of this region should depend on its size when the two cores get into contact, that is to say with the initial distance between the two cores.

ACKNOWLEDGEMENTS

J. Lega, E. Tirapegui and D. Walgraef are acknowledged for a number of fruitful discussions. We acknowledge the CCVR (Centre de Calcul Vectoriel pour la Recherche) where some numerical simulations presented in this paper have been performed, the NCAR, the CPAI (Centre Pilote d'Analyse d'Images) of the Observatoire de Nice, and the DRET (Direction des Recherches Etudes et Techniques) for a financial support under contract n° 86/1511.

REFERENCES

1. See for example L.D. Landau and E.M. Lifshitz, "Physical Kinetics" Pergamon Press (1981) and J.W. Cahn and J.E. Hilliard, J.Chem. Phys. **31**, 688 (1959).
2. J.S. Langer, Ann. Phys., **54** 258 (1969).
3. See for example "Physics of defects". Eds. R. Balian, M. Kléman and J.P. Poirier. North Holland (1980).
4. D. Mermin in Rev. Mod. Phys. **51**, 591 (1979).
5. Y. Pomeau, Physica **23D**, 45 (1986).
6. J. Guckenheimer and P. Holmes, "Nonlinear oscillations, dynamical systems, and bifurcations of vector fields", SpringerVerlag (1984).
7. P. Coullet and D. Repaux in "Instabilities and nonequilibrium structures". Eds. E. Tirapegui and D. Villarroel, Reidel (1987).
8. See for example K. Kawasaki and T. Ohta, Physica **116A**, 573 (1982).
9. P. Coullet, C. Elphick and D. Repaux, Phys. Rev. Lett., **58** 431 (1987).
10. For an example of such vortex core structure see N. Schopohl and T. J. Sluckin, Phys. Rev. Lett. **59**, 2582 (1987).

VORTEX VARIATIONS

FERNANDO LUND, ANDREAS REISENEGGER AND
CRISTIAN ROJAS
Departamento de Física
Facultad de Ciencias Físicas y Matemáticas
Universidad de Chile
Casilla 487-3, Santiago, Chile

ABSTRACT. We describe recent work on vortex mediated phase transitions in three dimensions, and on the relation of ultrasound scattering to two–point vorticity correlations.

1. Vortex Mediated Phase Transitions In Three Dimensions

Ever since the pioneering work of Onsager[1] and Feynman[2], from time to time one finds in the literature[3] the idea that the lambda transition in liquid helium might be driven by the unbinding of vortex loops, and it was shown by Langer and Fisher[4] that the behaviour of the critical velocity of a superfluid near the transition temperature could be understood in terms of thermally excited vortex rings. In spite of this result, no precise computation of the critical properties of a system of interacting vortex filaments had been done until recently, although the critical properties of liquid Helium have been computed to great accuracy by mapping the system of interacting bosons into an XY model, setting up the renormalization group equations for this system, and solving them in three dimensions using high order perturbation techniques[5]. Although this is a satisfactory computational tool, it offers no insight into the physics of the transition, and leaves open the question of what is its driving mechanism.

In two dimensions, one doesn't have of course vortex rings, but rather vortex–antivortex pairs with a logarithmic interaction that is formally equivalent to a Coulomb gas. Kosterlitz and Thouless[6] pioneered an approach to this system that has been highly succesful in computing its static and dynamic critical properties[7], as well as explaining the experimental results for two dimensional superfluids[8]. In its simplest form the Kosterlitz –Thouless transition may be understood as follows: at low temperatures there will be thermally excited dipoles. As the temperature is raised, the mean separation between the components of each dipole will increase and smaller dipoles will be created in between. The polarization of the medium due to the large dipole will facilitate the creation of the smaller dipoles, and they in turn will screen the interaction between the components of the larger dipoles, and this screening may, to a first approximation, be taken into account by introducing a scale–dependent dielectric constant. The crucial computation is to show that there is a

E. Tirapegui and D. Villarroel (eds.), Instabilities and Nonequilibrium Structures II, 207–220.
© *1989 by Kluwer Academic Publishers.*

finite (transition) temperature at which this dielectric constant diverges and allows the pair to break, as well as to compute the (critical) behaviour of the system near this transition temperature.

In three dimensions, the analogous problem is easily set up[8]: the energy of a system of interacting vortex loops is well–known to be

$$\frac{\rho}{8\pi} \sum_{i,j} \Gamma_i \Gamma_j \oint \oint \frac{d\ell_i \cdot d\ell_j}{|X_i - X_j|} \tag{1.1}$$

and all that would be needed would be to compute the thermodynamical properties of this system. Unfortunately, the phase space for the system is the space of all curves ("strings") that can be drawn in three dimensions, which is far too complicated an animal for exact calculations. Numerical simulations have been attempted[9] within the framework of superfluidity and of the XY model, and they suggest that vortex filaments do indeed play a crucial role at criticality.

Another approach has been followed by Williams[10], in which the phase space is drastically reduced by considering vortex filaments of circular shape only: vortex rings. Here, each state of the system is labeled not by an infinite number of parameters but only by six: three for position, two for orientation, and one, radius, for size. We have used a reasoning along these lines but with an approach that differs from that of Williams in several respects and we are led to a critical behaviour that is qualitatively similar but quantitatively different from his.

The energy associated with an isolated vortex ring of radius R and circulation Γ, both in classical and quantum fluids, is[11]

$$U_0(R) = \frac{\rho \Gamma^2 R}{2} (\ln(R/r) + C) \tag{1.2}$$

where ρ is the liquid density and C a constant whose exact value depends on the nature of the fluid and on what goes on inside the vortex core, of thickness r. For a superfluid, the value $C = 0.464$ has been proposed[11], and expression (1.2) is valid for a thin vortex ring, that is, for $R \gg r$. As we are interested in the critical behaviour of the system, what goes on at a microscopic scale should not matter and the precise values of r and C should be irrelevant as far as the universal quantities are concerned. A system of noninteracting vortex rings will clearly not have a phase transition, as the energy that it costs to get one grows monotonically with R. Interactions have to be taken into account, and now we follow the ideas of Kosterlitz–Thouless[6] whereby this interaction is approximately considered by way of a scale dependent screening. This makes sense, as a vortex ring behaves in many respects as a magnetic dipole, and we expect that the self–interaction of a large ring will be screened by smaller ones, whose creation will be facilitated by the presence of the larger one, and one might expect that as in two dimensions, it will cost a finite energy to form infinitely large rings. It has also been noted[12], that for a hypothetical system of vortices whose energy was proportional to length, allowing for arbitrarily shaped vortices would mean that at a finite temperature the gain in entropy would dominate over the cost in internal energy allowing for infinitely large vortices to occur. Indeed, embedding the system on a lattice of size r, and letting the vortex length be $L = nr$, one has that the number of states available is, for large n, of order $Aq^n n^{-h}$ [13], where $A > 0, q > 1$ and $h > 1$ are constants. The leading part of the free energy is then

$$\Omega = -k_B T \sum_{}^{\infty} A q^n n^{-h} e^{-\beta cn}$$

where c is the constant of proportionality between energy and length. This can be rewritten as

$$\Omega = -A k_B T \sum_{}^{\infty} n^{-h} e^{n(\ln q - \beta c)}$$

showing that there is a qualitative change in the thermodynamic behaviour at the critical temperature $\beta_c = (\ln q)/c$.

1.1. SCALE DEPENDENT MEAN FIELD APPROACH TO INTERACTING VORTEX RINGS

We wish to approximate the properties of a system of interacting vortex rings by way of something like a dielectric constant. An ideal incompressible fluid of density ρ, velocity $\vec{v}$ and vorticity $\vec{\omega}$ is described by the equations

$$\nabla \cdot \vec{v} = 0$$
$$\nabla \wedge \vec{v} = \vec{\omega}$$

and its energy is

$$E = \frac{\rho}{2} \int v^2 d\vec{r}.$$

These relations are clearly analogous to the magnetostatics equations

$$\nabla \cdot \vec{B} = 0$$
$$\nabla \wedge \vec{B} = \frac{4\pi}{c} \vec{j}$$
$$E = \frac{1}{8\pi} \int B^2 d\vec{r}$$

for a magnetic field $\vec{B}$ and current $\vec{j}$. This allows for a visualization of the fluid as a magnetic system with field

$$\vec{B} = \sqrt{4\pi\rho} \vec{v}$$

and current

$$\vec{j} = c \sqrt{\frac{\rho}{4\pi}} \vec{\omega}.$$

We now compute the energy that it costs to form a vortex ring with velocity field $\vec{v}_1$ in a prescribed uniform external flow $\vec{v}_0$ that may be produced, for instance, by a much larger vortex ring. The total energy of the fluid is

$$E = \frac{\rho}{2} \int (\vec{v}_0 + \vec{v}_1)^2 d\vec{r}$$

from which we extract the interaction energy

$$E_{int} \approx \vec{v}_0 \cdot \int \rho \vec{v}_1 d\vec{r} = \vec{v}_0 \cdot \vec{P},$$

where $\vec{P}$ is the vortex momentum which, for a vortex filament parametrized as $\vec{r} = \vec{r}(\sigma)$ is given by[14]

$$\vec{P} = \frac{\rho\Gamma}{2} \oint \vec{r}(\sigma) \wedge \frac{d\vec{r}(\sigma)}{d\sigma}.$$

Thus, within the magnetic analogy, the vortex momentum is related to the magnetic moment $\vec{m}$ of a current loop through

$$\vec{m} = \frac{\vec{P}}{4\pi\rho}.$$

There is however an important physical difference between the magnetic and fluid systems in the *sign* of the interaction. This is due to the abscence in the fluid of electromagnetic induction effects. These effects are responsible for the fact that in the magnetic system, for fixed current intensity, the force acting along some generalized coordinate ξ is

$$F_\xi = +\frac{\partial E}{\partial \xi},$$

whereas in the fluid it has the usual form

$$F_\xi = -\frac{\partial E}{\partial \xi}.$$

In order to study screening effects we need a "diamagnetic constant", for which we need a magnetic susceptibility, for which in turn we need the magnetic polarizability of a vortex system measuring its response to an external flow. This "magnetic polarizability" is

$$\alpha = \frac{\partial}{\partial v} \left\langle \frac{\vec{m} \cdot \vec{v}}{v} \right\rangle \Bigg|_{v=0},$$

where the statistical average $\langle\rangle$ is taken over all possible vortex orientations. One then has, for a dilute gas of vortices,

$$\alpha = \lim_{v \to 0} \frac{\int d\Omega \vec{m} \cdot \vec{v} \exp(-\beta\vec{P} \cdot \vec{v})}{v^2 \int d\Omega \exp(-\beta\vec{P} \cdot \vec{v})}$$

$$= -\frac{4\pi}{3}\beta\rho m^2.$$

For a vortex ring of radius R and circulation Γ the "magnetic moment is $m = \Gamma R^2/4$, giving for the scale–dependent "polarizability"

$$\alpha(R) = -\frac{\pi}{12}\beta\rho\Gamma^2 R^4.$$

The negative sign indicates that, contrary to a system of current loops, vortices are diamagnetic: they tend to orient themselves with their magnetic moment pointing *opposite* to the external flow.

The "magnetic susceptibility" is obtained by calculating $V^{-1} \sum_i \alpha_i n_i$, where V is the volume of the system, i labels the vortex states, α_i is the polarizability of a vortex in the ith state, and n_i is the number of vortices in that state. The sum over states is carried out replacing $\sum_i$ by $(V/r^6) \int R^2 dR d\omega$. The average number of vortices in a given sate is $e^{-\beta U}$ where U is the energy that it costs to put the vortex in that state. Consequently, the susceptibility χ is given by

$$\chi = \frac{4\pi}{r^6} \int R^2 dR \alpha(R) e^{-\beta U(R)}.$$

Since the energy of a thin isolated vortex ring is given by (1.2), it can be regarded as a string with a line tension

$$T_0(R) = \frac{dU_0}{dR} = \frac{\rho \Gamma^2}{2}(\ln(R/r) + C + 1)$$

due to its self–interaction. If, however, many more vortices are present, this tension will be modified with a scale–dependent "diamagnetic constant" $\varepsilon(R)$:

$$T_0(R) \longrightarrow T(R) = \frac{T_0(R)}{\varepsilon(R)},$$

and the energy of a vortex becomes

$$U(R) = \mu + \int_r^R T(R) dR = \mu + \frac{\rho \Gamma^2}{2} \int_r^R \frac{\ln(R'/r) + C + 1}{\varepsilon(R')} dR' \tag{1.3}$$

where μ is the energy needed to form a vortex of radius r. Under the assumption of a dilute gas, the self–energy of a ring of radius R is affected only by smaller vortices, acting as a magnetic medium with linear response. Using then $\varepsilon = 1 - 4\pi\chi$ we have

$$\varepsilon(R) = 1 + \frac{4\pi^3}{3}\beta\rho\Gamma^2 \int_r^R dR' \left(\frac{R'}{r}\right)^6 e^{-\beta U(R')}, \tag{1.4}$$

which, together with (1.3), contain all the relevant information for our model.

1.2. SCALING EQUATIONS AND CRITICAL BEHAVIOUR

We have derived a set of (coupled, integral) equations (1.3-1.4) for the effective potential $U(R)$ and the "diamagnetic constant" $\varepsilon(R)$. The relevant physical quantity is $\varepsilon(\infty)$ as a function of temperature. One looks for the asymptotic behaviour of ε as $R \to \infty$, and tries to find whether there is a temperature where this asymptotic behaviour changes qualitatively, signaling a phase transition. If such a critical temperature exists, the next task is to compute the behaviour of $\varepsilon(\infty)$ near the critical temperature.

A property of (1.3-1.4) that greatly simplifies the analysis is that it can be turned into a set of two coupled ordinary, nonlinear differential equations:

$$\frac{d\varepsilon}{dR} = \frac{4\pi^3}{3}\beta\rho\Gamma^2 \left(\frac{R}{\tau}\right)^6 e^{-\beta U(R)}$$

$$\frac{dU}{dR} = \frac{\rho\Gamma^2(\ln(R/\tau) + C + 1)}{2\varepsilon(R)},$$

(1.5)

with initial conditions $U(R = \tau) = \mu$ and $\varepsilon(R = \tau) = 1$. The first is the energy needed to create a vortex of minimum size, and the second is the statement that for a vortex of minimum size there are no smaller vortices and hence no screening. The critical behaviour of (1.5) should not depend on the precise values of the parameters or on the precise values of the initial conditions. For a numerical solution of (1.5) however, we do need numbers and for μ we choose the value given by Jones and Roberts[15] using a Landau–Ginzburg theory for an isolated vortex. This gives $\mu = 1.39\rho\Gamma^2\tau$ and for Γ, τ, and ρ we use the values appropriate to liquid Helium. A considerable simplification in the numerical search for a critical point of (1.5) is obtained through the change of variables

$$K = \frac{\beta\rho\Gamma^2 R(\ln(R/\tau) + C + 1)}{12\varepsilon(R)}$$

$$y = \frac{16\pi^3(R/\tau)^6 e^{-\beta U(R)}}{\ln(R/\tau) + C + 1}$$

$$l = \ln(R/\tau),$$

(1.6)

that gives the new set of equations

$$\frac{dK}{dl} = K\left(1 - Ky + \frac{1}{l + C + 1}\right)$$

$$\frac{dy}{dl} = 6y\left(1 - K - \frac{1}{6(l + C + 1)}\right),$$

(1.7)

with initial conditions

$$K(l = 0) = \frac{1}{12}\beta\rho\Gamma^2\tau(C + 1)$$

$$y(l = 0) = \frac{16\pi^3}{C + 1}e^{-\beta\mu},$$

(1.8)

which, upon alimination of the temperature β turn into the single condition

$$y(l = 0) = \frac{16\pi^3}{C + 1}\exp\left[-\frac{12\mu K(l = 0)}{\rho\Gamma^2\tau(C + 1)}\right].$$

We have solved numerically the system (1.7) using a fourth order Runge–Kutta method with adaptive step size. The trajectories are shown in Fig. 1, and it may be easily observed that there is a sharp change in the qualitative behaviour of the system at a critical value K_{0c} of $K(l = 0)$. It is given by

$$K_{0c} = 0.54222.$$

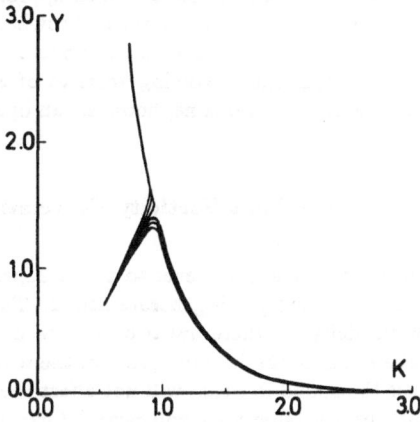

Figure 1. Trajectories for the system (1.7) for various initial conditions. There is a sharp transition at $K_{0c} = 0.54222$.

So, our first conclusion is that a system of interacting vortex rings in three dimensions has a phase transition. The next task is to find the critical behaviour. To do this, we define the critical temperature $t = 1 - (T/T_c)$ in terms of which one has

$$K_0 = \frac{K_{0c}}{1 - t},$$

and solve (1.7) for different values of K_0 in the vicinity of K_{0c}, corresponding to $t \in [10^{-5}, 10^{-2}]$. The resulting $\varepsilon(\infty)$ as a function of reduced temperature t is adjusted to the curve

$$\ln \varepsilon(\infty, t) = -\nu \ln t + \text{constant}$$

by way of a linear least–squares fit. This procedure yields the critical exponent

$$\nu - 0.57.$$

For $t < 0$ the quantity $\varepsilon(\infty)$ diverges.

As a check on the numerics, we redid the whole computation with the system (1.5) and the critical behaviour was confirmed.

1.3. COMMENTS

The main conclusion of this section is that a gas of vortex rings in three dimensions has a phase transition with a critical exponent that can be calculated, within the context of our model, to be $\nu = 0.57$. This compares fairly with the experimental value $\nu = 0.67$ for liquid Helium, and is quite better than the value $\nu = 0.50$ obtained on the basis of a $2 + \epsilon$ expansion[16] an is also an improvement on the value $\nu = 0.53$ claimed* by Williams[10].

---

*We have solved the scaling equations of Williams[10] and obtain $\nu = 0.50$, in disagreement with his claim.

The main drawback of the method is that it is an uncontrolled approximation in the sense that one does not have a quantitative estimate of what one is throwing away by considering a dilute gas of vortex rings only. The work of Banks, Myerson and Kogut[17], and of Wiegel[12] may be relevant for a calculation involving vortices of arbitrary shape. An accurate account of their mutual interaction remains, however, an open problem.

2. Ultrasound Scattering and Two–Point Vorticity Correlations

Consider a point vortex that it is hit by a sound wave: to a first approximation the vortex will oscillate with the wave, and in so doing will generate sound. This is the basic mechanism for sound scattering by vorticity. If there are two vortices present separated by a distance comparable to the wavelength of the incoming sound, the waves scattered off each will interfere significantly and if the sound is weak, will not affect the vortex dynamics. If there are many vortices present, the scattered wave will carry information about the vortex distribution, in analogy to electromanetic scattering and mass distribution, as well as neutron scattering and spin (Fig. 2). In the electromagnetic and neutron cases there is a well known[18] relation between scattering cross section and two–point correlations and as the motion of more than three vortices is in general chaotic, we would expect that interesting information about correlations in a system of vortices could be obtained through ultrasound scattering if, first, a precise relationship could be established, and second, the effect could be measured. Here we shall address the first problem. More generally, one could think of a target flow that, rather than having its vorticity confined to points, has it continuously distributed, and not just in two but also in three dimensions. Thus, we see that sound scattering might provide a nonintrusive direct way of probing turbulent flows.

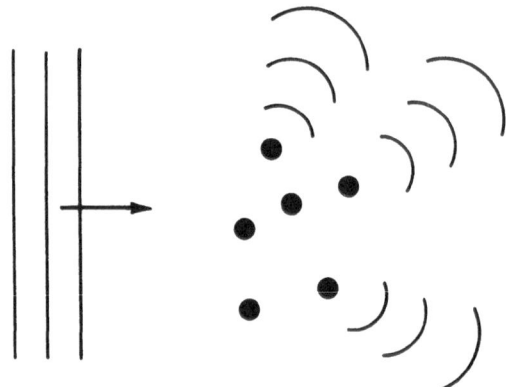

Figure 2. The interference pattern of the scattered ultrasonic waves carries information about the vortex distribution.

The generation of sound by nonmechanical means has been studied for many years[19], and one important point that has emerged from this study is that the source of such sound is associated with regions of unsteady vorticity, with the frequencies radiated being of course the frequencies associated with the source. Thus, we see that the reason for the point vortex we mentioned above has to radiate is that, being driven by the sound wave, it represents an unsteady vorticity. We also see here that in order for these ideas to work, it is necessary to have incoming sound of much higher frequency (i. e. ultrasound) than the frquencies of the target. Otherwise the scattered sound would be inextricably mixed with the spontaneously generated one.

The study of sound scattering by turbulence goes back to the work of Lighthill[20], Kraichnan[21], and Tatarskii[22]. The relation to vorticity, however, does not appear to have been noticed. This is not a mere detail, as vorticity is very basically related to turbulence. Also, having information, say, about the velocity of a flow, is a far cry from having information about its derivatives. Indeed, local measurements of velocity are routinely carried with hot–wire anemometry and laser Doppler velocimetry but similar measurements of vorticity are considerably more difficult[23]. Ultrasound would also provide a nonintrusive probe, and this is of course an important advantage, as is the possibility of exploring nontransparent fluids.

It has also been noted in the literature[24] that scattered pressure is linearly related to vorticity in Fourier space. This fact is however of limited usefulness in practice as one would need measurements at very many wavelengths to have information about the local values of vorticity. Vorticity correlations, on the other hand, appear to be a much more powerful tool to study the properties of disordered flows.

2.1. FORMALISM

We consider an adiabatic, slightly compressible fluid, and assume that viscosity will be irrelevant during the short time scales associated with the ultrasound. The idea is to have a target flow whose velocity in the abscense of the sound wave would be given by $\vec{u}$, with bounded vorticity $\vec{\omega} = \nabla \wedge \vec{u}$ and evolving with a typical frequency Ω, that is perturbed by a low amplitude sound wave of frequency $\nu \gg \Omega$. If $\vec{V}$ is the total velocity field, we associate the difference $\vec{v}_s = \vec{V} - \vec{u}$ with the high frequency sound. It is assumed that $v_s \ll u \ll c$, where c is the speed of sound, assumed uniform throughout the fluid. The density $\rho = \rho_0 + \rho_s$ will have fluctuations ρ_s obeying the equation[19]

$$\frac{\partial^2 \rho_s}{\partial t^2} - c^2 \nabla^2 \rho_s = \frac{\partial^2 (\rho V_i V_k)}{\partial x_i \partial x_k} \tag{2.1}$$

This is a wave equation with a source term. The strategy is to turn it into an integral equation and use a first Born approximation to obtain a linear relation between the scattered and scatterer fields. The first step is easily taken:

$$\rho_s = \rho_{inc} + G * S \tag{2.2}$$

where ρ_{inc} is the density associated with the incident wave and $G * S$ is a convolution of the source

$$S \equiv \frac{\partial^2 \rho V_i V_k}{\partial x_i \partial x_k}$$

with the Green function G of the wave equation.

Before taking the second step, we show that the source is indeed associated with the vorticity of the target, as was claimed at the beginning of this section. Note that, within our approximation scheme,

$$S \approx 2 \frac{\partial^2 ((\rho_0 + \rho_s) u_i v_{sk})}{\partial x_i \partial x_k}.$$

We have neglected terms quadratic in v_s because they are very weak, and terms quadratic in u because they contribute to the spontaneously generated sound, at frequency Ω, and not to the scattered sound at the much higher frequency ν. Using standard vector identities this turns, to leading order, into

$$\frac{S}{2} = \rho_0 \nabla \cdot (\vec{v}_s \wedge \vec{\omega}) + \rho_0 \nabla^2 (\vec{v}_s \cdot \vec{u}) - \rho_0 \partial_i (u_k \partial_i v_{sk}) + \rho_0 \partial_i (u_i \partial_k v_{sk})$$

and, using $u \ll c$ and $\nu \gg \Omega$ into

$$S = \rho_0 \nabla \cdot (\vec{v}_s \wedge \vec{\omega}) + \rho_0 \nabla^2 (\vec{v}_s \cdot \vec{u}) - \frac{\partial}{\partial t} (\vec{u} \cdot \nabla \rho).$$

Although the vorticity $\vec{\omega}$ does appear in the right hand side, there are also terms including $\vec{u}$ that at first sight are not related to it. However, the target flow is completely determined once $\vec{\omega}$ is specified, so it ought to be possible to express everything in terms of it. It will now be shown that this is so, at least within the first Born approximation.

Form the integral equation (2.2) we have that $G * S$ is the scattered pressure which, for an incident wave of small enough amplitude, will be very small, and we can solve the integral equation by iteration; that is, whenever v_s or ρ_s appear in $G * S$ we shall substitute for them the values for the incident, plane, wave. We have then

$$S = \rho_{s1} + \rho_{s2}$$

with

$$\rho_{s1} = \rho_0 G * (\nabla \cdot (\vec{\omega} \wedge \vec{v}_{inc}))$$

$$\rho_{s2} = \rho_0 G * \left(\nabla^2 (\vec{u} \cdot \vec{v}_{inc}) + \frac{1}{c^2} \frac{\partial}{\partial t} \left(\vec{u} \cdot \frac{\partial \vec{v}_{inc}}{\partial t} \right) \right)$$

where we have used that, to the order of accuracy we are using,

$$\frac{\partial \vec{v}_{inc}}{\partial t} = -\frac{c^2}{\rho_0} \nabla \rho_{inc}.$$

The subscript inc is used to label values associated with the incident plane wave. The first thing to note is that, for $\nu \gg \Omega$,

$$\int d^3 x' dt' \frac{\partial}{\partial t'} \left(\vec{u} \cdot \frac{\partial \vec{v}_{inc}}{\partial t'} \right) G = \int d^3 x' dt' \frac{\partial^2}{\partial t'^2} (\vec{u} \cdot \vec{v}_{inc}) G$$

$$= c^2 \int d^3 x' dt' \nabla'^2 (\vec{u} \cdot \vec{v}_{inc}) G,$$

where integration by parts and the equation obeyed by the Green function G have been used to get the second equality. Next, one has in the radiation zone

$$\nabla G \approx -\frac{\hat{r}}{c}\frac{\partial G}{\partial t},$$

and, for a plane incident wave

$$\partial_i \vec{v}_{inc} = -\frac{\hat{n}_i}{c}\frac{\partial \vec{v}_{inc}}{\partial t}.$$

where $\hat{n}_i$ is a unit vector pointing along the direction of incidence. Writing now ρ_{s1} in the form

$$\rho_{s1} = \rho_0 \int \left(\partial_i (v_k (\partial_k u_i)) - (\partial_i u_k)) \right) G,$$

using the next to last two equalities, $\nu \gg \Omega$, and integrating by parts leads to

$$\rho_{s1} = (\cos\theta - 1)\rho_{s2} \tag{2.3}$$

where θ is the scatteirng angle: $\cos\theta = \hat{r}\cdot\hat{n}$. Equations (2.2) and (2.3) show that the scattered field can be expressed as a convolution of the Green function with a source localized on the region of nonvanishing vorticity. We now use them to show that there is a linear relation between scattered pressure and vorticity in Fourier space, which leads to a linear relation between acoustic intensity and vorticity correlations. We shall study the two–dimensional case. Computations in three dimensions have been reported elsewhere[25].

2.2. TWO DIMENSIONAL CASE

The two dimensional case has the simplicity that vorticity is a scalar but is complicated by the Greens function having an infinitely long tail in time. Its Fourier transform (in time) is a Hankel function of the firs kind:

$$G(\vec{x} - \vec{x}', \nu) = \frac{i}{8\pi} H_0^{(+)}\left(\frac{\nu}{c}|\vec{x} - \vec{x}'|\right),$$

and we use its asymptotic behaviour to find the acoustic pressure in the radiation zone:

$$p_s = \left(\frac{\cos\theta}{1 - \cos\theta}\right) p_{s1}$$

with $p_{s1} = c^2 \rho_{s1}$. It is

$$p_{s1}(\vec{x}, t) = \frac{i\rho_0}{4\pi\sqrt{2\pi c|\vec{x}|}} \int d^2 x'\, dt'\, d\nu \sqrt{\nu}(\vec{\omega} \wedge \vec{v}_{inc}\cdot\hat{r}e^{i\left(\frac{\nu|\vec{x}|}{c} + \frac{\pi}{4}\right)} e^{-i\frac{\nu}{c}\vec{x}'\cdot\hat{r}} e^{-i\nu(t - t')}.$$

The incident plane wave is now explicitly written as

$$\vec{v}_{inc} = v_0 \hat{n}\cos\left(\vec{k}\cdot\vec{x} - \nu_0 t'\right)$$

so that the incident frequency is $\nu_0 \gg \Omega$. Calling ν the frequency of the scattered wave, our assumptions mean $\nu \sim \nu_0$ with the differences being of order Ω and a standard argument[18] shows that only frequencies $\nu \sim \nu_0$ contribute significantly to the integral on t'. Noting that the vorticity points along the direction perpendicular to the plane of the flow, one has

$$p_s(\vec{x}, \nu) = \left(\frac{\cos \theta \sin \theta}{\cos \theta - 1} \right) \left(\frac{\pi^3 \nu}{2c|\vec{x}|} \right)^{1/2} \rho v_0 e^{\left(\frac{\nu |\vec{x}|}{c} + \frac{3\pi}{4} \right)} \tilde{\omega}(\vec{q}, \nu - \nu_0) \tag{2.4}$$

where

$$\vec{q} = \frac{\nu}{c}\hat{r} - \vec{k}$$

is the momentum transfer and $\tilde{\omega}$ is the Fourier transform both in space and time of the vorticity. This is a linear relaton between acoustic pressure and vorticity that clearly points out that the measurement of scattered acoustic pressure may be used as a probe of vorticity. However, it is not likely that measurements of vorticity local in space can be carried out in this way, as an accuracy of order Δx in position would be obtained at the price of carrying out measurements over a range of wavelengths of order $(\Delta x)^{-1}$. A more likely candidate to give useful results is the acoustic intensity I, the total energy radiated away per unit area in a time interval τ:

$$I(\vec{x}) = \frac{1}{\rho c} \int_{-\tau/2}^{\tau/2} p_s^2 dt.$$

Assuming that the observation time is much larger than the inverse of the typical frequencies of the target, $\tau\Omega \gg 1$, standard arguments[18] again lead to

$$I(\vec{x}, \nu) = \frac{1}{4|\vec{x}|c^3} J_0 \pi \nu \cos^2 \theta \cot^2(\theta/2) \tilde{S}(\vec{q}, \nu - \nu_0) \tag{2.5}$$

where $J_0 = \rho c v_0^2$ is the energy flux of the incident wave, $\tau I(\vec{x}, \nu)$ is the energy per unit area and unit frequency radiated through $\vec{x}$ during the time interval τ defined through

$$I(\vec{x}) = \tau \int_0^\infty d\nu I(\vec{x}, \nu)$$

and $\tilde{S}$ is the Fourier transform, both in space and time of the two–point vorticity correlations $S(\vec{R}, T)$:

$$S(\vec{R}, T) = \int d^2 x \langle \omega(\vec{R} + \vec{x}, T)\omega(\vec{x}, 0)\rangle.$$

The brackets <> mean a time average:

$$\langle \omega(x', T)\omega(x'', 0)\rangle = \frac{1}{\tau} \int_{-\tau/2}^{\tau/2} dt \omega(x', T + t)\omega(x'', t).$$

Equation (2.5) is the main result of this section. Given the very nice experiments carried out recently[26] using soap films, it promises to be of more than just academic interest. The apparent divergence at forward scattering $\theta \sim 0$ is of course not physical but an artifact of our using plane waves of infinite extent. If, say, L is a measure af the width of the plane wave, our reasoning is valid for angles such that $qL \gg 1$ and breaks down for very small angles.

2.3. COMMENTS

From (2.4) we see that the ratio of scattered to incident pressure is inversely proportional to the square root of the acoustic wavelength, of the distance from target to receiver, to the speed of sound, and directly proportional to vorticity in Fourier space; that is

$$\frac{p_s}{p_{inc}} \sim \frac{1}{\sqrt{\lambda}\sqrt{|x|}}\frac{\tilde{\omega}}{c}.$$

In order to have a rough idea of what this means in practice, consider a target consisting of a single vortex Γ. In this case $\tilde{\omega} \sim \Gamma$ and, if say, $\Gamma \sim 10^3 \, cm2/sec$, a detector placed at a distance of tens of centimeters will pick up a pressure of about 1% of the incident pressure at a wavelength of millimiters. Similar estimates hold in three dimensions[25], and this should be well within the capabilities of current ultrasound technology.

ACKNOWLEDGMENTS

This work was supported by DIB Grant E–2854–8814 and Fondo Nacional de Ciencia Grant 0533–88.

REFERENCES

1. L. Onsager, *Nuovo Cimento* 6, Suppl. 2, 249 (1949).
2. R. P. Feynman, in *Progress in Low Temperature Physics*, edited by C. G. Gorter (North–Holland, Amsterdam, 1955).
3. P. W. Anderson, *Basic Notions of Condensed Matter Physics*, (Benjamin, Menlo Park, 1984). M. Rasetti and T. Regge, *Physica* A80, 217, (1975). V. N. Popov, *Sov. Phys. JETP* 37, 341 (1973).
4. J. S. Langer and M. E. Fisher, *Phys. Rev. Lett.* 19, 560 (1967).
5. J. C. Le Gillou and J. Zinn–Justin, *Phys. Rev.* B21, 3976, (1980). P. Hohenberg, *Physica* 109 & 110 B+C, 1436 (1982).
6. J. M. Kosterlitz and D. J. Thouless, *J. Phys.* C6, 1181 (1973). J. M. Kosterlitz, *J. Phys.* C7, 1046 (1974). A. P. Young, *J. Phys.* C11, L453 (1978). D. R. Nelson and J. M. Kosterlitz, *Phys. Rev. Lett.* 39, 1201 (1977). J. V. José, L. P. Kadanoff, S. Kirkpatrick and D. R. Nelson, *Phys. Rev.* B16, 1217 (1977). P. Minnhagen, *Rev. Mod. Phys.* 59, 1001 (1987).
7. D. R. Nelson, in *Phase Transitions*, Vol. 7 (Academic, London, 1983). V. Ambegaokar, B. I. Halperin, D. R. Nelson and E. D. Siggia, *Phys. Rev.* B21, 1806 (1980).
8. P. W. Adams and W. I. Glaberson, *Phys. Rev.* B35, 4633 (1987). D. Finotello, Y. Y. Yu and F. M. Gasparini, *Phys. Rev. Lett.* 57, 843 (1986) and references therein.

9. G. Kohring, R. E. Schrock and P. Wills, *Nucl. Phys.* **B288**, 397 (1987). C. Dasgupta and B. I. Halperin *Phys. Rev Lett.* **47**, 1556 (1981).

10. G. A. Williams, *Phys. Rev. Lett.* **59**, 1926 (1987).

11. P. H. Roberts and J. Grant, *J. Phys.* **A4**, 55 (1971).

12. F. W. Wiegel, *Physica* **65**, 321 (1973).

13. C. Domb, *Adv. Chem. Phys.* **15**, 229 (1969).

14. M. Rasetti and T. Regge, Ref. 3.

15. C. A. Jones and P. H. Roberts, *J. Phys.* **A15**, 2599 (1982).

16. D. S. Fisher and D. R. Nelson, *Phys. Rev.* **B16**, 4945 (1977).

17. T. Banks, R. Myerson and J. Kogut, *Nucl. Phys.* **B129**, 493 (1977).

18. See, for instance, S. W. Lovesey, *Condensed Matter Physics, Dynamic Correlations* (Benjamin, Reading, 1980).

19. F. Lund, in *Instabilities and Nonequilibrium Structures*, edited by E. Tirapegui and D. Villarroel (Reidel, Dordrecht, 1987). T. Kambe, *J. Fluid Mech.* **173**, 643 (1986). W. Mohring, E. A. Müller and F. Obermeier, *Rev. Mod. Phys.* **55**, 707 (1983). T. Kambe and T. Minota, *Proc. Roy. Soc. London* **A386**, 277 (1983). F. Obermeier, *Acoustica* **42**, 56 (1979). W. Mohring, *J. Fluid Mech.* **85**, 685 (1978). M. S. Howe, *J. Fluid Mech.* **71**, 625 (1975). A. Powell, *J. Acoust. Soc. Am.* **36**, 177 (1964). M. J. Lighthill, *Proc. Roy. Soc. London* **A222**, 1 (1954); **A211**, 564 (1952).

20. M. J. Lighthill, *Proc. Camb. Philos. Soc.* **49**, 531 (1953).

21. R. H. Kraichnan, *J. Acoust. Soc. Am.* **25**, 1096 (1953).

22. See A. S. Monin and A. M. Yaglom, *Statistical Fluid Mechanics* (MIT Press, Cambridge, 1980) and references therein.

23. E. Kit et. al., *Phys. Fluids* **30**, 3323 (1987). R. A. Antonia, D. A. Shah and L. W. B. Browne, *Phys Fluids* **30**, 3455 (1987).

24. M. S. Howe, *J. Sound and Vibration* **87**, 567 (1983). T. Kambe and U. Mya Oo, *J. Phys. Soc. Japan* **50ᵢ** 3507 (1981).

25. F. Lund and C. Rojas, to be published in *Physica D*.

26. M. Gharib and P. Derango, *Bull. Am. Phys. Soc.* **32**, 2031 (1987). Y. Couder and C. Basdevant, *J. Fluid Mech.* **173**, 225 (1986).

OSCILLATORY INSTABILITIES IN BENARD-MARANGONI CONVECTION IN A FLUID BOUNDED ABOVE BY A FREE SURFACE

R. D. BENGURIA
Departamento de Física, F.C.F.M
Universidad de Chile
Casilla 487/3, Santiago, Chile

M. C. DEPASSIER
Facultad de Física
Universidad Católica de Chile
Casilla 6177, Santiago 22, Chile

ABSTRACT. We study numerically the linear stability theory of a fluid bounded above by a free deformable surface and below by a rigid or free plane surface. Oscillatory instabilities are found which do not exist when surface deformation is neglected. Analytical results are given for a long wavelength oscillatory instability which we identify as gravity waves.

Introduction

In many analytical studies of convection it is assumed that the boundaries of the fluid are free but plane. This boundary condition is chosen since it is simpler to treat analytically than the rigid boundary condition.[1] A real free surface, however, is deformed due to the fluid motion, effect which is frequently neglected for the sake of simplicity. If the full boundary conditions on the free surface are used, the monotonicity principle is no longer valid in Rayleigh-Bénard convection and we may expect oscillatory instabilities. We have studied this problem numerically and analytically[2-3]; the results are described below.

Mathematical Formulation

Let us consider a two dimensional fluid bounded above by a thermally insulating passive gas and below by a plane stress-free perfect thermally conducting medium which, at rest, lies between $z = 0$ and $z = d$. Upon it acts gravity $\vec{g} = -g\hat{z}$. In the Boussinesq approximation the equations that describe the motion of the fluid are

$$\nabla \cdot \vec{v} = 0$$

$$\rho_o \frac{d\vec{v}}{dt} = -\vec{\nabla}p + \mu\nabla^2\vec{v} + \vec{g}\rho$$

$$\frac{dT}{dt} = \kappa\nabla^2 T$$

$$\rho = \rho_o[1 - \alpha(T - T_o)]$$

221

E. Tirapegui and D. Villarroel (eds.), Instabilities and Nonequilibrium Structures II, 221–225.
© *1989 by Kluwer Academic Publishers.*

where $d/dt = \partial/\partial t + \vec{v} \cdot \vec{\nabla}$ is the convective derivative; $\vec{v} = (u, 0, w)$ is the fluid velocity, p is the pressure, and T is the temperature. T_o and ρ_o are a reference temperature and density respectively. The viscosity, μ, thermal diffusivity, κ, and coefficient of thermal expansion, α are constant.

On the upper free surface $z = d + \eta(x, t)$ the boundary conditions are

$$\eta_t + u\eta_x = w$$

$$p - p_a - \frac{2\mu}{N^2}[w_z + u_x\eta_x^2 - \eta_x(u_z + w_x)] = 0$$

$$\mu(1 - \eta_x^2)(u_z + w_x) + 2\mu\eta_x(w_z - u_x) = 0$$

and

$$T_z - \eta_x T_x = -FN/k$$

Subscripts x and z denote derivatives with respect to the horizontal and vertical coordinates respectively. Here $N = (1 + \eta_x^2)^{1/2}$, F is the normal heat flux, k is the thermal conductivity, and p_a is a constant pressure exerted on the upper free surface.

We shall assume that the lower surface is either stress-free and plane or rigid, and it may be at constant temperature T_b or at constant heat flux. The boundary conditions on the lower surface $z = 0$ are then $w = u_z = 0$, if it is stress-free, or $w = u = 0$ if it is rigid; and $T = T_b$ if it is held at constant temperature or $dT/dz = -F/k$ if the heat flux F is held fixed.

The static solution to these equations is given by $T_s = -F(z - d)/k + T_o$, $\rho_s = \rho_0[1 + (\alpha F/k)(z - d)]$, and $p_s = p_a - g\rho_0[(z - d) + (\alpha F/2k)(z - d)^2]$. It is convenient to adopt d as unit of length, d^2/κ as unit of time, $\rho_o d^3$ as unit of mass, and $T_s(0) - T_s(d)$ as unit of temperature. Then only three dimensionless parameters are involved in the problem, the Prandtl number $\sigma = \mu/\rho_o\kappa$, the Rayleigh number $R = \rho_0 g\alpha(T_s(0) - T_s(d))d^3/\kappa\mu$ and the Galileo number $G = gd^3\rho_o^2/\mu^2$. Introducing a stream function $\psi(x, z, t)$ in terms of which the velocity is given by $\vec{v} = (\psi_z, 0, -\psi_x)$, the linear equations for the perturbations to the static state may be reduced to

$$(D^2 - a^2)(D^2 - a^2 - \lambda/\sigma)\psi = iaR\theta$$

$$(D^2 - a^2 - \lambda)\theta = ia\psi$$

where $D = d/dz$, θ is the perturbation to the static temperature profile, and where we have assumed that all perturbations evolve in time as $\exp(\lambda t)$ and in the horizontal variable as $\exp(iax)$. The linearized boundary conditions become

$$\lambda(D^2 - 3a^2 - \lambda/\sigma)D\psi - a^2(\sigma G + a^2/C)\psi = 0$$

$$(D^2 + a^2)\psi - \frac{R\Gamma}{\sigma G + a^2/C}(D^2 - 3a^2 - \lambda/\sigma)D\psi + iaR\Gamma\theta = 0$$

The dimensionless numbers that have appeared are the Capillary number $C = \mu\kappa/\tau_o d$ and $\Gamma = \gamma/\rho_o g\alpha d^2$. The Marangoni number is given by $M = \Gamma R$, we have chosen to use Γ as an independent parameter instead of M. The thermal boundary conditions are either $\theta = 0$ or $D\theta = 0$. We have solved the linear equations for different boundary conditions on the lower surface in search for oscillatory instabilities.

Results

The effect of surface deformation on the eigenvalues λ and R is measured by the size of $\sigma G + a^2/C$. When this coefficient tends to infinity we recover the simpler stress-free boundary conditions. In the case when the lower surface is rigid, an oscillatory instability is found at finite wavenumber. The critical R for overstability decreases with increasing Γ and increases with G (figs.1-2). This shows that surface tension is a driving mechanism for this instability. The critical Rayleigh number remains higher than the corresponding value for the onset of steady convection within the range of validity of the Boussinesq approximation. When the lower surface is stress-free an oscillatory instability with similar features to the case just described is present. In addition we have found a long wavelength instability at critical R considerably lower than the corresponding value for the onset of steady convection (figs.3-4). The driving mechanism for this instability is buoyancy alone. An asymptotic analysis of this long wave instability shows that the leading order stream function and temperature perturbations are given by $\psi = z$ and $\theta = a(z^3 - 3z)/6$. The critical Rayleigh number is given by $R_c = 30/(1 - 5\Gamma/2)$ and the frequency at criticality is given by $\omega = a\sqrt{\sigma^2 G - \sigma \Gamma R_c}$. This instability corresponds to gravity waves in a shallow viscous fluid.

Acknowledgments

This work was financed by FONDECYT, by DIB U. de Chile and by DIUC U. Católica.

References

1. S. Chandrasekhar, Hydrodynamic and Hydromagnetic Stability, Oxford, Clarendon Press 1961.
2. R. D. Benguria & M. C. Depassier, Phys. Fluids 30, 1678 (1987).
3. R. D. Benguria & M. C. Depassier, to appear in Phys. Fluids.

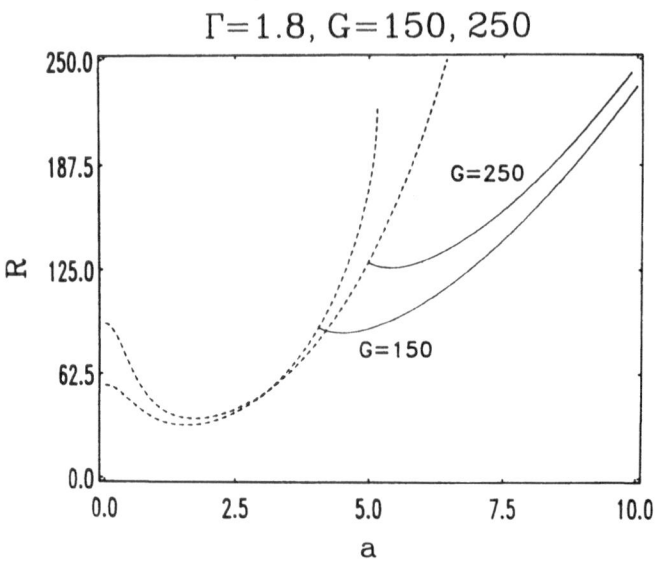

Figure 1. Rayleigh   number vs. wavenumber for different values of
the Galileo number, with Γ = 1.8. The dashed lines correspond to
the marginal curves.

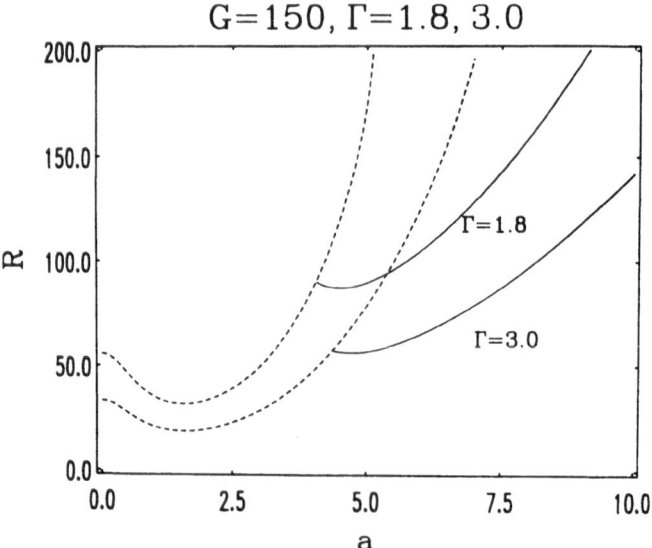

Figure 2. Rayleigh number vs. wavenumber for G = 150 and different
values of Γ.

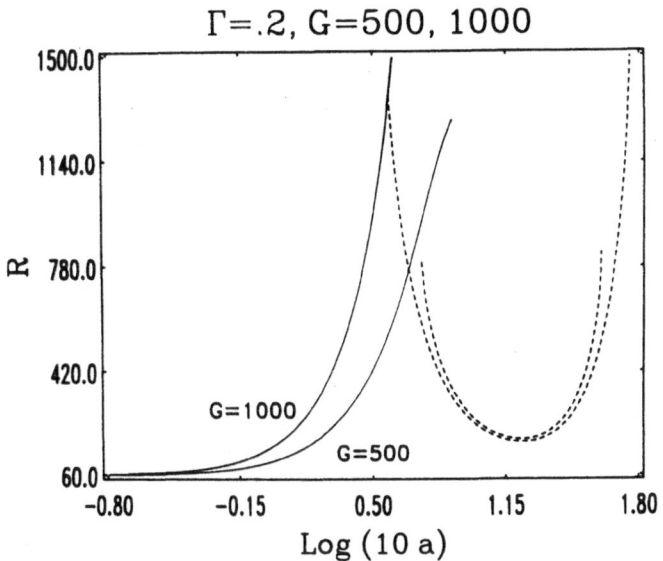

Figure 3. Rayleigh number vs. wavenumber for a small value of Γ (= 0.2) and different values of G.

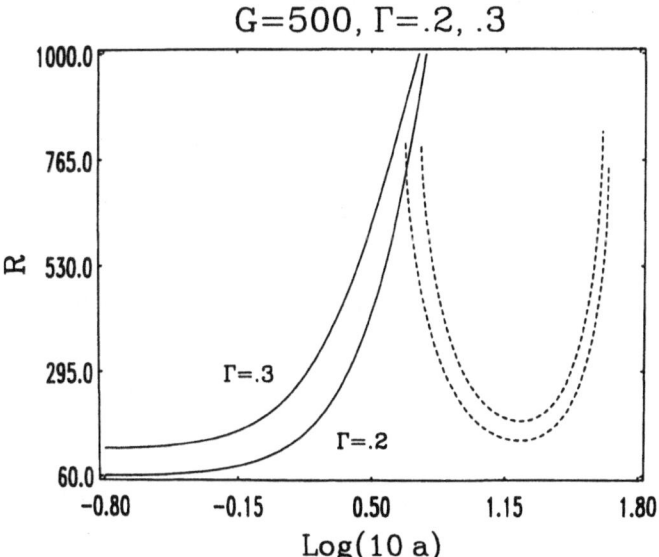

Figure 4. The same as in Fig. 3 for two different (small) values of Γ and fixed G.

PARAMETRIC INSTABILITIES

O. Thual
CERFACS
42, avenue Coriolis
31057 Toulouse
France

S. Douady and S. Fauve
Ecole Normale Supérieure de Lyon
46, allée d'Italie
69364 Lyon
France

ABSTRACT. We consider systems parametrically driven by an external time periodic forcing, and derive the evolution equations that govern the amplitude of the unstable modes. We emphasize the main differences with systems driven far from equilibrium by a stationary forcing. The period doubling bifurcation is understood as a parametric instability, and the corresponding amplitude equation is derived. In spatially extended systems, we show that a parametric forcing always generates a standing wave, and we study its secondary instabilities.

1. Introduction

Pattern forming instabilities in nonlinear dissipative systems driven far from equilibrium by a constant forcing are well documented [1]. We consider here the case of parametric forcing ; simple examples can be found in mechanics [2, 3]. In spatially extended continuous media, a well known example is the Faraday instability [4], i. e. the generation of waves on the free surface of a vertically oscillating fluid layer. We do not present here a review of parametric instabilities, but emphasize their characteristic features. First, the existence of two relevant parameters, the external forcing amplitude and frequency, make the bifurcation diagram richer than for a Hopf bifurcation. In the limit of small dissipation, the parametric instability can be supercritical or subcritical, depending on the forcing frequency, and one thus have a tricritical point in the bifurcation diagram. In the bifurcated state, the external temporal forcing quenches the phase of the parametric oscillation which

E. Tirapegui and D. Villarroel (eds.), Instabilities and Nonequilibrium Structures II, 227–237.

thus strongly differs from a limit cycle generated through a Hopf bifurcation . This affects the dynamics of waves generated by parametric instabilities in spatially extended systems. At instability onset, one always get standing waves instead of propagating waves. The secondary instabilities and the defects of these waves are also characteristic of the parametric forcing. This paper is organized as follows : in section 2 we give elementary results about the parametric oscillator, its amplitude equation and the corresponding bifurcation diagram. In section 3, we show that the period doubling instability of a limit cycle in an autonomous system can be understood as a parametric instability. Section 4 concerns spatially extended systems, where parametric instabilities always generate standing waves ; we study their secondary instabilities in connection with an experiment on surface waves.

2. The parametric oscillator

We consider a parametric pendulum governed by the Mathieu equation

$$\frac{d^2u}{dt^2} + 2\Lambda\frac{du}{dt} + \omega_o^2(1 + F sin\omega_e t)sin u = 0 \tag{1}$$

where ω_o is the pendulum eigenfrequency, Λ represents the dissipation, F is the external forcing at frequency ω_e. The linear stability of the state $u = 0$ is given by the diagram of figure 1 [2]. Without dissipation ($\Lambda = 0$), the instability occurs for vanishing excitation amplitude at frequencies $\omega_{e,n} = 2\omega_o/n$, with n = 1, 2, ... In the small dissipation limit, the critical amplitudes $F_{c,n}$ are finite, and increase with n like $\Lambda^{1/n}$.

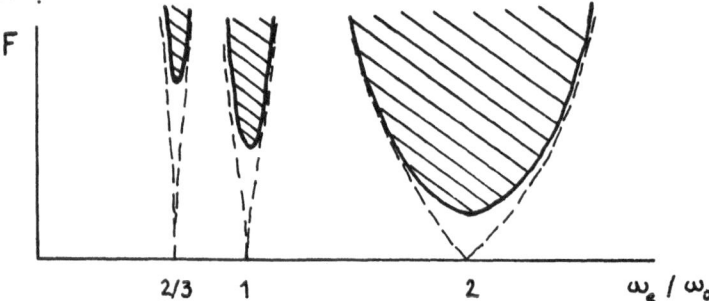

Figure 1. Linear stability diagram of the parametric oscillator.

Let us consider the case $\omega_e = 2\omega_o$ that corresponds to the smallest critical amplitude. We write $\omega_e = 2\omega$ and define the detuning, $\Delta = \frac{\omega_o^2}{\omega_e^2} - \frac{1}{4}$. In the limit of small detuning and small dissipation, we write $\Delta = \delta\epsilon^2$, $\Lambda = \lambda\epsilon^2$, $F = f\epsilon^2$, and look for an asymptotics expansion in ϵ under the form

$$u(t) = \epsilon[u_o(s, T) + \epsilon^2 u_1(s, T) + h.o.t.],$$

where $\frac{d}{dt} = \frac{\partial}{\partial s} + \epsilon^2 \frac{\partial}{\partial T}$. Using standard asymptotic methods [5], we get at leading order

$$u_o(s, T) = A(T) \exp iws + c.c. + \dots \tag{2}$$

The solvability condition at the next order gives the evolution equation for A

$$\frac{dA}{dT} = (-\lambda + i\nu)A + \mu\bar{A} + i\beta A^2 \bar{A}, \tag{3}$$

where ν represents the detuning, $\epsilon^2 \nu = 2\omega(\frac{\omega_e^2}{\omega^2} - 1)$, μ is the scaled external forcing, $\mu = f\omega/4$, and the term $i\beta A^2 \bar{A}$ with $\beta = -\omega/4$, corresponds to the nonlinear frequency correction of the pendulum versus its amplitude. This term is stabilizating at large oscillation amplitude as it nonlinearly increases the detuning between the pendulum frequency and the external forcing frequency. However, when ν and β have opposite signs, the nonlinear detuning first decreases when the pendulum amplitude increases from 0, and the nonlinear term enhances the parametric instability at small oscillation amplitude ; in that parameter range the instability is thus subcritical [6] (see below).

The form of equation (3) follows from symmetry constraints. In the absence of time dependent external forcing, the system is invariant under continuous translations in time, $t \rightarrow t + \theta$, thus from equation (2) one has the invariance $A \rightarrow A \exp iw\theta$. Up to the third order, the evolution equation for A reads

$$\frac{dA}{dT} = (-\lambda + i\nu)A + (\beta_r + i\beta_i)A^2 \bar{A}.$$

This is the normal form for a Hopf bifurcation [7]. For dissipationless systems we have the symmetry constraint $t \rightarrow -t$, $A \rightarrow \bar{A}$, which implies that the coefficients are pure imaginary. When an external forcing is applied, the continuous translation symmetry in time is broken, and the system is invariant under the discrete symmetry $t \rightarrow t + 2\pi/\omega_e$, and thus from (2), $A \rightarrow -A$. Therefore the external forcing allows the new terms, $\bar{A}$, A^3, $\bar{A}^2 A$ and $\bar{A}^3$. If the parametric instability occurs for a small value of the external forcing, i.e. when the the dissipation is small enough, we have to keep only the leading order term $\mu\bar{A}$, where μ is proportional to the external forcing and can always be chosen real with the appropriate origin of time. In the limit of small dissipation, the leading order correction is $-\lambda A$, and the amplitude equation reads

$$\frac{dA}{dT} = (-\lambda + i\nu)A + \mu\bar{A} + i\beta_i A^2 \bar{A}.$$

The nonlinear dissipation, $\beta_r A^2 \bar{A}$, can be kept in a model and modifies the discussion we give below ; in particular it can make the transition always supercritical[8].

The linear part of (3) gives the marginal stability curve,

$$\mu_c(\nu) = \sqrt{\lambda^2 + \nu^2}.$$

Writting $A = R \exp i\theta$, equation (3) becomes

$$\frac{dR}{dT} = (-\lambda + \mu \cos 2\theta)R \tag{4a}$$

$$\frac{d\theta}{dT} = \nu - \mu \sin 2\theta + \beta R^2. \tag{4b}$$

Stationary solutions A_o of (4) represent limit cycles of frequency $\omega_e/2$. Eliminating θ_o, we obtain for the amplitude, $\mu^2 - \lambda^2 = (\nu + \beta R_o^2)^2$. As β is negative we have

$$if \ \nu \leq 0, \ \ R_o^2 = -\sqrt{\mu^2 - \lambda^2} - \nu.$$

$$if \ \nu \geq 0, \ \ R_o^2 = \pm\sqrt{\mu^2 - \lambda^2} - \nu$$

The domains of existence of these solutions are represented in the diagram of figure 2a. The linear stability analysis of the bifurcated solutions shows that they are stable if $\nu + \beta R_o^2 \leq 0$ (see figure 2b). For each value of R_o, equation (4) shows that the phase θ_o is quenched by the external forcing and we have two possible limit cycles with different phases θ_o.

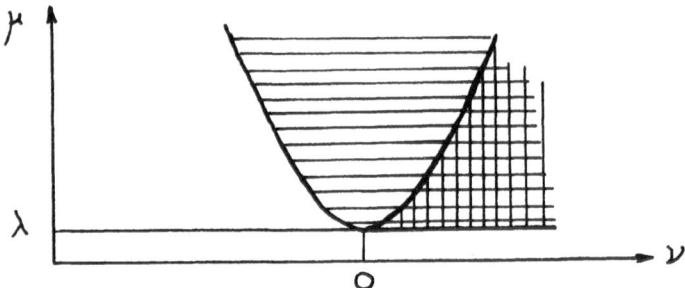

Figure 2a. Nonlinear stability diagram of the $\omega_e/2$ resonance tongue.

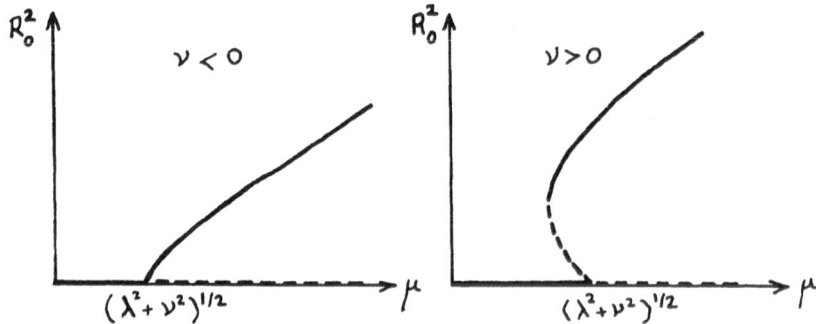

Figure 2b. Amplitude R_o of the $\omega_e/2$ parametric resonance.

3. The period doubling bifurcation

The period doubling instability is one of the three generic bifurcations of a limit cycle. It occurs when a real eigenvalue of the linearized Poincaré map gets off the unit circle by the value -1, or, equivalently, when a Floquet exponent crosses the imaginary axis at the value $i\omega/2$, where ω is the limit cycle pulsation. The bifurcated state is a new limit cycle the period of which is twice the basic cycle period [3, 7]. We want to show, on a simple example, that this bifurcation can be understood as a parametric instablity, and we will give the associated normal form. One of the simplest dynamical system undergoing a period doubling bifurcation is the following "oscillator":

$$\frac{d^3u}{dt^3} + \mu_2 \frac{d^2u}{dt^2} + \mu_1 \frac{du}{dt} + \mu_0 u = u^2. \tag{5}$$

This equation describes the dynamics of a system in the vicinity of a codimension three bifurcation, and the equilibrium state $u = 0$ can undergo a transition to chaos through a Hopf bifurcation and a period doubling cascade when the real parameters μ_0, μ_1, μ_2 are varied around the codimension 3 surface $\mu = (\mu_0, \mu_1, \mu_2) = 0$ [9]. Here we shall restrict our attention to the Hopf bifurcation of the equilibrium $u = 0$, and to the first period doubling instability of the bifurcated limit cycle.

The stability of the state $u = 0$ is given by the real parts of the roots of the charateristic polynomial :

$$P_\mu(s) = s^3 + \mu_2 s^2 + \mu_1 s + \mu_0.$$

A Hopf bifurcation occurs when $\mu_0 = \mu_1 \mu_2$, with a frequency at onset $\omega = \sqrt{\mu_1}$. Numerical simulation of (5) has shown that the period doubling cascade occurs in the vicinity of the equilibrium state $u = 0$ when $\mu = (\eta^3 \nu_0, \eta^2 \nu_1, \eta \nu_2)$ with $\eta \to 0$[9]. We thus consider this scaling and first compute the amplitude of the limit cycle created by the Hopf bifurcation. Then, we consider the stability of this limit cycle with respect to the period doubling bifurcation and compute its approximate onset value. Finally, using symmetry arguments, we derive the normal form for the period doubling bifurcation, and show how the temporal phase of the limit cycle is coupled with the amplitude of the unstable mode at the period doubling bifurcation.

In the vicinity of the Hopf biburcation, we note $-\lambda, \epsilon \pm i\omega$ the three roots of P_μ, and we write :

$$u(t) = a(t) + \overline{a}(t) + \sum_{n+m \geq 2} v^{(n,m)} a(t)^n \overline{a}(t)^m. \tag{6}$$

At leading order, the complex amplitude $a(t)$ is governed by the equation

$$\frac{da}{dt} = (\epsilon + i\omega)\, a + \alpha |a|^2 a, \tag{7}$$

where the coefficient α can be calculated by standard methods [3, 5, 7]. The sign of its real part tells us whether the bifurcation is supercritical ($\alpha_r < 0$) or subcritical ($\alpha_r > 0$), whereas its imaginary part corresponds to the nonlinear frequency correction of the limit cycle. We have,

$$\alpha_r = \frac{-1}{\omega^2 \lambda} \frac{8\omega^2 + \lambda^2}{4\omega^6 + 5\lambda^2\omega^2 + \lambda^4} \quad \text{and} \quad \alpha_i = \frac{-1}{3\omega^3} \frac{26\omega^2 + 5\lambda^2}{4\omega^6 + 5\lambda^2\omega^2 + \lambda^4},$$

and, as $\lambda > 0$, the Hopf bifurcation is always supercritical ($\alpha_r < 0$), and the limit cycle amplitude is $\sqrt{\epsilon/-\alpha_r}$.

We now study the period doubling destabilization of the limit cycle $v_0(t)$ created by the Hopf bifurcation. Writting $u(t) = v_0(t) + w(t)$, equation (5) becomes :

$$\frac{d^3w}{dt^3} + \mu_2\frac{d^2w}{dt^2} + \mu_1\frac{dw}{dt} + \mu_0 w = 2v_0(t)w + w^2. \tag{8}$$

We have thus a situation where the limit cycle $v_0(t)$ acts as a parametric forcing on the perturbation $w(t)$. To analyse the stability of the state $w = 0$ we must use the Floquet theory [3, 7]. However, as the period doubling instability occurs not too far from the Hopf bifurcation with our scaling, an approximate evaluation of this secondary instability onset can be found, using the leading order approxiation of the limit cycle and identifying leading order Fourier components of the linearized part of (8). We write the double period cycle as a Fourier series :

$$w(t) = \sum_{n=-\infty}^{+\infty} w_{\frac{n}{2}} e^{i\frac{\Omega}{2}t}$$

where $w_{-\frac{n}{2}} = \overline{w}_{\frac{n}{2}}$. Projecting on the Fourier basis we obtained a infinite linear system for the w_m's, with m integer, decoupled from a linear system for the half integer indices. The explicit calculation of the Floquet coefficients would require the manipulation of infinite derterminant [10], but as the forcing oscillation $v_0(t)$ is weakly nonlinear and small we will only consider the order 1 in ϵ :

$$P_\mu\left(i\frac{\Omega}{2}\right) w_{\frac{1}{2}} = 2\left(\overline{v}_1 w_{\frac{3}{2}} + v_0 w_{\frac{1}{2}} + v_1 \overline{w}_{\frac{1}{2}}\right)$$

$$P_\mu\left(3i\frac{\Omega}{2}\right) w_{\frac{3}{2}} = 2\left(\overline{v}_0 w_{\frac{3}{2}} + v_1 w_{\frac{1}{2}}\right)$$

The elimination of $w_{\frac{3}{2}}$ leads to :

$$\left[P_\mu\left(i\frac{\Omega}{2}\right) - v_0 - \frac{4|v_1|^2}{P_\mu(3i\frac{\Omega}{2}) - v_0}\right] w_{\frac{1}{2}} = 2v_1 \overline{w}_{\frac{1}{2}}$$

The onset of instability is reached when the two coefficients of this equation have the same modulus. Let

$$f(\epsilon) = \left| \left[P_\mu \left(i\frac{\Omega}{2} \right) - v_0 \right] \left[P_\mu \left(3i\frac{\Omega}{2} \right) - v_0 \right] - 4|v_1|^2 \right|^2 - 4|v_1|^2 \left| P_\mu \left(3i\frac{\Omega}{2} \right) - v_0 \right|^2$$

the fonction of ϵ when ω and λ are fixed. The root of this function gives an approximate value of the threshold ϵ_c of the period doubling. For $\omega = 1$ and $\lambda = 1$, the graphical calculation of this root gives $\epsilon_c^{(g)} = .080$ to be compared with the result of the numerical integration of (5), $\epsilon_c^{(m)} = .098$. The error is as expected of order of magnitude ϵ^2 (see figure 3). We see that the limit $\frac{\omega^2}{\lambda} \to 0$ is a good limit for our analysis, but is hard to reach numerically.

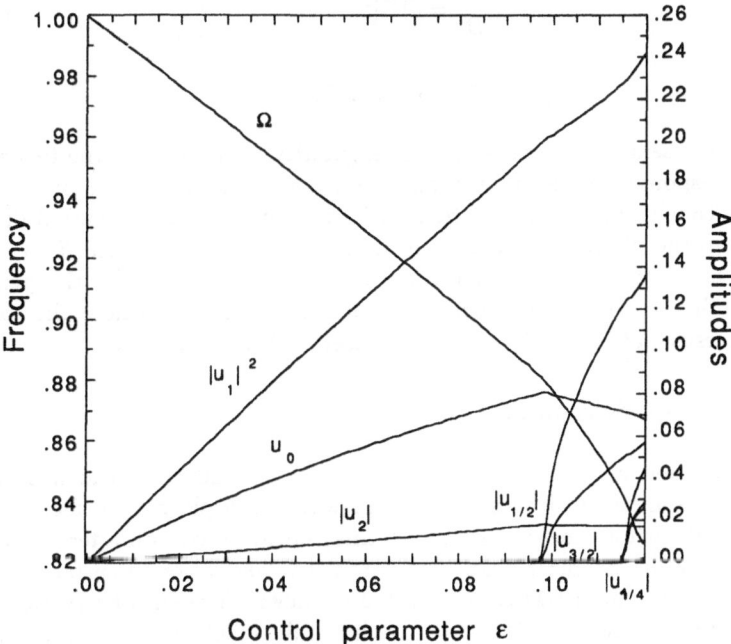

Figure 3. Period doubling bifurcation for the model described by (5), with $\omega = 1$, $\lambda = 1$ and ϵ varying. $u_{m/2^n}$ is the complex amplitude of the Fourier component at frequency $m\Omega/2^n$.

The above analysis of a period doubling bifurcation understood as a parametric instability, can be extended to the nonlinear regime. However one must take into account the existence of the neutral mode that results from translational invariance in time of equation (5). Indeed, for $\epsilon > 0$, equation (5) has a family of solutions,

$$u(t) = v_0(A, s),$$

of period 2π in $s = \omega(A)t$, which represent the limit cycles of period $T(A) = 2\pi/\omega(A)$. A parametrizes the limit cycle amplitudes, and the period T depends on it as usual for nonlinear oscillations. As equation (5) represents an autonomous sytem, $\frac{\partial v_0}{\partial s}$ is a neutral mode, i.e. a mode with a zero growthrate connected with translational invariance in time. At the onset of the period doubling instability, we thus write,

$$u(t) = v_0[s + \phi(T)] + B(T)\exp is/2 + c.c. + h.o.t.$$

and look for an evolution equation for ϕ and B. From symmetry constraints, we get evolution equations of the form,

$$\frac{d\phi}{dT} = \zeta|B|^2 \tag{9a}$$

$$\frac{dB}{dT} = \alpha B + \beta\bar{B} + \gamma|B|^2 B \tag{9b}$$

Equations (9) represents a codimension two singularity [3, 7]. One of the two zero eigenvalues corresponds to the limit cycle amplitude instability. The other one comes from the time translational invariance of equation (5). The absence of linear coupling between the phase and the amplitude is due to the invariance $B \rightarrow -B$. Equation (9a) describes the renormalisation of the frequency Ω after the period doubling instability onset ; this effect is visualized on figure 3 where the slope increase of Ω versus ϵ at the period doubling onset is clearly observed.

4. Parametric amplification of waves

Parametric generation of waves by a spatially uniform oscillatory field has been observed in various areas of physics : Langmuir waves in a plasma, surface waves on a molten metal with a time-dependent heating, surface waves on a ferrofluid in an alternating horizontal magnetic field. The parametric excitation of surface waves on a horizontal layer of fluid vertically vibrated is known since the observations of Faraday [11], and provides the simplest experimental model to study parametric amplification in spatially extended systems, i.e. in systems with a large number of degrees of freedom.

Surface waves are characterized by their dispersion relation,

$$\omega_0(k) = k\tanh(kh)(g + \frac{T}{\rho}k^2), \tag{10}$$

where h is the fluid depth, g is the acceleration of gravity, T is the surface tension and ρ is the fluid density. When the fluid layer is vertically vibrated, g is the time dependent effective gravity $g_e(t)$, and the first unstable mode is the one of wavevector $\vec{k}$ with $\omega(|\vec{k}|)$ as close as possible of $\omega_e/2$. In large containers, and particularly in square containers, several modes with different wavevectors are simultaneously critical, and there is a nonlinear selection mechanism even at instability onset [12].

We consider here the case of an annular cell (diameters 105-115 mm, height 8 mm) which can be modelled as a one dimensional system with periodic boundary conditions. The experiment is performed with a filled-up cell in order to pin the meniscus at the edge of the lateral wall of the container, and thus to avoid lateral perturbations [8]. The dispersion relation (10) and the dissipation are modified by the lateral boundary conditions [13], but this does not change the qualitative behaviors in this geometry.

Close to the the instability onset, the surface deformation can be written under the form

$$\xi(x,t) = A \exp i(\omega t - kx) + B \exp i(\omega t + kx) + c.c. + h.o.t., \qquad (11)$$

where A and B are the slowly varying amplitudes of the right and left waves at frequency $\omega = \omega_e/2$. The evolution equations for A and B can be derived by symmetry considerations. In the absence of time dependent external forcing, the system is invariant under continuous translations in time, $t \to t + \theta$, and in space, $x \to x + \phi$, thus from equation (11) one has the invariance :

$$A \to A \exp i\omega\theta, \ B \to B \exp i\omega\theta$$

$$A \to A \exp -ik\phi, \ B \to B \exp ik\phi,$$

and up to the third order, the evolution equations read

$$\frac{\partial A}{\partial T} = (-\lambda + i\nu)A + c\frac{\partial A}{\partial X} + \alpha\frac{\partial^2 A}{\partial X^2} + [\beta|A|^2 + \gamma|B|^2]A$$

$$\frac{\partial B}{\partial T} = (-\lambda + i\nu)B - c\frac{\partial B}{\partial X} + \alpha\frac{\partial^2 B}{\partial X^2} + [\beta|B|^2 + \gamma|A|^2]B$$

For dissipationless systems, we have the symmetry constraint $t \to -t$, and thus from (11), $A \to \bar{B}, B \to \bar{A}$, which implies that $\lambda = 0$, ν and c are real, and α, β, γ are pure imaginary. In the limit of small dissipation, the leading order correction are respectively, $-\lambda A$ and $-\lambda B$. When an external forcing is applied, the continuous translation symmetry in time is broken, and the system is invariant under the discrete symmetry $t \to t + 2\pi/\omega_e$, and thus from (11), $A \to -A$, $B \to -B$. Keeping only the leading order terms, we obtain [14],

$$\frac{\partial A}{\partial T} = (-\lambda + i\nu)A + \mu\bar{B} + c\frac{\partial A}{\partial x} + \alpha\frac{\partial^2 A}{\partial x^2} + [\beta|A|^2 + \gamma|B|^2]A \qquad (12.a)$$

$$\frac{\partial B}{\partial T} = (-\lambda + i\nu)B + \mu\bar{A} - c\frac{\partial B}{\partial X} + \alpha\frac{\partial^2 B}{\partial X^2} + [\beta|B|^2 + \gamma|A|^2]B. \qquad (12.b)$$

The first point to notice without any computation is the fact that propagative waves cannot be parametrically amplified. Indeed, the forcing term for one wave, say A, is proportional to the amplitude of the other, $\mu\bar{B}$. The physical explanation is obvious : the parametrically amplified wave must be locked on the excitation frequency, and this is not possible for a propagative wave [15]. Therefore A and B must grow together at instability onset, and only standing waves can be amplified parametrically.

This is observed experimentally in the annulus cell. At the instability onset, a perfectly ordered standing wave pattern bifurcates supercritically. Three types of secondary instability are observed :

-when the driving frequency is increased (decreased), a wavelength changing instability occurs, and the system bifurcates abruptly toward another standing wave pattern by nucleation (anihilation) of one wavelength. These instabilities involve large hysteresis, and are the transitions between the different resonance tongues corresponding to different wavenumbers. They have been observed by numerical integration of equations (12). Analytically, they corresponds to saddle-node bifurcations.

-When the driving amplitude is increased, a drift instability occurs, and the pattern begins to move uniformly at a constant velocity. This bifurcation is not described by equations (12) because of the form chosen for (11). We have to consider a standing wave drifting at velocity V, and find how V is coupled to the standing wave amplitude S. The form of the evolution equation is at leading order,

$$\frac{dV}{dt} = V(-\Upsilon + \kappa|S|^2)$$

Above a critical amplitude $\sqrt{\Upsilon/\kappa}$, the stationary pattern undergoes a pitchfork bifurcation and begins to drift at constant velocity.

-When the driving amplitude is increased further, a time-dependent compression mode of the pattern is observed, and its wavelength oscillates in space and time at low frequency (compared to the driving frequency). This instability has been observed by numerical integration of equations (12), and corresponds to a Hopf bifurcation. The experimentally observed phenomena can thus be understood in the framework of the amplitude equation formalism up to the onset of quasiperiodic regimes.

References

[1] - see for instance, Propagation in systems far from equilibrium J. E. Wesfreid, H. Brand, P. Maneville, G. Albinet and N. Boccara (editors), Springer Series in Synergetics **41**, (1988).

[2] - L. D. Landau and E. M. Lifschitz, Mechanics(third edition) , Pergamon Press (1982).

[3] - V. Arnold Geometrical methods in the theory of ordinary differential equations, Springer Verlag (1982).

[4] - T.B. Benjamin and F. Ursell, 'The stability of the plane free surface of a liquid in vertical periodic motion', Proc. Roy. Soc. London **A225**, 505 (1954).

[5] - A. H. Nayfeh, Perturbation Methods, Wiley Interscience (1973).

[6] - E. Meron, 'Parametric excitation of multimode dissipative systems', Phys. Rev. **A35**, 4892 (1987).

[7] - J. Guckenheimer and P. Holmes, Nonlinear Oscillations, Dynamical Systems and Bifurcations of Vector Fields, Springer Verlag (1983).

[8] - S. Douady, 'Experimental study of the Faraday instability', submitted to J. Fluid Mech. (1988).

[9] - A. Arnéodo, P. H. Coullet, E. A. Spiegel, C. Tresser, 'Asymptotic chaos', Physica **14D**, 327 (1985).

[10] -W. Magnus, S. Winkler, Hill's equation, Interscience publishers, Wiley and Sons (1966).

[11] - M. Faraday, 'On the form and states assumed by fluids in contact with vibrating elastic surfaces', Phil. Trans. Roy. Soc. London, **52**, 319 (1831).

[12] - S. Douady and S. Fauve, 'Pattern selection in the Faraday instability', Europhysics Letters **6**, 221 (1988).

[13] - S. Douady, 'Capillary-gravity surface wave modes in a closed vessel with edge constraint : eigenfrequency and dissipation', WHOI Tech. Report (1988).

[14] - These equations have been derived by, A.B. Ezerskii, M.I. Rabinovich, V.P. Reutov and I.M. Starobinets, 'Spatiotemporal chaos in the parametric excitation of a capillary ripple', Sov. Phys. JETP **64**, 1228 (1986).

[15] - For the same reason, a propagative spatial pattern undergoes a transition to a stationary pattern when a temporal forcing is applied.

SPATIO-TEMPORAL RADIATION PATTERN OF A HIGH ENERGY ELECTRON

Danilo Villarroel

Departamento de Física,

Universidad Técnica Federico Santa María,

Casilla 110-V, Valparaíso, Chile

ABSTRACT. Synchrotron radiation along the radial direction has an interesting structure that depends on the distance to the electron orbit. Thus for detection very close to the orbit all the radiation of a high energy electron tends to be concentrated into a line giving rise to a very high density of energy per unit area. This paper deals with the study of the distance dependence of the field energy that it is radiated away as well as with that part that remains bound to the electron. Fortunately, for a high energy electron the structure of the radiation as a function of the distance can be described by relatively simple formulas, which are valid even very close to the electron orbit.

INTRODUCTION

It is well-known that the radiation emitted by a high energy electron is mainly concentrated around a cone whose apex and axis coincide with the instantaneous position and velocity of the electron respectively, and where the angular aperture of the cone is of the order of $\gamma^{-1} \equiv (1 - v^2/c^2)^{1/2}$. From an experimental point of view the radiation that is measured is, of course, not the instantaneous one, but that emitted by the electron during a very small interval of time, as it happens in the circular accelerators, where the radiation that reaches the detector is generated in a tiny arc of the orbit located at the tangential point. The radiation emitted by a monoenergetic electron in circular orbit has been studied in two spatial directions: the radiation emitted in the tangential direction, that is, in the direction of the electron motion; and the radiation emitted in the radial direction, that is, in the direction defined by means of a straight line that goes through the center of the electron orbit. For a high energy electron, the radiation emitted in the tangential

239

E. Tirapegui and D. Villarroel (eds.), Instabilities and Nonequilibrium Structures II, 239–253.
© 1989 by Kluwer Academic Publishers.

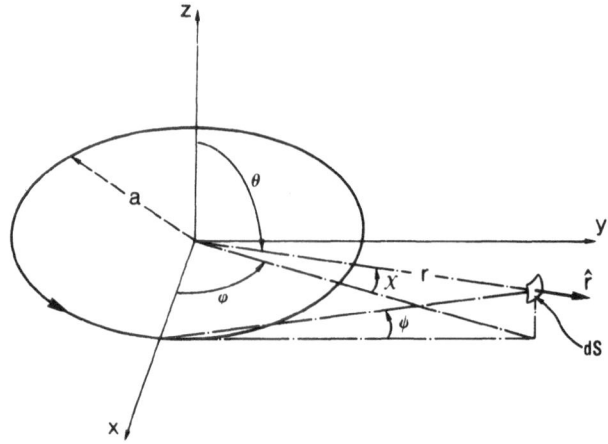

Figure 1. Coordinate system used to describe radiation along the radial direction which is defined by the unit vector $\hat{r}$.

direction is described by a well-known formula due to Schwinger [1]; this formula has been tested in numerous experiments, and plays a central role in the great diversity of practical applications that synchrotron radiation has nowadays.

The radiation emitted in the radial direction has been described by means of and old formula due to Schott [2], which, as was shown by Ivanenko and Sokolov [3-4], is considerably simplified for a high energy electron, reducing to

$$\frac{dI_m}{d\Omega} = \left(\frac{e^2}{3\pi^2 a}\right) m^2 \eta^2 \left\{ K^2_{2/3}\left(\frac{m}{3}\eta^{3/2}\right) + \chi^2 \eta^{-1} K^2_{1/3}\left(\frac{m}{3}\eta^{3/2}\right) \right\} \qquad (1.1)$$

Here $dI_m/d\Omega$ represents the energy associated with the harmonic number m that is radiated in the solid angle $d\Omega$ during an electron period; a is the radius of the orbit; χ is the angle between the radial direction and its projection in the orbit plane; $K_{2/3}$ and $K_{1/3}$ are modified Bessel functions of the second kind (see [4] for details), and

$$\eta = \gamma^{-2} + \chi^2. \qquad (1.2)$$

Equation (1.1) is formally identical to Schwinger's formula, but in this last one instead of the angle χ appears the angle ψ defined as the angle between the line of observation with its projection in the orbit plane tangent to the orbit (see Fig. 1). Even though far from the orbit the angles χ and ψ are practically equal, the difference between them became significant for detection at distances of the order of the orbit radius, because the radiation is concentrated around the orbit plane in a small angle of the order of γ^{-1}.

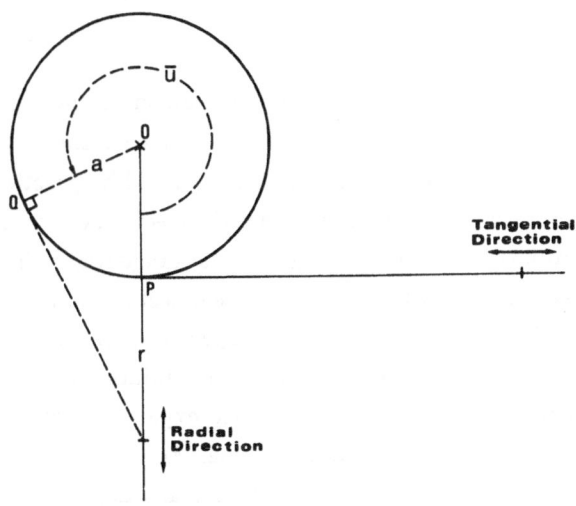

Figure 2. The radiation that is detected along the tangential direction comes from a vicinity of the point P, whose location does not depend on the distance to the detector. On the other hand, in the radial direction the radiation comes mainly from a vicinity of the point Q whose location is changing as the observer approaches to the electron orbit.

The geometry associated with the radial direction presents advantages over the tangential one, for the purpose of the experimental study of the spatio-temporal pattern of the radiation in the vicinity of the electron. Figure 2 represents the detection process along both directions in the orbit plane. Due to the high directionality of the radiation of a relativistic electron, the radiation that is detected along the tangential direction comes from a tiny arc located around the point P. In this direction it is impossible to approach a physical detector to a vicinity of P, because the electron will colide with the detector before it reaches the tangential point. On the other hand, the radiation detected along the radial direction is originated in a vicinity of the point Q, which has a location that is changing in the orbital ring as the detector approaches to the orbit, tending to coincide with P for detection very close to the orbit. In the radial direction there are not, at least in principle, difficulties in order to put a detector at small distances from the orbit. This paper deals with the radiation along the radial direction, and as in other works on the radiation in this direction [1-5], the distances are measured from the center of the orbit (see Figs. 1 and 2).

At first sight it may appear that in the radial direction there is a difficulty when the detector is very close to the orbit, since in this case the point Q is almost in front of the detector, and then the radiation cone would be practically parallel to it. However, as the observation point approaches the electron orbit, the distance to the orbit goes to zero faster than the distance to the tangential point Q. In fact, from fig. 2 it can be see that the distance between the detector and the point P is $r(1-\xi)$, where $\xi = a/r$; while the distance to the tangential point is $r(1 - \xi^2)^{1/2}$. So, no matter how small is the distance to the orbit, the radiation cone is never parallel to the detector, allowing in consequence a detailed exploration of the spatio-temporal pattern of the whole apex of the radiation cone. According to this description the radiation that is measured must have an important dependence on the distance near the orbit. In particular, the height of the radiation spot (measured along the perpendicular to the orbit plane) must decrease in a significant way in the proximity of the electron orbit, since in this case it represents the structure of the apex of the radiation cone. In fact, for a detector located almost touching the electron orbit, all the emitted radiation tends to be concentrated into a line producing a very high density of energy per unit area [6-8].

As Fig. 2 shows, the lenght of the orbit arc that contributes to the radiation that is measured increases as the detector approaches the orbit, especially if it is considered that near the orbit it looks like a straight line. Fortunately, the directionality of the radiation of a high energy electron and the singularity of the electron field are such that even very close to the orbit the lenght of the arc that contributes remains small. This property allows to obtain simple analytical representations for the total intensity of radiation as well as for the different spectral components of the radiation as functions of the distance to the orbit. Representations which are valid even very near to the orbit [8].

2. LOCAL CHARACTERIZATION OF THE RADIATION

In order to study the radiation near the orbit it is necessary to identify the radiation at arbitrary distances from the electron, and not merely at large distances from it as it is usually done in the literature. In Schott's work [2] the radiation is studied far from the orbit, and for this reason the distance dependence which must be present according to the analysis of the previous section does not appear in Eq. (1.1). The radiation is analized in Schwinger's paper [1] with the help of a superposition of retarded and advanced fields, introduced for the first time in the well-known paper of Dirac [9]. But in the radial direction Schwinger reproduces Schott's formula. Subsequently Sokolov, Galstov, and Kolesnikova [5] studied the radiation in the radial direction without resort to the asymptotic form of the fields for large distances from the electron orbit, obtaining once again Schott's formula.

The radiation emitted by an electron can be characterized locally at arbitrary distances from the electron using only the physical fields, which are the retarded ones [10]. This description is physically more satisfactory than the one based on a superposition of advanced and retarded fields, because the radiation is then defined only after its emission, and moreover it has a nonvanishing source on the electron world-line. These two properties are not shared by the description that uses advanced fields, which brings in serious difficulties of physical interpretation. As a purely technical device, the introduction of advanced fields may present certain advantages in the derivation of the equations of motion of a point particle within the framework of the energy-momentum tensor [9], [11], but it appears unnecessary when one works directly with the coupled equations of motion of the particle and the field [12], [15]. Even more, it is only the use of retarded fields which allows a clear physical interpretation of the different terms of the equation of motion [10]. Thus, for example, the origin of the puzzling "Schott term" [16] may be traced back to that part of the energy which is not radiated but remains bound to the electron.

The local characterization of the electron radiation is achieved by means of a splitting of the energy-momentum tensor $T_{\mu\nu}$ of the retarded electromagnetic field of the electron in two parts, which are dynamically independent outside the electron.

$$T_{\mu\nu} = T_{\mu\nu}^{(r)} + T_{\mu\nu}^{(b)}. \tag{2.1}$$

The radiation tensor $T_{\mu\nu}^{(r)}$ is constructed using only the piece of the retarded Liénard-Wiechert field that contains the electron acceleration. The energy-momentum constructed with the velocity fields is amalgamated with the interference between the acceleration and velocity fields in order to produce the bound energy-momentum tensor $T_{\mu\nu}^{(b)}$. The splitting (2.1) is such that both parts $T_{\mu\nu}^{(r)}$ and $T_{\mu\nu}^{(b)}$ of $T_{\mu\nu}$ are separately conserved off the electron world-line, that is,

$$\partial^\nu T_{\mu\nu}^{(r)} = 0, \tag{2.2}$$

$$\partial^\nu T_{\mu\nu}^{(b)} = 0. \tag{2.3}$$

Apart from satisfying (2.2), the radiation tensor $T_{\mu\nu}^{(r)}$ has the supplementary property of having no flux through the future light cones emerging from the electron world-line. The properties of $T_{\mu\nu}^{(r)}$ are such that they allow to picture the energy contained in a small three-dimensional volume element as having detached itself completely from the electron and travelling radially outwards from it with the speed of light, just as a free photon would do. Thus, it can be concluded without any ambiguity whatsoever that $T_{\mu\nu}^{(r)}$ represents that part of the energy-momentum of the electron field that is emitted in an irreversible way by the electron. The identification of $T_{\mu\nu}^{(r)}$ as describing the radiation is achieved without having to go to large distances from the electron. The avoidance of asymptotic limits is satisfactory, since a local treatment is necessary if one wants to picture radiation as an emission of something (photons) by the electron: that "something" begins to exist immediately after emission and it must be possible to identify it without waiting for it to arrive at the wave zone. The splitting (2.1) together with the properties of $T_{\mu\nu}^{(r)}$, are the underlying basis of Rohrlich's local radiation criterion [17].

Contrary to the case for $T_{\mu\nu}^{(r)}$, the flux across the light cones with apex on the electron world-line of the tensor $T_{\mu\nu}^{(b)}$ is not zero; which is an indication that $T_{\mu\nu}^{(b)}$ describes energy-momentum which remains bound to the electron. This interpretation becomes transparent when one considers the four-momentum associated to $T_{\mu\nu}^{(b)}$, that is, the integral of $T_{\mu\nu}^{(b)}$ over a three-dimentional spacelike surface. The remarkable property of the bound four-momentum is that, in spite of the fact that in the integrand appears the whole history of the electron, the integral is an state function, since it depends only on the instantaneous values of the velocity and acceleration of the electron associated with the point where the spacelike surface intercepts the electron world-line [10]. So, the energy-momentum of $T_{\mu\nu}^{(b)}$ can be visualized as being emitted and subsequently reabsorbed by the electron.

In the laboratory reference frame of Fig. 1, equation (2.1) reduces itself to the following form for the Poynting vector.

$$\mathbf{S} = \mathbf{S}^{(r)} + \mathbf{S}^{(b)}, \tag{2.4}$$

where $\mathbf{S}^{(r)}$ is the Poynting vector $(c/4\pi)\mathbf{E}_a \times \mathbf{B}_a$ constructed with the acceleration part of the electric and magnetic field $\mathbf{E}_a$ and $\mathbf{B}_a$ respectively. On the other hand equation (2.2) for $\mathbf{S}^{(r)}$ takes the form

$$\nabla \cdot \mathbf{S}^{(r)} + \partial U^{(r)}/\partial t = 0, \tag{2.5}$$

where $U^{(r)}$ is the energy density of the radiation field, that is, $(\mathbf{E}_a^2 + \mathbf{B}_a^2)/8\pi$.

If $\hat{\mathbf{r}}$ denotes the unit vector in the radial direction (see Fig. 1), then the radiation intensity in this direction is

$$\frac{dI^{(r)}}{d\Omega} = r^2 < \hat{\mathbf{r}} \cdot \mathbf{S}^{(r)} >, \tag{2.6}$$

where $< \hat{\mathbf{r}} \cdot \mathbf{S}^{(r)} >$ is the energy flux along $\hat{\mathbf{r}}$ during an electron period, namely,

$$< \hat{\mathbf{r}} \cdot \mathbf{S}^{(r)} >= \int_{t=0}^{t=T} (\hat{\mathbf{r}} \cdot \mathbf{S}^{(r)}) dt. \tag{2.7}$$

In order to see in a clear way the differences of the present treatment with other works on this subject [1], [5], it is convenient to represent (2.7) as a power series in the parameter $\xi = a/r < 1$

$$< \hat{\mathbf{r}} \cdot \mathbf{S}^{(r)} >= (e^2/16a^3) \sum_{m=2}^{\infty} a_m \xi^m. \tag{2.8}$$

The coefficients a_m can be evaluated in an explicit way for arbitrary values of the electron energy [6]. In the case of a high energy electron, the first three coefficients of (2.8) turn out to be

$$a_2 = \gamma^5 Z^5 (12 - 5Z^2),$$
$$a_4 = \frac{1}{2}\gamma^5 Z^5 (-48 + 90Z^2 - 35Z^4), \tag{2.9}$$
$$a_6 = \frac{1}{8}\gamma^5 Z^5 (96 - 600Z^2 + 840Z^4 - 315Z^6),$$

where

$$Z = \{1 + (\gamma\chi)^2\}^{-1/2} . \tag{2.10}$$

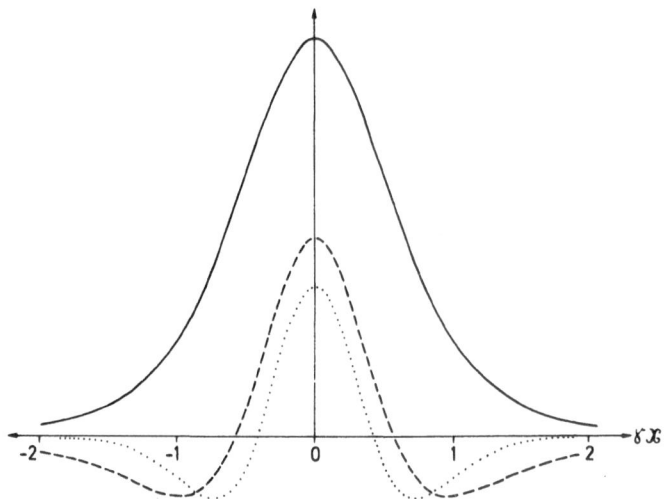

Figure 3. Curves related to the energy flux in the radial direction (see Eq. (2.9)). The coefficient a_2 is represented by the full line, a_4 by the dashed line, and a_6 corresponds to the dotted line.

These coefficients have been drawn in figure 3 without the overall factor γ^5. Excepting a_2 the coefficients of (2.8) have in general a complicated oscillatory behaviour with total area equal to zero. This property, which is quite clear in the graphs of a_4 and a_6, is valid for any energy of the electron. In fact, using Eq. (2.5) and the periodicity of the electromagnetic field it can be shown that

$$\int_0^{2\pi} \int_0^{\pi} a_m \sin\theta d\theta d\varphi = 0 \qquad \text{if} \qquad m > 2. \tag{2.11}$$

According to (2.11) the energy flux across the whole surface of a sphere enclosing the electron orbit is determined exclusively by the coefficient a_2 of (2.8); and it turns out to be equal to the instantaneous power of emission, given by Larmor formula, multiplied by the electron period. Nevertheless, this does not mean that the remaining coefficients are of no physical relevancy, since the intensity of radiation in a given direction, defined in (2.6) is, just as the instantaneous power radiated in all directions, a physical measurable quantity. These remaining coefficients are precisely the ones that give the information about the distance dependence of the radiation along the radial direction.

Even if in a separate way each coefficient of (2.8) has not physical meaning, they can be used in order to see in a qualitative manner the structure that will have the intensity of radiation as a function of the distance to the orbit. Thus, far from the orbit the only significant contribution will be that of a_2; in this case, as it is shown in Fig. 3, the radiation will be concentrated in an angle of the order $2\gamma^{-1}$ around the orbit plane, which is also the result contained in Schott's formula [4]. But as the observation point approaches the orbit, the contribution of the remaining coefficients begins to play a role, and because of their oscillatory behaviour, the angular spread of the radiation will decrease at the same time as the intensity of the radiation in the orbit plane increases due to energy conservation.

If instead of the power series representation (2.8) one considers the corresponding power series of $S^{(b)}$, then it turns out that all the coefficients of such expansion satisfy Eq. (2.11). In particular, the energy associated with $S^{(b)}$ that goes across a surface enclosing the electron orbit during an electron period is zero, which is in agreement with the fact that $S^{(b)}$ represents energy-momentum that remains linked to the electron.

3. DISTANCE DEPENDENCE OF THE RADIATION.

The power series representation (2.8) is not appropriate for the analysis of the distance dependence of the radiation, since near the orbit it is necessary to consider a great number of terms. Fortunately the integral (2.7) can be evaluated in a simple manner for a high energy electron. Introducing in (2.6) the explicit form of the acceleration fields for a monoenergetic electron in circular orbit (see Fig. 1), it can be written as

$$\frac{dI^{(r)}}{d\Omega} = (7e^2\gamma^5/16a)A_1,\tag{3.1}$$

where

$$A_1 = \frac{4\gamma^{-5}}{7\pi}\left\{\int_0^{2\pi}\frac{(1-\xi\cos\chi\cos u)du}{\kappa^3\rho^3}\right.$$

$$\left.-\gamma^{-2}\int_0^{2\pi}\frac{(1-\xi\cos\chi\cos u)(\cos\chi\cos u-\xi)^2 du}{\kappa^5\rho^5}\right\}.\tag{3.2}$$

The variables κ and ρ that appear in the integrals are the following:

$$\kappa = 1 + \rho^{-1}\beta\cos\chi\sin u,\tag{3.3}$$

$$\rho = (1 - 2\xi\cos\chi cosu + \xi^2)^{1/2},\tag{3.4}$$

where β is, as usual, the quotient between the electron velocity and the velocity of light. For a high energy electron the term $\kappa\rho$ in the integrals of (3.2) become significant only in a small vicinity of the tangential point defined by (see Fig. 2).

$$\sin\bar{u} = -(1-\xi^2)^{1/2} \equiv -\varepsilon.\tag{3.5}$$

In this vicinity $\kappa\rho$ can be approximated by

$$\kappa\rho = \frac{\varepsilon}{2}[\gamma^{-2}(1+\lambda^2) + (u-\bar{u})^2],\tag{3.6}$$

where

$$\lambda = \frac{\gamma\chi}{\varepsilon}.\tag{3.7}$$

With the help of (3.6) the integrals (3.2) can be easily carried out, to obtain

$$A_1 = \frac{4}{7\varepsilon}\left\{\frac{3}{(1+\lambda^2)^{5/2}} - \frac{5}{4(1+\lambda^2)^{7/2}}\right\}.\tag{3.8}$$

This result is remarkable for its simplicity, because, aside from the overall factor ε^{-1}, the three variables that define A_1, namely, the electron energy γ, the elevation angle with respect to the orbit plane χ, and the distance to the orbit ξ, appear only in the combination (3.7). Numerical studies show [7], [8] that the expression (3.8) is a very good approximation of (3.2), even at distances of the orbit as small as $1 - \xi = 10^{-6}$ for electrons over 500 Mev. Figure 4 shows that equation (3.8) describes a combination of two effects, which are the decrease of the angular spread coming together with the increase of the radiation intensity in the orbit plane as the observer approaches the orbit.

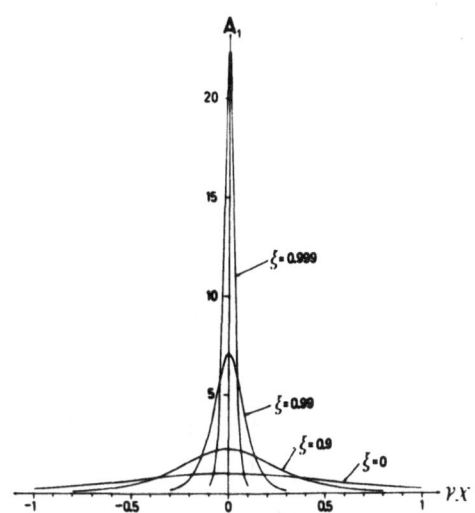

Figure 4. Plots of the radial intensity of radiation as a function of the elevation angle χ measured in γ^{-1} units for four different distances from the electron orbit.

From Eq. (3.8) it is easy to see that even if the intensity (3.1) has an important dependence on the distance to the electron orbit, that is, on the parameter ξ, the energy radiated in all directions does not depend on this parameter. This independence is reflected in the fact that the area of all the curves of Fig. 4 is the same. According to (3.8) the electron radiation tends to be concentrated into a line for detection almost touching the electron orbit, giving rise to a very high density of energy per unit area. In this case the radiation intensity may be several orders of magnitude higher than the intensity that usually exists in the applications of synchrotron radiation.

Considering the bound piece of the Poynting vector $\mathbf{S}^{(b)}$ instead of $\mathbf{S}^{(r)}$ in Eq. (2.6), the associated "radiation intensity" can be written as

$$\frac{dI^{(b)}}{d\Omega} = (7e^2\gamma^3/16a)A_2,\tag{3.9}$$

where, for a high energy electron, A_2 in given by

$$A_2 = \frac{\xi^2}{7\varepsilon^3}\left\{\frac{16}{(1+\lambda^2)^{5/2}} - \frac{50}{(1+\lambda^2)^{7/2}} + \frac{35}{(1+\lambda^2)^{9/2}}\right\}.\tag{3.10}$$

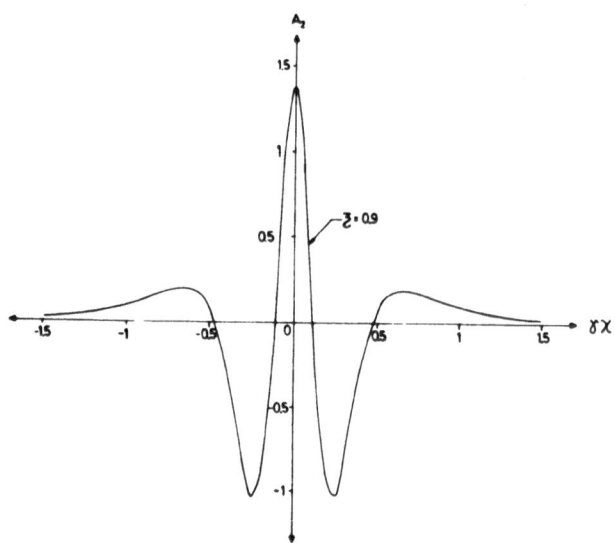

Figure 5. Plot of the energy flux that remains bound to the electron for $\xi = 0.9$ (see Eq. (3.10)).

Like A_1, the angular spread of A_2 decreases as the observer approaches the electron orbit. But due to the different physical nature of $\mathbf{S}^{(b)}$, A_2, in constrast with A_1, becomes negative for certain angles as it is shown in Fig. 5. The values of χ for which A_2 is negative can be easily determined from (3.10), and they depend, of course, on the electron energy and on the distance to the orbit. As (3.10) shows, $dI^{(b)}/d\Omega$ has an important dependence on the distance to the orbit, being even more singular than $dI^{(r)}/d\Omega$ in the vicinity of it. However, as it is easy to show, the energy flux radiated in all directions is not only independent of ξ, but even more, it vanishes indentically; in agreement with the fact that $\mathbf{S}^{(b)}$ represents the energy-momentum that remains bound to the electron. This property can also be seen in Fig. 5, where the area under the curve is zero.

A photographic plate that measures the electron radiation will of course react upon the whole Poynting vector. But if the detection process is not carried out in the immediate vicinity of the electron orbit, the contribution due to $\mathbf{S}^{(b)}$ is negligible with respect to the one of $\mathbf{S}^{(r)}$ in the case of a high energy electron, since the contribution of $\mathbf{S}^{(r)}$ is proportional to γ^5, in contrast with that of $\mathbf{S}^{(b)}$ which is proportional to γ^3. Besides to make clear the bound nature of the energy-momentum associated with $\mathbf{S}^{(b)}$, the form of the curve of Fig. 5 is directly linked with the circular motion of the electron. The energy that is radiated away from the electron in the angles where A_2 is positive, returns to the electron along the angles where A_2 becomes negative. In particular, a physical detector placed in the vicinity of the electron orbit will tend to disturb the energetic balance illustrated in Fig. 5, giving rise to instabilities on the electron orbit.

4. DISTANCE DEPENDENCE OF THE SPECTRAL COMPONENTS

Using the acceleration terms of the Liénard-Wiechert fields, the energy radiated away by the harmonic number m can be written in the form

$$\frac{dI_m}{d\Omega} = 2T\mathbf{a}_m \cdot \mathbf{a}_m^*, \tag{4.1}$$

where $\mathbf{a}_m^*$ is the complex conjugate of

$$\mathbf{a}_m = \frac{e}{c}\left(\frac{c}{4\pi}\right)^{1/2}\frac{1}{2\pi}\int\limits_0^{2\pi}\frac{(\hat{n}\cdot\hat{r})^{1/2}}{\rho}\frac{\hat{n}\times[(\hat{n}-\boldsymbol{\beta})\times\dot{\boldsymbol{\beta}}]}{\kappa^2}\,e^{im(u+\beta\rho\xi^{-1})}\,du \tag{4.2}$$

In this equation $\hat{n}$ is a unit vector that points from a variable point of the electron orbit to the detection point. As in the integral (3.2), the integrand of (4.2) is significant only in a vicinity of the angle $\bar{u}$ defined in equation (3.5), which allows to do some approximations that lead up to the following expression for the distance dependence of the spectrum

$$\frac{dI_m}{d\Omega} = \left(\frac{e^2}{3\pi^2 a}\right)\frac{m^2\bar{\eta}^2}{\varepsilon}\left\{K_{2/3}^2\left(\frac{m}{3}\bar{\eta}^{3/2}\right) + \frac{\gamma^{-2}\lambda^2}{\bar{\eta}}K_{1/3}^2\left(\frac{m}{3}\bar{\eta}^{3/2}\right)\right\}, \qquad (4.3)$$

where

$$\bar{\eta} = \gamma^{-2} + \chi^2\varepsilon^{-2} \qquad (4.4)$$

Far from the orbit, that is, for $\xi = 0$, equation (4.4) is reduced to equation (1.2) and therefore the spectral distribution given by (4.3) becomes identical to the Ivanenko-Sokolov formula (1.1). The distinction between radial and tangential directions is irrelevant for $\xi = 0$, and for this reason (4.3) also reduces in this case to Schwinger's formula [1]. The consistency between the approximations used in the derivation of (4.3) with those used in the derivation of (3.8) can be easily verified using the practically continuous nature of the spectrum given by (4.3). On general grounds it can be said that each one of the spectral components of the radiation has a behaviour similar to that of the total intensity of radiation as a function of the distance to the orbit. Thus, as the distance to the orbit decreases, the radiation intensity in the orbit plane of the corresponding harmonic increases, while its angular spread decreases.

ACKNOWLEDGMENTS

This work was supported in part by Fondo Nacional de Investigación Científica y Tecnológica, Chile, under Grant N° 88-0609.

REFERENCES

1. J. Schwinger, Phys. Rev. **75**, (1949), p. 1912.
2. G. A. Schott, "Electromagnetic Radiation," Cambridge University Press, Cambridge, England, 1912.
3. D. Ivanenko and A.A. Sokolov, DAN (U.S.S.R.) **59** (1948), p. 1551.
4. A. A. Sokolov and I.M. Ternov, "Synchrotron Radiation," Akademie, Berlin, 1968.

5. A.A. Sokolov, D. V. Galstov, and M.M. Kolesnikova, Izv. Vyssh. Uchebn. Zaved, Fiz. (USSR) **4** (1971), p. 14.

6. D. Villarroel and V. Fuenzalida, J. Phys. A **20** (1987), p. 1387.

7. D. Villarroel, Phys. Rev. A **36** (1987), p. 2980.

8. D. Villarroel and C. Millán, Phys. Rev. D **38** (1988), p. 383.

9. P.A.M. Dirac, Proc. R. Soc. London Ser. A **167** (1938), p. 148.

10. C. Teitelboim, D. Villarroel and Ch. G. van Weert, Riv. Nuovo Cimento **3** (1980), p. 1.

11. H.J. Bhabha and Harish-Chandra, Proc. R. Soc. London Ser. A **187** (1946), p. 250.

12. A. Barut, Phys. Rev. D **10** (1974), p. 3335.

13. A. Barut and D. Villarroel, J. Phys. A **8** (1975), p. 156.

14. A. Barut and D. Villarroel, J. Phys. A **8** (1975), p. 1537.

15. M.E. Brachet and E. Tirapegui, Nuovo Cimento A **47** (1978), p. 210.

16. G.A. Schott, Philos. Mag. **29** (1915), p. 49.

17. F. Rohrlich, "Classical Charged Particles," Reading, Mass, 1965.

STABILITY OF TAYLOR-COUETTE FLOW WITH RADIAL HEATING

P. D. Weidman and M. E. Ali
Department of Mechanical Engineering
University of Colorado
Boulder, CO 80309   USA

1. Introduction

Beginning with Taylor's [1] classical study of the motion of a viscous fluid in an annular gap driven by differential rotation of the coaxial cylinders, circular Couette flow has provided a rich testing ground for both linear and nonlinear stability theory. The present study employs linear stability theory to investigate the more complex problem of viscous fluid motion in a vertical annulus driven by the *combined* agencies of cylinder rotation and radial heating. With one exception, all previous infinite aspect ratio theoretical attacks on this problem have neglected the crucial role of gravity. These include studies by Yih [2], Becker and Kaye [3], Walowit, Tsao and DiPrima [4], Bahl [5], and Soundalgekar, Takhar, and Smith [6]. These analyses showed that isothermal Taylor cells are destabilized (stabilized) by positive (negative) radial heating gradients across the annular gap. Roesner [7] properly included the effect of gravity in the Boussinesq approximation but, like most his predecessors, considered only axisymmetric disturbances. Roesner's results contrast those neglecting gravity: isothermal Talylor cells are stabilized by both positive and negative radial heating and the stability boundary is *symmetric* with respect to the sense of radial heating.

The present investigation pursues the infinite aspect ratio idealization by taking proper account of gravity in the Boussinesq approximation. The stability of the full linearized perturbation equations are tested with respect to both axisymmetric and nonaxisymmetric disturbances. Preliminary remarks given in §2 lead to the mathematical formulation and outline of the solution procedure in §3. A presentation of results in §4 is followed by a comparison with the finite aspect ratio, narrow gap experiments of Snyder and Karlsson in §5. The discussion of results and concluding remarks are given in §6.

2. Preliminary Remarks

It is presumed that the reader is familiar with the problem of stability of isothermal circular Couette flow. With the inner cylinder rotating and the outer cylinder stationary the flow is driven by *centrifugal* forces to an axisymmetric instability consisting of counter-rotating toroidal cells. Excellent reviews of the field are given by DiPrima and Swinney [8] and Stuart [9].

255

E. Tirapegui and D. Villarroel (eds.), Instabilities and Nonequilibrium Structures II, 255–268.
© *1989 by Kluwer Academic Publishers.*

The stability of natural convection in the gap formed between vertical concentric cylinders may be less familiar. Of importance for the present study is the fact that the convection base flow falls into three identifiable regimes: conduction, transition and convection. The base flow for each of these regimes may be regarded as a single closed cell of fluid rising along the hot wall and descending along the cold wall. Thomas and De Vahl Davis [10] place the transition regime in the range

$$400 \ A < Ra < 3000 \ A \qquad (A > 5) \tag{1}$$

where **Ra** is the Rayleigh number and A = L/d is the aspect ratio, with L the annulus length and d the gap width. Rayleigh numbers below 400 A correspond to the conduction regime and Rayleigh numbers above 3000 A are in the convection or boundary layer regime. Although this result is reported strictly valid only for unity Prandtl number, the authors claim that the boundaries (1) delineating the transition regime are relatively insensitive to both Prandtl number **Pr** and radius ratio η.

In the limit $A \to \infty$, analytical solution for the base flow in the conduction regime is readily obtained and a linear stability analysis for axisymmetric disturbances has been carried out by Choi and Korpela [11]. They found that the flow may be driven to instability by *shear* forces for **Pr** < 15 and by *buoyancy* forces for **Pr** > 15, approximately. McFadden, Coriell, Boisvert and Glicksman [12] extended these results to include the first nonaxisymmetric disturbance. They reported that for air with Prandtl number **Pr** = 0.71 the most dangerous disturbance is nonaxisymmetric for radius ratios 0 < η < 0.44 but axisymmetric for 0.44 < η < 1, and in either case the instability is due to the action of the shear forces. For water with **Pr** = 3.5 it was shown that the asymmetric shear mode at **Pr** = 0.71 is superseded by an axisymmetric buoyancy-driven instability when 0.03 < η < 0.16.

An analytical solution for the base flow in a differentially heated vertical annulus at high Rayleigh number in the convection regime is not available. Weidman and Mehrdadtehranfar [13] reported an experimental study of the stability of this problem at A = 64 and and η = 0.62. Their visualization studies revealed an unstable flow consisting of two separate progressive wave systems, one ascending the hot inner wall and the other descending the cold outer wall for Prandtl numbers in the range 20 < **Pr** < 110.

3. Mathematical Formulation and Numerical Solution Procedure

3.1 EQUATIONS OF MOTION

The motion of a thermally active viscous fluid is governed by the equation of continuity, the Navier-Stokes equations, and the energy equation. Gravity is taken into account in the Boussinesq approximation. The equations are made dimensionless by scaling lengths with the gap width d, time with d^2/ν, temperature with the radial temperature contrast $\Delta T = T_1 - T_2 > 0$, axial and radial velocities with the convection speed $U_0 = g\beta d^2 \Delta T/\nu$, the azimuthal velocity with $\Omega_1 R_1$, and the reduced pressure (thermodynamic pressure less the hydrostatic head) with $\rho_2 U_0^2$. Here g is gravity and β, ν and ρ are the volume coefficient of thermal expansion, the kinematic viscosity and the fluid density, respectively. Subscripts 1 and 2 denote inner and outer cylinder. The nondimensional conservation equations governing the velocities (u,v,w) in coordinate directions (r,ϕ,z), the pressure p and the temperature θ may be written

$$\frac{\partial u}{\partial t} + G\left[u\frac{\partial u}{\partial r} + S\frac{v}{r}\frac{\partial u}{\partial \phi} + w\frac{\partial u}{\partial z}\right] - GS^2\frac{v^2}{r} = -G\frac{\partial p}{\partial r} + \nabla^2 u - S\frac{2}{r^2}\frac{\partial v}{\partial \phi} - \frac{u}{r^2} \tag{2a}$$

$$S\left[\frac{\partial v}{\partial t} + G\left(u\frac{\partial v}{\partial r} + S\frac{v}{r}\frac{\partial v}{\partial \phi} + w\frac{\partial v}{\partial z} + \frac{uv}{r}\right)\right] = -G\frac{1}{r}\frac{\partial p}{\partial \phi} + S\nabla^2 v + \frac{2}{r^2}\frac{\partial u}{\partial \phi} - S\frac{v}{r^2} \tag{2b}$$

$$\frac{\partial w}{\partial t} + G\left[u\frac{\partial w}{\partial r} + S\frac{v}{r}\frac{\partial w}{\partial \phi} + w\frac{\partial w}{\partial z}\right] = -G\frac{\partial p}{\partial z} + \theta + \nabla^2 w \tag{2c}$$

$$\frac{\partial \theta}{\partial t} + G\left[u\frac{\partial \theta}{\partial r} + S\frac{v}{r}\frac{\partial \theta}{\partial \phi} + w\frac{\partial \theta}{\partial z}\right] = \frac{1}{Pr}\nabla^2\theta \tag{2d}$$

$$\frac{\partial u}{\partial r} + \frac{u}{r} + S\frac{1}{r}\frac{\partial v}{\partial \phi} + \frac{\partial w}{\partial z} = 0; \qquad \nabla^2 = \frac{\partial^2}{\partial r^2} + \frac{1}{r}\frac{\partial}{\partial r} + \frac{1}{r^2}\frac{\partial^2}{\partial \phi^2} + \frac{\partial^2}{\partial z^2}. \tag{2e,f}$$

Dimensionless ratios that appear in (2) are the Grashof number $G = U_0 d/\nu$, the Prandtl number $Pr = \nu/\alpha$, and the swirl parameter $S = \Omega_1 R_1/U_0$. Here α is fluid thermal diffusivity. Computed stability results will be presented in terms of the Taylor number defined by

$$Ta = \frac{2\eta^2}{1-\eta^2}\left[\frac{\Omega_1 d^2}{\nu}\right]^2 = \frac{2(1-\eta)}{1+\eta}S^2 G^2, \tag{3}$$

in lieu of S. Here $\eta = r_1/r_2$ is the radius ratio of the annulus with nondimensional inner radius $r_1 = \eta/(1-\eta)$ and outer radius $r_2 = 1/(1-\eta)$. The swirl parameter S is just a convenient parameter which allows easy reduction of the full equations to the limiting equations describing flows driven solely by rotation or radial heating. The classical isothermal Taylor-Couette equations are obtained by assigning S, G, and θ the values $S = 1$, $G = [Ta(1+\eta)/2(1-\eta)]^{1/2}$ and $\theta = 0$. The equations for natural convection in the absence of rotation are obtained by simply setting $S = 1$.

3.2 BASE FLOW SOLUTION

The conduction regime spiral base flow in an infinitely tall annulus is found by assuming the flow to be steady $(\partial/\partial t = 0)$, axisymmetric $(\partial/\partial \phi = 0)$ and independent of the axial coordinate $(\partial/\partial z = 0)$. Under these conditions the equation of continuity is satisfied identically and conservation of mass is insured by the integral constraint

$$\int_{r_1}^{r_2} r\overline{w}(r)\, dr = 0. \tag{4}$$

The solution of equations (2a) – (2d) satisfying (4) with boundary conditions $v = \theta = 1$, $w = 0$ at $r = r_1$ and $v = \theta = w = 0$ at $r = r_2$ yields

$$\bar{u}(r) = 0; \qquad \bar{v}(r) = \frac{\eta}{1+\eta}\left[\frac{1}{(1-\eta)^2 r} - r\right]; \qquad \bar{\theta}(r) = \frac{1}{\ln(\eta)}\ln(r/r_2) \tag{5a,b,c}$$

$$\bar{w}(r) = \frac{1}{(1-\eta)^2}\left\{\left[\frac{A}{B}\right]\left[(1-\eta)^2 r^2 - 1 + (1-\eta^2)\frac{\ln(r/r_2)}{\ln(\eta)}\right] - \frac{1}{4}[(1-\eta)^2 r^2 - \eta^2]\frac{\ln(r/r_2)}{\ln(\eta)}\right\} \tag{5d}$$

$$\frac{\bar{p}}{S^2} = \frac{\eta^2}{(1+\eta)^2}\left\{\frac{r^2}{2} + \frac{1}{2r^2(1-\eta)^4} - \frac{1}{2(1-\eta)^2}\left[\frac{\eta^4+1}{\eta^2}\right] - \frac{2}{(1-\eta)^2}\ln\left[\frac{(1-\eta)r}{\eta}\right]\right\} \tag{5e}$$

where $A = (1-\eta^2)(1-3\eta^2) - 4\eta^4\ln(\eta)$, and $B = 16[(1-\eta^2)^2 + (1-\eta^4)\ln(\eta)]$ are the constants in (5d). The azimuthal velocity is identical to that given by Chandrasekhar [14] for the classical Taylor problem. The axial velocity and temperature solutions are the same as those reported by Choi and Korpela [11] for conduction regime base flow in the absence of rotation. The base flow pressure solution is that reported by Roesner [7] when $S = 1$.

3.3 PERTURBATION EQUATIONS

The independent variables are written as the sum of a base flow and a perturbed flow as follows

$$\begin{Bmatrix} u \\ v \\ w \\ \theta \\ p \end{Bmatrix} = \begin{Bmatrix} 0 \\ v \\ w \\ \theta \\ p \end{Bmatrix} + \begin{Bmatrix} U(r) \\ V(r) \\ W(r) \\ \Theta(r) \\ P(r) \end{Bmatrix} \exp[i(Kz + n\phi + \sigma_i t) + \sigma_r t]. \tag{6}$$

This modal function admits toroidal $(n = 0)$ or spiral $(n \neq 0)$ disturbances with axial wavenumber K, frequency $-\sigma_i$ and growth rate σ_r. The radial eigenfunctions $U(r)$, $V(r)$, $W(r)$, $\Theta(r)$ and $P(r)$ are complex quantities. Substituting (6) into (2), subtracting the base flow (5) and linearizing furnishes the linear stability equations

$$\frac{d^2U}{dr^2} = (\sigma + iKG\bar{w})U + G\frac{dP}{dr} + \frac{inSG\bar{v}}{r}U - \frac{1}{r}\frac{dU}{dr} + \left[K^2 + \frac{1+n^2}{r^2}\right]U + \left[\frac{2inS}{r^2} - \frac{2S^2G\bar{v}}{r}\right]V \tag{7a}$$

$$\frac{d^2V}{dr^2} = (\sigma + iKG\bar{w})V + \frac{inG}{Sr}P + \frac{inSG\bar{v}}{r}V - \frac{1}{r}\frac{dV}{dr} + \left[K^2 + \frac{1+n^2}{r^2}\right]V - \frac{2in}{Sr^2}U$$

$$+ G\left[\frac{\bar{v}}{r} + \frac{d\bar{v}}{dr}\right]U \tag{7b}$$

$$\frac{d^2W}{dr^2} = (\sigma + iKG\bar{w})W + G\frac{d\bar{w}}{dr}U + iKGP + \left[inSG\frac{\bar{v}}{r}\right]W - \theta - \frac{1}{r}\frac{dW}{dr} + \left[K^2 + \frac{n^2}{r^2}\right]W \tag{7c}$$

$$\frac{d^2\theta}{dr^2} = Pr\ (\sigma+iKG\overline{w})\theta + PrG\frac{d\overline{\theta}}{dr}U + PrSG in\frac{\overline{v}}{r}\theta - \frac{1}{r}\frac{d\theta}{dr} + \left[K^2+\frac{n^2}{r^2}\right]\theta \tag{7d}$$

$$\frac{dU}{dr} = - \frac{U}{r} - in\frac{SV}{r} - iKW. \tag{7e}$$

The phase function $\gamma(z,\phi,t) = (Kz + n\phi + \sigma_i t)$ in (6) with solution values for n, K, and σ_i completely determines the kinematics of the disturbance flow patterns at onset of instability. For example, the axial propagation speed C of the disturbances, the wavelength λ normal to the phase lines and the inclination ψ of phase lines with respect to the horizontal are given by

$$C = \frac{-\sigma_i}{K}, \qquad \lambda = \frac{2\pi}{(n^2/r^2 + K^2)^{1/2}}, \qquad \psi = - \tan^{-1}\left[\frac{n}{rK}\right]. \tag{8a,b,c}$$

Note that the wavelength and inclination angle of the nonaxisymmetric spiral disturbances depend on the radial coordinate; for a given mode of instability, the spiral wavelength (inclination) will be shorter (steeper) if observed at the inner wall than if observed at the outer wall.

3.4 NUMERICAL SOLUTION PROCEDURE

The usual method of eliminating the pressure by cross-differentiation and reducing the higher order equations to a set of first-order equations for the real and imaginary components of the primiteive variables is extremely tedious. It is far simpler to follow the procedure suggested by Garg and Rouleau [15] and Garg [16] and introduce new complex velocities f(r) and g(r) defined by the equations

$$f(r) = U(r) + iV(r), \qquad g(r) = U(r) - iV(r). \tag{9a,b}$$

Using this technique it is fairly straightforward to write down the set of real ordinary differential equations describing the real and imaginary parts of the radial functions. The system was solved subject to zero disturbances at $r = r_1$ and r_2 using the linear boundary-value problem software package **SUPORT** [17] in combination with the nonlinear equation solver **SNSQE** (SLATEC [8]). In order to avoid singular matricies, we followed a procedure described by Keller [19] wherein one of the complex boundary conditions, say $W = 0$ at $r = r_2$, is replaced by the boundary condition $dW/dr = 1$ at $r = r_2$. For given values of the parameters, a solution of the linear system can be determined which generally will not satisfy $W = 0$ at $r = r_2$. The two parameters chosen to be eigenvalues are then varied until both the real and imaginary parts of $W(r)$ are driven to zero at the outer boundary. The object of this study is to determine curves of neutral stability for which $\sigma_r = 0$. Hence the eigenvalue problem may be written symbolically as

$$\mathscr{F}(G, Ta, Pr, K, \sigma_i, n, \eta) = 0. \tag{10}$$

Ta, Pr, K, n and η are generally fixed and solution of the ordinary differential equations are obtained by iteration on G and σ_i. At fixed mode number n a search is then conducted to find the minimum value G_m over all wavenumbers K. Critical conditions are then determined by searching for the minimum G_m over all n modes and the critical values so obtained are denoted Ta_c, G_c and K_c. In some instances solutions were more readily obtained by fixing G, Pr, K, n and η, and iterating on initial guesses for Ta and σ_i. Minimum values for either G or Ta were determined by a search incrementing the wavenumber in steps $\Delta K = 0.01$. The error incurred in the stepping procedure depends on the shape of the neutral stability curve in the vicinity of the minimum. An analysis of results shows that typical errors for the critical values of either G or Ta are on the order of 0.05% with a worst case of about 0.2%. All computations were performed in double precision on a Pyramid 90X or in single precision on a Cyber 205.

4. Presentation of Results

Stability results for a wide gap with $\eta = 0.6$ at three Prandtl numbers Pr = 4.35, 15 and 100 is presented. A narrow gap calculation at $\eta = 0.959$ for comparison with the experimental results of Snyder and Karlsson [20] will be reported in §5.

4.1 NEUTRAL STABILITY CURVES

In the process of locating the critical stability boundaries, neutral stability curves of Ta versus K or G versus K were generated for given parameter conditions at various mode numbers n. We note a persistent feature observed during this search. Computations revealed that open loop Ta-K neutral stability curves were often accompanied by a set of closed loop branches in a limited range of overlapping values of G as shown in Fig.1(a) for Pr = 15 and n = -2.

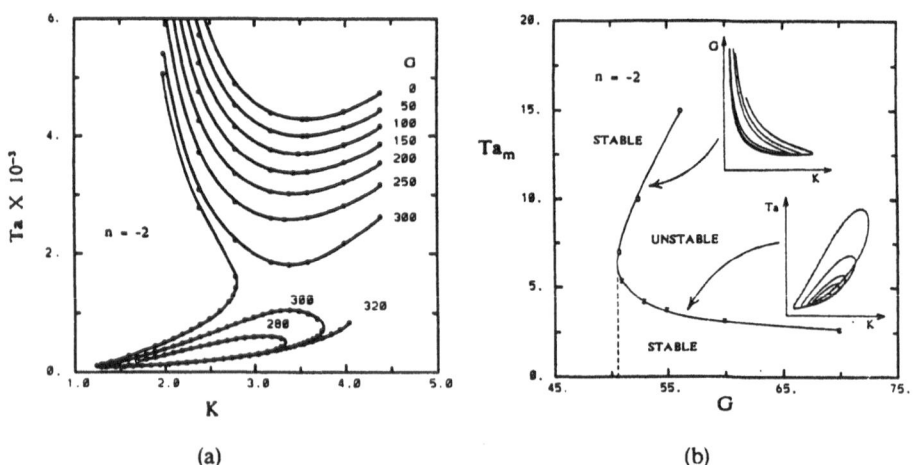

(a)   (b)

Figure 1. (a) Neutral stability curves for Pr = 15. (b) Critical stability boundary for Pr = 100, showing the turning point separating contributions from open and closed loop stability branches.

The closed loops evolve to smaller loops and finally degenerate to a limit point at $G \cong 254$ where $Ta \cong 217$ and $K \cong 2.33$. Beyond this point we were not able follow solutions in Ta-K space but did locate them in G-K space by changing the eigenvalue pair from (Ta, σ_i) to (G, σ_i). These limit points were found to be associated with a turning point of the Ta-G stability curve as shown in Fig. 1(b) for a different case with $Pr = 100$ and $n = -2$. The lower branch of the stability boundary is formed from the values of Ta_m determined at fixed values of G from the closed loops shown in the inset, while the upper branch is formed from the values of G_m determined at fixed values of Ta from the open loops. A curve fit to the stability boundary in Fig. 1(b) gives a turning point value of $G \cong 50.3$, virtually identical to the value obtained from an estimate of the limit point for the closed loops. This feature is a manifestation of the complicated topology of the stability curves in three-dimensional Ta-G-K space. Similar closed loop stability maps have been reported by Hart [21] for convection in an inclined heated box and by McFadden, Coriell, Boisvert, Glicksman and Fang [22] in a study of the stability of crystal morphology driven by fluid flow in the melt.

4.2 CRITICAL STABILITY RESULTS

Critical stability boundaries for $\eta = 0.6$ at three Prandtl numbers are given in Fig. 2(a). Corresponding axial phase speeds determined from (8a) at the two extreme Prandtl numbers are presented in Fig. 2(b).

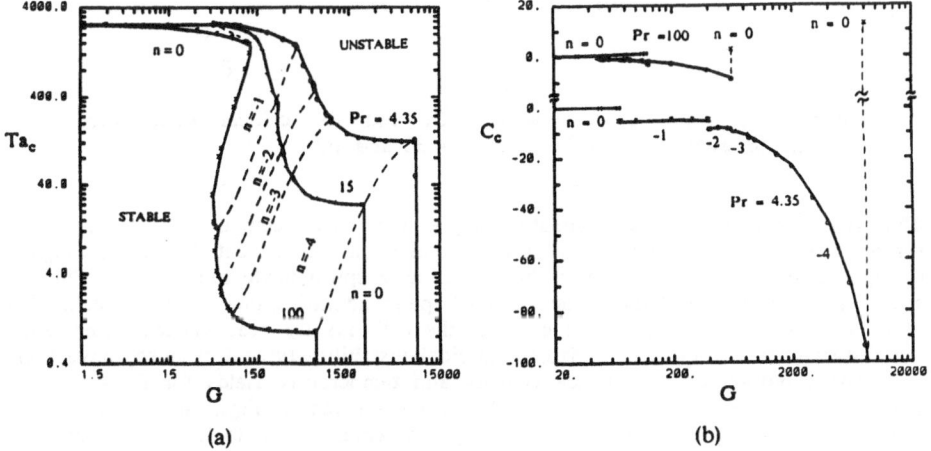

(a)   (b)

Figure 2. (a) Critical stability boundaries for $Pr = 4.35, 15, 100$. (b) Critical axial phase speed for $Pr = 4.35$ and 100.

Each stability curve emanates from the same axisymmetric centrifugally-driven Taylor instability at $G = 0$ and evolves through four spiral modes to terminate on separate axisymmetric convection mode branches. A comparison with the work of Choi and Korpela [11] and McFadden, et al [12] shows that the terminal convection branch corresponds to a shear-driven instability for $Pr = 4.35$, but becomes a buoyancy-driven instability for $Pr = 15$ and 100. Since both Ta and G are independent of the thermal diffusivity, the destabilization of the flow with increasing Pr is tied directly to changes in the thermal properties of the fluid. Fig. 2(b) shows that the effect of radial heating is to impart a weak upward drift to the Taylor cells, that spiral modes propagate downward with phase speed increasing with increasing spiral mode number and that the terminal axisymmetric cells drift vertically upward.

The variation of the wavelength λ and the spiral inclination angle ψ defined in (8b,c) for the two extreme values of Pr are presented in Figs. 3(a,b), respectively.

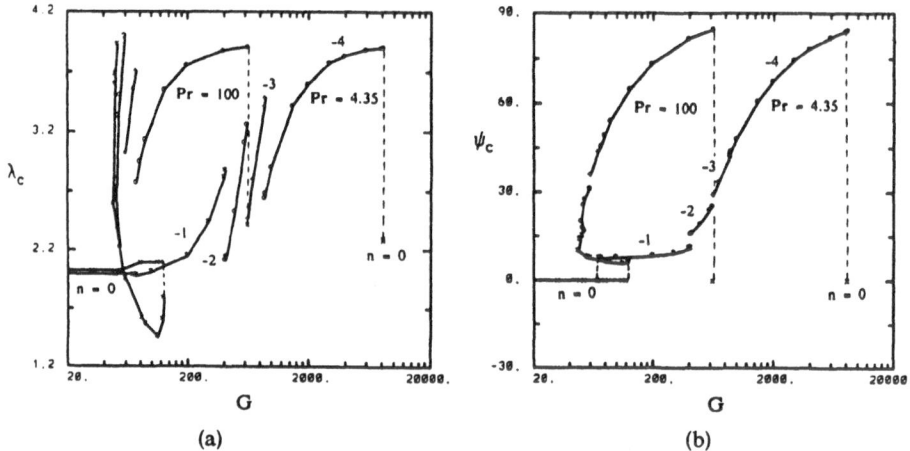

Figure 3.   (a) Critical wavelength and (b) critical disturbance inclination angle for $\eta = 0.6$ at $Pr = 4.35$ and 100.

The wavelengths and spiral inclination angles have been computed at $r = r_2$ giving results one would observe in a flow visualization experiment looking through a transparent outer cylinder. Fig. 3 shows that as a spiral mode tilts upward, the cells stretch significantly until they reach a break point admitting one additional cell. The terminal ($n = -4$) spiral mode elongates as the cell tilts up near vertical after which the system reverts back to a horizontal toroidal instability typical of pure convection.

The evolution of disturbance velocity and temperature fields for $Pr = 15$ along the critical stability boundary in Fig. 2(a) are presented in Figs. 4(a,b), respectively. The axial wavelengths have been scaled to be of equal height to facilitate comparison of the cell morphologies. In Fig. 4(a) the sequence commences with a centrifugally-driven Taylor instability at $G = 0$ and evolves through four spiral modes to the axisymmetric shear-driven instability at $Ta = 0$. The corresponding sequence for the isotherms given in Fig. 4(b) has the first frame omitted since the Taylor instability at $G = 0$ is isothermal.

Ali [23] has demonstrated that radial heating may either stabilize or destabilize isothermal Taylor cells. A study of the stability boundaries for axisymmetric disturbances in the neighborhood $G = 0$ for $\eta = 0.6$ shows that Taylor cells are stabilized in the region $4.35 < Pr < 63.5$ and it is inferred that destabilization persists for all $Pr > 63.5$.

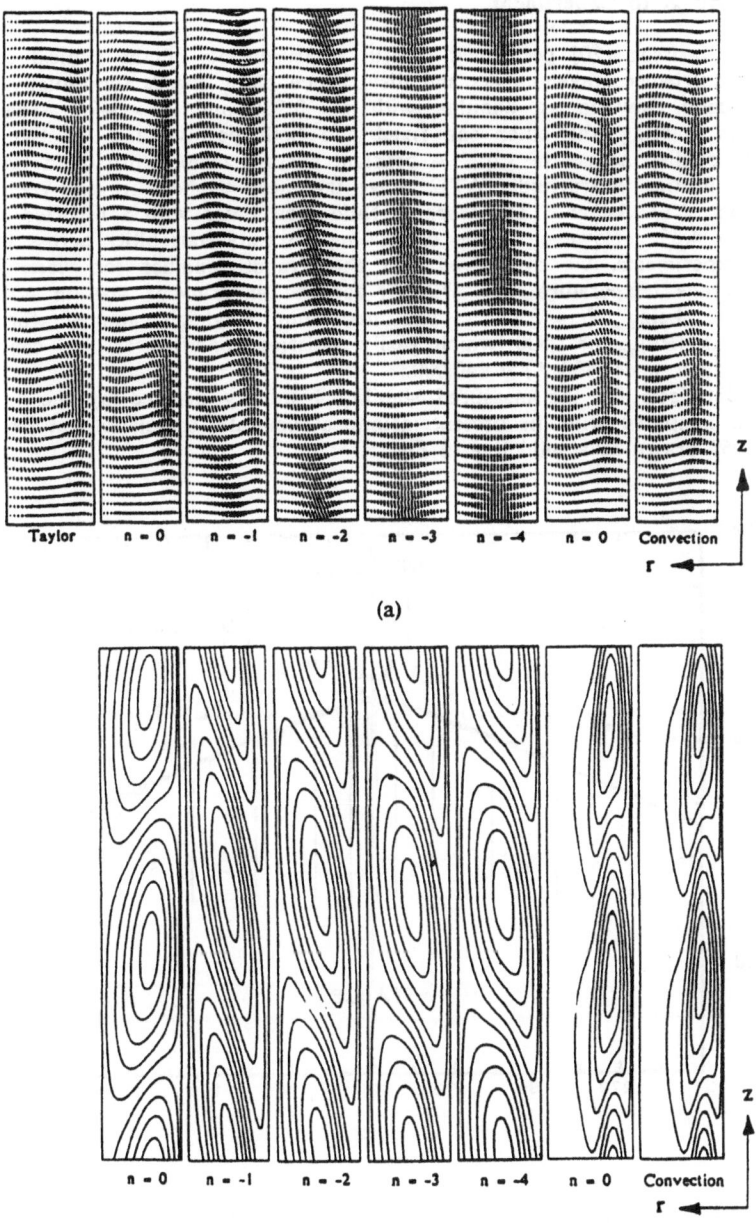

(a)

(b)

Figure 4. Evolution of (a) disturbance vector velocity fields and (b) temperature contours along the stability boundary for $\eta = 0.6$ and $\mathbf{Pr} = 15$.

5. Comparison With Experiment

Snyder and Karlsson [20] performed careful measurements of the critical stability boundary, disturbance wavenumbers and disturbance phase speeds for radially heated circular Couette flow in an annulus with $\eta = 0.959$ and $A = 349$. The inner cylinder was rotating, the outer one fixed and both positive and negative radial heating was applied across the gap. A constant average temperature in the gap was achieved by heating one cylinder $\Delta T/2$ above the mean temperature and cooling the other cylinder $\Delta T/2$ below. The working fluid was water with a reported Prandtl number $Pr = 4.35$. However, while calculating viscosity variability across the gap for these experiments, we discovered that the Prandtl number for water at the reported average temperature ($T = 27.5\ ^0C$) for the experiments was in error, the correct value being $Pr = 5.77$. The maximum Rayleigh number for these experiments is $(Ra)_{max} \cong 1,800$ which easily satisfies condition (1) for conduction regime base flow since $(400\ A) \cong 140,000$.

Two data sets for the measured stability boundary denoted by the + and * symbols are compared with the present infinite aspect ratio calculations at $Pr = 5.77$ and $\eta = 0.959$ drawn as a solid line in Fig. 5.

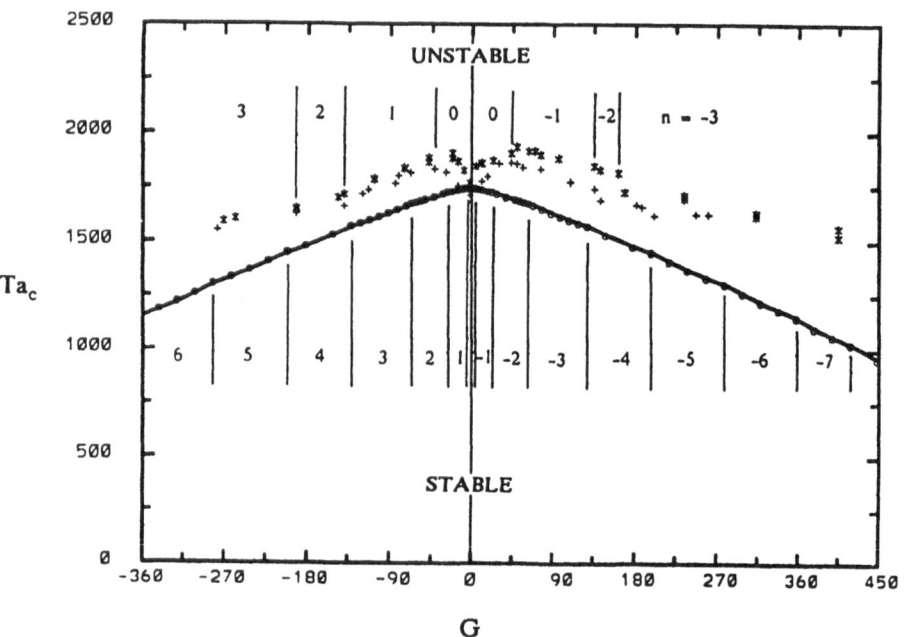

Figure 5.   Comparison of computed infinite aspect ratio stability boundary (solid curve) with experimental data (+,*) obtained by Snyder and Karlsson [20] at $\eta = 0.959$, $Pr = 5.77$ and $A = 349$.

The experimentally observed transition boundaries between different spiral modes are indicated above the measured data points while the numerically computed mode transition boundaries are indicated below the theoretical curve.   Computed axial wavenumbers K_c are compared with the axial wavenumbers measured at slightly supercritical conditions in Fig. 6.

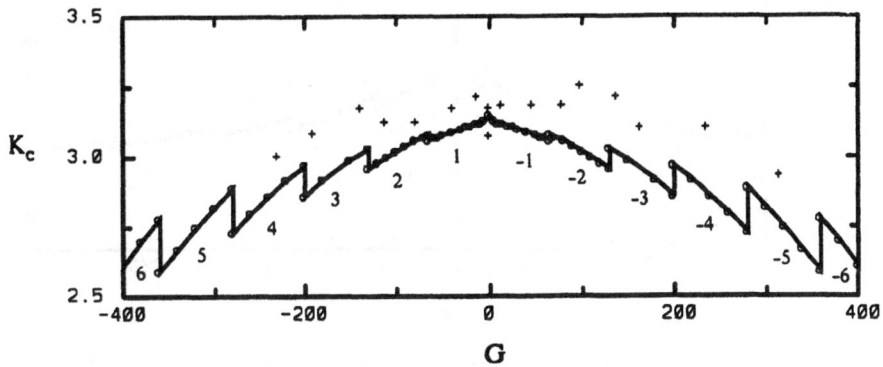

Figure 6.   Comparison between computed and measured critical axial wavenumbers. + Measurements of Snyder and Karlsson [20]; solid line respresents   the   computed   wavenumbers   showing   discontinuous transitions between successive modes.

The dimensional angular rotation rate of spiral disturbances may be calculated from the equation

$$\omega \; = \; \frac{-\,\sigma_i}{n\,(d^2/\nu)}. \tag{11}$$

In Fig. 7 we compare computed and measured results for the critical angular velocity ω_c of the helical waves normalized with the angular velocity Ω_1 of the inner cylinder.   The near zero velocities for the axisymmetric disturbances are omitted to facilitate   a   comparison   on   an   enlarged   vertical   scale.   Both   theoretical   and experimental results show that over a broad range of Grahsof numbers the spiral disturbances rotate at nearly the average angular speed of the inner and outer cylinder, namely $\Omega_1/2$.

Fundamental differences exist between the theory and experiment which need mentioning.   First, the present calculations are for an infinite aspect ratio geometry while the experiment corresponded to A = 349.   Second, the experiments exhibited a small but finite density-stabilizing axial temperature gradient, whereas the stability calculations are for zero axial temperature gradient.   In addition, Snyder and Karlsson [20] observed   an   early   transition   to   weak   cat's-eye   cells;   bifurcation   to   fully-developed toroidal or spiral cells filling the gap were observed as a transition from the cat's-eye flow regime.   Furthermore, Snyder and Karlsson observed spirals composed of counter-rotating cells of nonuniform width (cell width ratio about 3:1). The disturbance function selected for the present calculations enforces uniform cell width in each counter-rotating vortex pair.

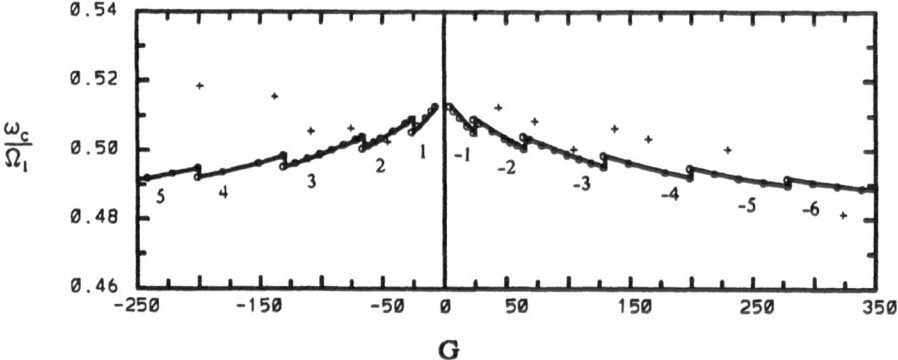

Figure 7. Comparison between computed and measured critical angular rotation rates of the spiral disturbances. + Measurements of Snyder and Karlsson [20]; solid line respresents the computed rotation rates showing discontinuous transitions between successive modes.

In spite of the above differences, the measured and computed stability boundaries exhibit similar behavior. Snyder and Karlsson observed stabilization of the axisymmetric Taylor vortex mode up to $G \cong 49$ for negative radial temperature gradients and $G = -30$ for positive gradients. The present calculations also predict stabilization, but only up to $|G| = 5.5$. The enhanced stabilization in the experiments is attributed to the density-stabilizing axial temperature gradient observed in the finite aspect ratio apparatus. The calculated rms percent deviation between theory and experiment for the critical Taylor numbers is 13.8%. It is clear in Fig. 5 that increasingly higher mode numbers come into play with increasing values of $|G|$. However, the rate of increase of mode number in the numerical results is almost exactly double that observed in the experiments. Even though the azimuthal mode numbers differ by a factor of two, good agreement between theoretical and experimental axial wavenumbers K_c is observed in Fig. 6. The rms deviation for the wavenumber comparison is 4.4%. Good agreement is also found in the comparison of normalized spiral rotation rates ω_c/Ω_1 given in Fig. 7 where the rms deviation is only 2.0%. It is further noted that the orientation of the helical disturbances determined from the present computations agree with those observed in the experiments as listed in figure 9 of Snyder and Karlsson [20].

Except for the linear dependence of density on temperature, the numerical results are for constant property fluids. Calculations of fluid property variability in the experiments of Snyder and Karlsson have been made at the highest temperature contrast for both positive and negative radial heating. At positive Grashof number the rms variabilities in Pr and G across the gap are 15% and 45%, respectively. At negative Grashof number the corresponding variabilities are 11% and 30%.

6. Discussion and Conclusion

A linearized analysis of the stability of circular Couette flow in the annular gap between vertical, coaxial differentially-heated cylinders of infinite aspect ratio has been performed. Gravitational effects are included in the Boussinesq approximation and constant diffusion coefficients for momentum and heat transport are assumed. Positive Grashof number computations have been carried out for a wide gap at $\eta = 0.6$ for Prandtl numbers 4.35, 15 and 100. In each case the flow evolves from an axisymmetric centrifugal instability (Taylor cells) at $G = 0$ and bifurcates through four spiral modes to an axisymmetric shear-driven or buoyancy-driven instability at $Ta = 0$. The intervening spiral modes tilt upwards from horizontal at $G = 0$ to nearly vertical $[(\psi_c)_{max} > 81°]$ before plummeting back to horizontal at $Ta = 0$ corresponding to flow driven by radial heating across the gap of a stationary annulus. Stability of the system is greatly affected by the Prandtl number. A plot of the stability curves in Ta-G space shows that flow destabilization with increasing Prandtl number is linked solely to changes in fluid thermal diffusivity. It is further found that radial heating imparts a vertical drift to Taylor cells. The weak axial phase speed increases with both increasing Grashof number and increasing Prandtl number.

Positive and negative Grashof number computations for a narrow gap geometry have been carried out at $\eta = 0.959$ and $Pr = 5.77$ to compare with the measurements of Snyder and Karlsson [20]. Results for the critical stability boundary, axial wavenumber and angular rotation rate of spiral disturbances presented in Figs. 5, 6 and 7 exhibit perfect even symmetry about the $G = 0$ axis. This symmetry property with respect to the sense of radial heating may be noted in the n = 0 results of Roesner [7], but here we see that it extends to the nonaxisymmetric modes as well and is evidently a fundamental property of the linearized system. The agreement between computational and experimental results is very good considering the infinite aspect ratio idealization in the linearized theoretical model. Rms deviations between theory and experiment are 14% for the critical Taylor numbers, 4% for the wavenumbers and 2% for the azimuthal phase speeds compared over a significant range of positive and negative Grashof numbers.

Topics for future work include a continuation of the present linear stability analysis to include low Prandtl number fluids such as liquid metals where $Pr = O(10^{-2})$ or even air for which $Pr = 0.7$. An investigation of the effects of variable fluid properties on system stability offers another avenue for research on Taylor-Couette flow with radial heating.

Acknowledgments

We are indebted to Professor Snyder who kindly supplied his original laboratory data to facilitate a determination of the critical Grashof, Taylor and Prandtl numbers as defined in the present investigation. Appreciation is extended to the Von Neumann Center for awarding free supercomputer time on the Princeton Cyber 205.

References

[1]   Taylor, G. I. *Phil. Trans. Roy. Soc.*, **A223**, 289, 1923.
[2]   Yih, C.-S. *Phys. Fluids*, **4**, 806, 1961.

[3]   Becker, K. M. and Kaye, J. *Trans. ASME, J. Heat Trans.*, **80**, 106, 1962.
[4]   Walowit, J., Tsao, S. and Diprima, R. C. *Trans. ASME, J. of Appl. Mech.*, **31E**, 585, 1964.
[5]   Bahl, S. K. *J. Appl. Mech.*, **39**, 593, 1972.
[6]   Soundalgekar, V. M., Takhar, H. S. and Smith, J. T. *Warme und Stoff.*, **15**, 233, 1981.
[7]   Roesner, K. G. *Arch. Mech.*, **30**, 619, 1978.
[8]   DiPrima, R. C. and Swinney, H. L. *Topics in Applied Physics*, **45**, 139, 1981.
[9]   Stuart, J. T. *SIAM*, **26**, 315, 1986.
[10] Thomas, R. W. and de Vahl Davis, G. *Proc. 4th Int. Heat Transfer Conf.*, Paris, paper NC 2.4, 1970.
[11] Choi, I. G. and Korpela, S. A. *J. Fluid Mech.*, **99**, 725, 1980.
[12] McFadden, G. B., Coriell, S. R., Boisvert, R. F. and Glicksman, M. E. *Phys. Fluids*, **27**, 1359, 1984.
[13] Weidman P. D. and Mehrdadtehranfar G. *Phys. Fluids*, **28**, 776, 1985.
[14] Chandrasekhar, S. *Hydrodynamic and Hydromagnetic Stability*, Oxford University Press, London, 1961.
[15] Garg, V. K. and Rouleau, W. T. *J. Fluid Mech.*, **54**, 113, 1972.
[16] Garg, V. K. *J. Fluid Mech.*, **110**, 209, 1981.
[17] Scott, M. R. and Watts. H. A. *SIAM J. Num. Anal.*, **14**, 40, 1977.
[18] SLATEC Common Math Library, National Energy Software Center, Argonne National Laboratory, Argonne, IL, written by K. L. Hiebert.
[19] Keller, H. B. *SIAM Regional Conference Series in Applied Mathematics*, **24**, Philadelphia, PA, 1976.
[20] Snyder, H. A. and Karlsson, S. K. F. *Phys. Fluids*, **7**, 1696, 1964.
[21] Hart, J. E. *J. Fluid Mech.*, **47**, 547, 1971.
[22] McFadden, G. B., Coriell, S. R. and Boisvert, R. F., Glicksman, M. E., Fang, Q. T. *Metallurgical Transactions A*, **15A**, 2117, 1984.
[23] Ali, M. E. PhD thesis, University of Colorado, Boulder, Colorado, USA, 1988.

FLOW FIELD EFFECTS ON DYNAMICAL INSTABILITIES

D.WALGRAEF †
Laboratoire de Physique Théorique,
Université de Nice, Parc Valrose,
Nice Cedex 06034, France.

ABSTRACT. Flow field effects on pattern forming instabilities are studied in the framework of amplitude equations and phase dynamics. In thin horizontal fluid layers, it is shown that the vertical profile of imposed fluid velocity fields induces anisotropy effects which determine the selection and stability properties of spatio-temporal patterns associated with Turing and Hopf bifurcations. In deformed metals, the role of plastic flows in the formation of dislocation microstructures is discussed.

1. INTRODUCTION.

The effect of external fields on pattern forming systems suscited recently much interest. For example, the Lowe-Gollub experiment [1] shows that a periodic field imposed on a system which undergoes a patterning instability may induce transitions between structures with incommensurate wavelengthes . The theoretical aspects of this problem have been discussed in the framework of amplitude equations and phase dynamics [2][3]. In the case of Rayleigh-Benard convection, an imposed flow field leads to a spatial variation of the wavelength of the patterns [4], while in chemically active media, such a flow is able to disorganize spiral waves and leads to spatial disorder [5] . It has also been shown that these effects may be interpreted within the phase dynamical analysis of the structures [6][7]. Furthermore Brand [6] recently discussed how flow fields can affect the phase dynamics and proposed minimal model equations able to describe qualitatively several experimental observations.

---

† Senior Research Associate, National Fund for Scientific Research (Belgium), permanent address: Faculté des Sciences, Université Libre de Bruxelles, CP 231, Brussels, Belgium.

E. Tirapegui and D. Villarroel (eds.), Instabilities and Nonequilibrium Structures II, 269–283.
© 1989 by Kluwer Academic Publishers.

Since his discussion was restricted to flow field effects in one-dimensional systems, I would like to show here that, when dynamical instabilities of the Turing or Hopf type occur in thin horizontal fluid layers in the presence of an imposed flow, one has to take into account the vertical profile of the associated fluid velocity field in the analysis of the pattern selection and stability properties which are strongly affected by the anisotropy effects induced by the flow.

Another problem I would like to discuss here is the effect of plastic flow on the spatio-temporal organization of dislocation populations in deformed metals. Effectively, the evidence of instabilities and pattern formation is overwhelming in the case of plasticity. It is now well-known for about fourty years that the elementary carriers of plastic deformation of metals are dislocations. In spite of their complex shapes and interactions it has been observed, e.g. via transmission electron microscopy that they form well organized spatial structures. At high temperatures these regular structures mainly consist of planar networks which form the walls of cellular planforms. At lower temperatures these cells become thicker and more diffuse. Under cyclic loading the tendency to form ordered structures is even more pronounced. Especially the so-called persistent slip bands (PSB), a regular ladderlike structure embedded in a less regular vein or matrix structure has been observed in crystals oriented for single slip [8]. In the case of multiple slip, PSB with different orientations may coexist leading to "labyrinth structures"[9].

The instabilities of plastic flow and the associated development of slip localisation manifest themselves in the stress-strain response of the sample. In particular a plateau is associated to the development of PSB leading to the analogy with phase transitions formulated by Winter [10]. Since other transitions of this kind, related for example to the formation of kink bands, coarse slip bands or shear bands which appear during monotonous loading tests also lead to dramatic changes in the stress-strain curves, there should be an underlying principle and a unified framework for the interpretation of these different phenomena. Our point here is that plastic deformation is a nonlinear dissipative process and, as such, should be described by the collective behavior of the dislocation populations under stress. Plastic strain rates should then be related to dislocation densities via phenomenological equations such as the Orowan law for example. In fact it will be shown in the following that a physically based set of nonlinear differential equations of the reaction-diffusion type derived for the dislocation densities is able to qualitatively describe the various aspects of dislocation patterning during deformation processes and provides the framework where quantitative aspects may be obtained via specific analysis and numerical simulations. Furthermore, several properties of the dislocation microstructures will appear to be strongly dependent on the characteristics of the plastic flow associated with the deformation process.

2. CHEMICAL INSTABILITIES AND HYDRODYNAMIC FLOWS.

It is well known that, in a solution, the presence of convective flows strongly modifies the diffusive transport of passive solutes. Since the classical work of Taylor on dispersion in a shear flow, the way how the fluid velocity profile modifies the diffusion coefficients of advected solutes has been studied in various limits and for various flows [11][12]. As the flow fields induce anisotropy effects in the transport coefficients, it is also likely to affect the stability and selection properties of the spatio-temporal patterns associated to dynamical instabilities in chemically active media.

Let me illustrate this point in the case of pattern forming or Hopf instabilities occuring in reaction-diffusion systems in the presence of a passive convective field. For example, when nonlinear chemical reactions take place in a fluid layer, the concentration inhomogeneities are coupled to the fluid velocity field and the dynamics of such a system may be written as:

$$\dot{X}_i = D_i \nabla^2 X_i + F_i(\{X_j\}, b) - \vec{v}\vec{\nabla}X_i \tag{1}$$

where the X_i represent the concentrations of the reacting species, the F_i are nolinear functions of the X_i, D_i the Fick's diffusion coefficients, ν the kinematic viscosity, g the gravity field (parallel to $0z$), and β the mass expansion coefficient (considered to be equal for all X_i to simplify the notations). The reactions are supposed here to be isothermal for simplicity, but couplings with temperature gradients could also have been considered. Near a Turing-like instability which occur, for the chemical system alone, in the absence of hydrodynamic motion, at $b = b_c$ and for a critical wavelength $\lambda = \lambda_c = 2\pi q_c^{-1}$, the time and space scale separation leads to a reduction of the dynamics to normal forms or amplitude equations [13-15]. If, furthermore, this instability is well separated from any hydrodynamical instability in the layer, the asymptotic dynamics of the isolated chemical system may be written as:

$$\tau_0 \dot{\sigma} = [\epsilon - d(q_c^2 + \nabla^2)^2]\sigma - v\sigma^2 - u\sigma^3 \tag{2}$$

where $\epsilon = (b - b_c)/b_c$, and where τ_0, d, u, v may be explicitly calculated for specific models [14]. As a result of the adiabatic elimination of the stable modes, the nonlinearities of the velocity equations are now σ dependent.

In the case of passive convection, where a velocity field $\vec{v}$ is established in the medium when the bifurcation parameter b reaches its critical value, the amplitude of the order parameter-like variable σ is sufficiently small to have no relevant feedback effect on the velocity field. The dynamics will then be restricted to:

$$\tau_0 \dot{\sigma}(\vec{r}, t) = [\epsilon - d(q_c^2 + \nabla^2)^2]\sigma(\vec{r}, t) - v\sigma(\vec{r}, t)^2 - u\sigma(\vec{r}, t)^3 - \vec{v}(\vec{r}).\vec{\nabla}\sigma(\vec{r}, t) \tag{3}$$

Due to the spatial dependence of the fluid velocity field we are now confronted to a nonautonomous problem. However, when the active medium is restricted to a thin horizontal layer (such that its thickness l is smaller than the critical wavelength λ_c), the solute concentrations may be represented as a Fourier sum, and in the case of divergence free boundary conditions, the order parameterlike variable is written as $\sigma(\vec{r}, t) = \sum_n \sigma_n(x, y, t) \cos n\pi z/l$ (a similar results may of course be performed in the case of other boundary conditions). Since $\pi/l \ll q_c$ the only unstable component of σ corresponds, in the absence of velocity field, to $n = 0$ while the other ones remain stable unless $\epsilon \geq d(n\pi/l)^4$. As a result the only stable patterns given by eq.(3) correspond to rolls, hexagons or triangles in the horizontal plane [14]. In the presence of a passive convective flow however, the slow mode dynamics (3) may be rewritten as:

$$\tau_0 \dot\sigma_0 = [\epsilon - d(q_c^2 + \nabla_\perp^2)^2]\sigma_0 - v\sigma_0^2 - u\sigma_0^3 - \sum_n (v_n + u_n\sigma_0)\sigma_n^2$$

$$- \sum_n (\vec{v}\vec{\nabla})_{0n}\sigma_n$$

$$\tau_0 \dot\sigma_n = [\epsilon - d(q_c^2 - (\frac{n\pi}{l})^2 + \nabla_\perp^2)^2]\sigma_0 - (\vec{v}\vec{\nabla})_{n0}\sigma_0 + \sum_{m \neq 0} (\vec{v}\vec{\nabla})_{nm}\sigma_m$$

$$+ \text{nonlinear terms} \tag{4}$$

where $\nabla_\perp^2 = \nabla_x^2 + \nabla_y^2$ and $(\vec{v}\vec{\nabla})_{nm} = \int_0^l dz \cos(n\pi z/l)(v_{0x}(\vec{r})\nabla_x + v_{0y}(\vec{r})\nabla_y + v_{0z}(\vec{r})\nabla_z) \cos(m\pi z/l)/\int_0^l dz \cos^2(n\pi z/l)$.

In the case of thin layers, the $n = 0$ mode remains the critical one and the other ones may be adiabatically eliminated since near the instability, one has, for low Peclet numbers,

$$\sigma_n \simeq \frac{l^2}{dn^2\pi^2}(\vec{v}\vec{\nabla})_{n0}\sigma_0 \tag{5}$$

leading to:

$$\tau_0 \dot\sigma_0 = [\epsilon - d(q_c^2 + \nabla_\perp^2)^2]\sigma_0 - v\sigma_0^2 - u\sigma_0^3 - \sum_n (v_n + u_n\sigma_0)\frac{l^4}{d^2 n^4 \pi^4}((\vec{v}\vec{\nabla})_{n0}\sigma_0)^2$$

$$- (\vec{v}\vec{\nabla})_{00}\sigma_0 + \sum_n \frac{l^2}{n^2\pi^2}(\vec{v}\vec{\nabla})_{0n}(\vec{v}\vec{\nabla})_{n0}\sigma_0 \tag{6}$$

Hence the effect of an imposed space-dependent flow field on the patterning instability is twofold. On one side there appears an advection term with a material derivative associated with the mean value of the fluid velocity, the mean being taken over the layer thickness. The other effect corresponds to an additional anisotropic diffusion term. The influence of these terms on the pattern selection and stability properties of course depend on the characteristics of the imposed flow.

For example, in the case of plane Poiseuille or shear flows, $\vec{v} = v_0(z)\vec{1}_x$ with, respectively, $v_0(z) = v_M 4\frac{z}{l}(1 - \frac{z}{l})$ and $v_0(z) = az$. Equation (6) then becomes:

$$\tau_0\dot{\sigma}_0 = [\epsilon - d(q_c^2 + \nabla_\perp^2)^2]\sigma_0 - \bar{v}\sigma_0^2 - \bar{u}\sigma_0^3 - \bar{v}_0\nabla_x\sigma_0 + D_x\nabla_x^2\sigma_0 \qquad (7)$$

where the nonlinear couplings $\bar{v}$ and $\bar{u}$ correspond to the renormalization of v and u resulting from the elimination of the noncritical modes and where D_x is proportional to $v_M l^2$ for Poiseuille flows and to $a^2 l^4$ for shear flows, while $\bar{v}_0$ is the vertical mean of the velocity field.

As a consequence of the anisotropic diffusion effect triggered by the flow, the roll patterns which appear first have their wavevector perpendicular to the flow velocity (in the present case $\vec{q} = \vec{q}_c\vec{1}_y$) and their amplitude equation may be written as ($\sigma_0 = Aexpiq_cy + A^*exp - iq_cy$):

$$\tau_0\dot{A} = (\epsilon + D_y\nabla_y^2 + D_x\nabla_x^2)A - \bar{v}_0\nabla_x A - 3\bar{u}|A|^2 A \qquad (8)$$

The phase dynamics associated to a pattern having a wavenumber q slightly different from critical is then:

$$\tau_0\dot{\phi} = (D_\parallel\nabla_y^2 + D_\perp\nabla_x^2)\phi - \bar{v}_0\nabla_x\phi \qquad (9)$$

where

$$D_\parallel = D_y\frac{\epsilon - 3D_y(q - q_c)^2}{\epsilon - D_y(q - q_c)^2}, \quad \text{and} \quad D_\perp = D_x + D_y\frac{q - q_c}{q_c}$$

We see that at the leading order the Eckhaus instability is not affected by the flow while the zig-zag instability is shifted towards smaller values of the wavenumber, and the stability domain of the pattern is increased compared to the isotropic case. Furthermore, if the zig-zag instability triggers modulations along the axis of the rolls, these modulations will propagate due to the presence of the advection term $\bar{v}_0\nabla_x\phi$. As in other symmetry-breaking instabilities the nucleation of defects is expected. From the phase dynamics (9) we see that as a matter of fact dislocations may be triggered by phase singularities and that the advection term should induce a climbing motion for these dislocations.

The patterns with hexagonal symmetry which may appear via a first orderlike transition are also affected by the anisotropic diffusion terms triggered by the flow field and are either deformed or destabilized as discussed in [16]. On the other hand, if the system is very narrow in the y direction and cannot accomodate rolls with wavevectors in that direction, we may consider the system as as nearly one-dimensional . In that case the phase dynamics of the patterns may be written, at the leading order, as:

$$\tau_0 \dot{\phi} = D_\parallel \nabla_x^2 \phi - \bar{v}_0 (\bar{q} + \nabla_x \phi) \tag{10}$$

where $\bar{q}$ is the critical wavenumber in the x direction. We recover here the expression analyzed by Pocheau [17] to interpret his experiment where a through flow induces wavelength variations of Rayleigh-Benard rolls in an annulus.

Consider now imposed flows corresponding to spatially periodic patterns associated to forced or natural convection and represented, for simplicity, by a velocity field of the form $\vec{v} = v_0 f(z) \sin q_0 x \vec{1}_z + (v_0/q_0) f' \cos q_0 x \vec{1}_x$ with $f(0) = f(l) = 0$. In this case, the mean flow is zero and there is no advection term associated to it. Hence, if there is no strong resonance between the periodicities of the unstable modes and of the imposed flow, the slow mode dynamics corresponding to eq.(6) becomes:

$$\tau_0 \dot{\sigma}_0 = [\epsilon - d(q_c^2 + \nabla_\perp^2)^2]\sigma_0 - \bar{v}\sigma_0^2 - \bar{u}\sigma_0^3 + D_x \nabla_x^2 \sigma_0 \tag{11}$$

where $D_x = (1/2) \sum_n (v_0 \alpha_n/q_0)^2$ with $\alpha_n = (2/l) \int_0^l dz f(z) \sin(n\pi z/l)$. Here also the anisotropy effect induces a selection principle favoring patterns with wavevectors oriented along $0y$ and their phase dynamics is given by the usual expression:

$$\tau_0 \dot{\phi} = (D_\parallel \nabla_y^2 + D_\perp \nabla_x^2)\phi \tag{12}$$

If strong resonances occur between the two periodicities and if the system is nearly one-dimensional, commensurate-incommensurate transitions may occur as discussed for example by Coullet[2].

In the case of a Hopf bifurcation, a similar analysis leads to the following type of amplitude equation for the order parameter-like variable in the presence of an imposed flow:

$$\tau_0 \dot{A} + \bar{v}_0 \nabla_x A = \epsilon A + (1 + i\alpha)\nabla_\perp^2 A + \frac{1 - i\alpha}{1 + \alpha^2} D \nabla_x^2 A - (1 + i\beta)|A|^2 A \tag{13}$$

where $D = \sum_n \frac{l^2}{n^2 \pi^2}(\vec{v})_{0n}(\vec{v})_{n0}$, the flow being parallel to the x direction and the mean flow velocity is supposed to be sufficiently small to remain in the case of an absolute instability [18].

Besides the advection term induced by the mean flow velocity the anisotropy effect will affect the shape of the spiral or concentric phase waves able to devlop in the medium but it also increases the stability domain of the oscillations with respect to inhomogeneities in the x direction. Effectively, the phase turbulence criterion is, for perturbations of wavevector oriented along $0x$ and $0y$ respectively:

$$\alpha\beta < -\frac{1+D}{1-D}, \qquad \text{and} \qquad \alpha\beta < -1$$

Furthermore, for increasing velocity the anisotropic advection should modify the threshold of convective instability and it would be interesting to study the spatio-temporal behavior of such systems.

The conclusion of this discussion is that the effect of an imposed flow in a thin horizontal fluid layer where patterning or Hopf instabilities occur is twofold. Besides the advection term associated to the mean flow, an anisotropic renormalization of the diffusion coefficients is induced by the vertical velocity profile. As a result the pattern selection is modified since roll structures tend to be oriented with the roll axis orthogonal to the flow. Cellular patterns and two-dimensional waves are deformed or even destabilized. On the other hand the above analysis may provide useful tools for the understanding of chemical patterning in fluid solutions where the separation between chemical and hydrodynamical effects is usually not obvious [19]. Passive convection was only considered here but, from eq.(1) it is clear that in the absence of preexisting convection, solute concentration inhomogeneities in the horizontal plane induce hydrodynamic motion eventually able to modify the characteristics of pattern forming instabilities induced by nonlinear chemical reactions. This case, and the more complex one where Turing-like and convective instabilities are coupled and lead to bifurcations of higher codimension, requires further analysis.

3. PLASTIC FLOW AND DISLOCATION MICROSTRUCTURES.

The example of chemical instabilities is typical of the fact that in non equilibrium pattern forming systems, as in phase transitions, the global properties of the structures do not depend on microscopic properties but rather on the symmetries of the problem and on the characteristics of the bifurcation. This seems also to be the case for many patterns and microstructures which form in driven or metastable materials and it is expected that the methods developed for the study of instabilities and bifurcations of complex dynamical systems should be useful in the understanding of materials behavior as I will try to illustrate it below.

Let me first emphasize that during the last years the whole field of materials science and related technologies has experienced a complete renewal. Effectively, by using techniques corresponding to strong non equilibrium conditions, it is now possible to escape from the constraints of equilibrium thermodynamics and to process totally new materials structures including different types of glasses, nano- and quasicrystals, superlattices, ... These techniques include ion implantation, laser beam surface melting as well as electron beam heating. For example in laser annealing, after the extremely rapid melting of a shallow layer of material followed by a comparably fast recrystallization, microstructures with superior resistance to friction, corrosion, ... may be frozen into place. Ultra-rapid solidification of alloys may trigger the formation of quasicrystalline structures. Point defect and void patterning in irradiated systems are related to the precipitation of solid solutions. Finally the self-organisation of dislocation populations in stressed or irradiated materials is associated to plastic instabilities and deformation localization. As materials with increased resistance to fatigue and fracture are sought for actual applications, a fundamental understanding of the collective behavior of dislocations during cyclic and monotonous loading and their role in crack initiation and propagation is highly desirable. Since usual thermodynamical and mechanical concepts are not adapted to describe such highly nonequilibrium situations, progress in this field is expected from nonlinear dynamics and instability theory [20].

Hence I would like to show in this framework that the competition between plastic flow and dislocation interactions may lead to the destabilisation of random dislocation distributions and triggers the formation of microstructures whith well defined orientations and symmetries.

3.1. Dislocation Dynamics in Fatigue.

Let us discuss first a dynamical system of coupled nonlinear differential equations describing the collective behavior of dislocations populations during fatigue experiments and presenting patterning instabilities leading to the ladder-like structure of persistent slip bands . This as they are experimentally observed [21] .

Consider a monocrystal oriented for single slip, the primary slip direction being parallel to the x axis, and submitted to cyclic loading of frequency ν. After the hardening period the forest of immobile dislocations is already well developped and we hence consider two types of dislocations: "trapped" or nearly immobile ones, of density ϱ_i , and "free" or mobile ones gliding on the primary slip plane and of density ϱ_m . In this regime, the dislocation density is sufficiently high $(10^{13}, 10^{14}m^{-2})$ to be represented by a continuous concentration field on a space scale larger than a few lattice spacings.

For each family of dislocations a balance equation is set up. The source and flux terms are determined by the following basic processes:

• the creation of dislocations under an applied stress is described by means of internal sources. In the case of PSB formation, a large number of dislocation clusters is already present in the crystal. Hence the majority of the newly created dislocations will be almost immediately pinned by the forest of nearly immobile ones and these mechanisms induce source terms in the kinetic equation for ρ_i;

• when thermal, or stress-induced activation dominates local energy barriers, dislocations may break free and move rapidly with a stress-dependent velocity in their glide planes. The freeing processes of trapped dislocations then lead to a linear source term $b\rho_m$ in the free dislocation kinetics and to the corresponding sink term in the trapped dislocation kinetics. The parameter b, specifying the freeing rate, is vanishingly small below threshold and suddenly reaches a finite value beyond the corresponding stress level. This parameter will be taken as bifurcation parameter and its variation depends, as discussed in [22], on the strain rate, the temperature, the nature of the crystal, the early stages of the deformation process, ... ;

• the pinning of mobile dislocations by dipoles or multipoles induces a sink term in their evolution equation . This term is generally nonlinear and depends on the elementary interactions between the two dislocation families. It is of the form $\sum_{n>1} c_n \rho_i^n \rho_m$ where the index n models the order of the cluster participating in the pinning process.

• on time scales larger than the period of the fatigue process and on space scales larger than the mean distance between obstacles, the plastic flow is effectively diffusive. As discussed in [23], this is a consequence of the combined effect of the back and forward motion of dislocations during each fatigue cycle and to the presence of a large number of pinning clusters. The effective diffusion coefficient is then proportional to the square of the mean velocity (which may be related to the stress intensity τ via phenomenological laws of the type $v \simeq v_0 exp(-(\tau/\tau_0)^m)$) and inversely proportional to the total pinning rate. For the trapped dislocations, the motion in the slip and cross-slip directions is diffusive as well. Their diffusion coefficients, however, are much smaller since their mobility is mostly due to thermal effects or to the coupling with vacancies or other defects.

By taking into account these dynamical processes, the balance equations for the two dislocation populations are written as [22]:

$$\dot{\rho}_i = \nabla_i (D_{ij}^{(0)} - D_{ijk}^{(1)} \nabla_k^2) \nabla_j \rho_i + f(\rho_i) - b\rho_i + \sum_n c_n \rho_i^n \rho_m$$

$$\dot{\rho}_m = D_M \nabla_x^2 \rho_m + b\rho_i - \sum_n c_n \rho_i^n \rho_m \qquad (14)$$

where $f(\rho_i)$ represents creation and annihilation of trapped dislocations. Due to the attractive character of the elastic interactions between dislocations, the diffusion coefficients $D^{(0)}$ may be negative in high density regime while the $D^{(1)}$ remain posistive. Hence a diffusive instability may occur for high creation rates or high stress levels [22][24]. The anisotropy of the underlying crystal structure is reflected in the diffusion tensors.

By performing the linear stability analysis of the uniform steady state, $f(\rho_i^0) = 0$, $\rho_m^0 = b\rho_i^0/\sum_n \rho_i^{0n}$, one obtains, in Fourier transform, the following evolution matrix:

$$L = \begin{pmatrix} \beta - a - q_i q_j (D_{ij}^{(0)} + q_k^2 D_{ijk}^{(1)}) & \gamma \\ -\beta & -\gamma - q_x^2 D_M \end{pmatrix} \quad (15)$$

where $a = -f'(\rho_i^0)$ is positive as a result of the stability of the uniform steady state at low stress intensities and where $\gamma = \sum_n c_n \rho_i^{0n}$, and $\beta = b(\rho_i^0 \frac{\partial \ln \gamma}{\partial \rho_i^0} - 1)$.

The computation of the eigenvalues of this evolution matrix shows that, in the absence of diffusional instability $(D_{ij}^{(0)} > 0)$, by increasing the stress intensity and since the motion of mobile dislocations in the primary slip direction is much faster than any other, the uniform steady state becomes unstable first versus density modulations along the x direction at a bifurcation point given by :

$$\beta = \beta_c = (\sqrt{a} + \sqrt{\gamma D_{xx}^{(0)}/D_M})^2, \quad \vec{q} = q_c \vec{1}_x, \quad q_c = \frac{2\pi}{\lambda_c} = (\frac{a\gamma}{D_{xx}^{(0)} D_M})^{1/4} \quad (16)$$

Hence, beyond this pattern forming instability, ladder-like structures are expected to develop with a wavelength λ_c which is a material property depending on the dislocation mobilities, and on their multiplication and pinning rates. By expressing the links between the parameters of the model and experimental quantities, it may be shown [25] that the wavelength satisfies usual phenomenological relations such as $q_c \propto \sqrt{\rho_i^0} \propto \sqrt{\tau_s}$, where τ_s is the resolved shear stress.

When $D_{ij}^{(0)} < 0$, a diffusional instability may occur leading first to the formation of cellular structures which may be associated to the vein structure of the matrix. By increasing the stress intensity, the freeing of trapped dislocations becomes more and more efficient and, as a result of the high anisotropy of the plastic flow, this cellular structure is destabilized versus ladder-like structures with wavevectors parallel to the primary slip direction [22.23].

The selection and stability properties of the patterns near the instability may be discussed with the slow mode dynamics :

$$\tau_0 \partial_t \sigma = [\epsilon - d_x(q_c^2 + \nabla^2)^2 + d_y \nabla_y^2]\sigma - v\sigma^2 - u\sigma^3 \quad (17)$$

where $\epsilon = (\beta - \beta_c)/\beta_c$, $d_x = D^{(0)}/\beta_c q_c^2$, $d_y = D^{(0)}/\beta_c(1 + \gamma/q_c^2 D_M)$, $\nabla^2 = \nabla_x^2 + \nabla_y^2$, while u and v may be computed explicitly in terms of the parameters of the model [25].

In isotropic cases $(d_y = 0)$, the only stable steady states beyond threshold $(\epsilon > 0)$ correspond to hexagonal or layered patterns while in highly anisotropic media the cellular structures become unstable [16] . It turns then out that in our case, layered structures with walls perpendicular to the primary slip direction will be the only possible ones.

When the bifurcation parameter increases far above threshold, the weakly non-linear approximation breaks down and numerical analysis is needed to study the behavior of the patterns and to test the permanence of the results obtained in the near-bifurcation regime. The one-dimensional aspects of the reaction-diffusion model (14) , corresponding to the patterning in the primary slip direction have been tested numerically [25]. The various kinetic rates were obtained from ex-perimentally accessible quantities such as the mean dislocation density, the mean dipole annihilation length, typical plastic deformation rates, ... as extensively dis-cussed in [27]. Stable periodic structures were found above the predicted threshold with wavelengthes between $0.5\mu m$ and $1.5\mu m$ and maximum amplitudes for ρ_i of the order of $2.10^{15}m^{-2}$, the wall fraction being around 0.09. Different nucleation processes have been tested, from slight bulk inhomogeneities in the dislocation density to surface stress variations.

3.2. Multiple Slip and Orientational Effects.

3.2.1. *Double Slip and Labyrinth Structures.* The experimental observations of dislocation patterning in fatigued crystals oriented for double slip show interesting additional features [26]. Effectively, according to the interactions between the dislocations associated to each slip system, different types of structure emerge. For example, if the interaction is strong, leading to locking effects of the Lomer-Cottrel type, PSB and ladder-like structures associated to the two primary slip systems develop in separated domains. On the other hand, if the interaction produces other dislocations with different orientations for their Burgers vector, cellular structures may emerge.

The first situation may be described via a dynamical model such as (14) with two families of mobile dislocations corresponding to each slip system. For the simplicity of the presentation, let us consider these primary slip directions as orthogonal. The balance equations may then be written as:

$$\dot{\rho}_i = \nabla_i (D_{ij}^{(0)} - D_{ijk}^{(1)} \nabla_k^2) \nabla_j \rho_i + f(\rho_i) - (b_1 + b_2)\rho_i + \sum_n c_n \rho_i^n (\rho_{m1} + \rho_{m2})$$

$$\dot{\rho}_{m1} = D_1 \nabla_x^2 \rho_{m1} + b_1 \rho_i - \sum_n c_n \rho_i^n \rho_{m1}$$

$$\dot{\rho}_{m2} = D_2 \nabla_y^2 \rho_{m2} + b_2 \rho_i - \sum_n c_n \rho_i^n \rho_{m2} \qquad (18)$$

b_1 and b_2 are the individual freeing rates and the two families of free dislocations interact via the trapped ones. Direct interaction between them could be considered without any qualitative change in the results. One gets, from the linear stability analysis around the uniform steady state, the following bifurcation point :

$$\beta = \beta_c = (\sqrt{a} + \sqrt{\gamma D_s/D_m})^2, \quad \vec{q} = q_c \vec{1}_x \text{or} \vec{q} = q_c \vec{1}_y, \quad q_c = \frac{2\pi}{\lambda_c} = (\frac{a\gamma}{D_s D_m})^{1/4}$$
$$(19)$$

As a result, the slow mode dynamics, in the three-dimensional space, is of the form:

$$\tau_0 \partial_t \sigma = [\epsilon - d(q_c^2 + \nabla^2)^2 - \kappa \nabla_x^2 \nabla_y^2 + d_z \nabla_z^2]\sigma - v\sigma^2 - u\sigma^3 \qquad (20)$$

Hence, steady state solutions corresponding to modulations of uniform amplitude are:

$$\sigma_x = 2R_0 \cos(q_c x + \phi_0), \quad \sigma_y = 2R_0 \cos(q_c y + \psi_0), \quad R_0 = \sqrt{\frac{\epsilon}{3u}}, \quad \phi_0, \psi_0 = \text{cst}$$
$$(21)$$

Since the two slip directions are equivalent as far as the pattern forming instability is concerned, one may also expect the formation of square patterns . However, the linear stability analysis of the uniform amplitude steady state, $|A_x^0| = |A_y^0| = \sqrt{\epsilon/9u}$, shows that such square planforms are unstable. Hence the only possible structures correspond to PSB associated with one or the other slip systems, or to mixed structures corresponding to regions where the two slip systems dominate alternatively. But as a consequence of the instability of square patterns, when one wall structure nucleates in some region of the crystal it empedes the other one to develop in the same region. For example, if one considers amplitude variations in the x direction only, chains of alternating kink-antikink solutions of the amplitude equation

$$\tau_0 \partial_t A_x = \epsilon A_x + \xi_\parallel^2 \nabla_x^2 A_x - 3u A_x(|A_x|^2 + 2|A_y|^2)$$
$$\tau_0 \partial_t A_y = \epsilon A_y + \xi_\perp^2 \nabla_x^2 A_y - 3u A_y(|A_y|^2 + 2|A_x|^2) \qquad (22)$$

exist, corresponding to a succession of domains where A_x and A_y are alternatively zero and nearly equal to $\sqrt{\epsilon/3u}$. As recently discussed by Coullet et al. [27], these solutions are dynamically unstable with very long transients. But random pinning of these kink and antikink solutions may occur as a result of the interactions of these "defects" with inhomogeneities of the medium. Since the wall structures are initiated within the relatively regular vein structure of the matrix, the experimentally observed alternating domains with PSB associated to one or the other slip system are consistent with their description within the amplitude equation formalism. A similar analysis may justify the existence of labyrinth structures when secondary slip is activated [9.26].

3.2.2. *Plastic Flow and Orientational Effects.* As I mentioned it in preceeding sections the attractive character of the dislocation interactions may induce diffusionnal instabilities in the dislocation dynamics and this led Holt to formulate his theory of stress induced cell formation [24]. In his approach, the mobility tensor of the initially randomly distributed dislocations is almost isotropic leading to the development of regular cells, while the presence of slight anisotropies would induce cell distortions. I would like to show here that the effect of higly anisotropic plastic flow strongly influences the pattern selection even in the early stages of the deformation process. Consider a crystal oriented for single slip submitted to monotonous loading. At the beginning of the plastic deformation process, the mobile dislocations are dominant. Due to their interactions with other defects and crystal imperfections, they have small mobilities in the glide and climb directions and a large glide current in the primary slip direction associated with the plastic flow. By writing separate balance equations for the density of dislocations with positive and negative Burgers vectors and by adding source and sinks terms representing dislocation creation and annihilation, we get the following dynamical system:

$$\partial_t \rho_+ = -D\nabla^2 \rho_+ - E\nabla^4 \rho_+ - \vec{v}\vec{\nabla}\rho_+ - NL_+(\rho_+,\rho_-)$$
$$\partial_t \rho_- = -D\nabla^2 \rho_- - E\nabla^4 \rho_- + \vec{v}\vec{\nabla}\rho_- - NL_-(\rho_+,\rho_-) \qquad (23)$$

where D and E are positive as discussed in [18] as a consequence of the attractive character of dislocation interactions, while $\vec{v}$ is the mean glide velocity. The linear stability analysis around the uniform steady state leads to the following evolution matrix, in Fourier space:

$$L = \begin{pmatrix} -\alpha + q^2 D - q^4 E - i\vec{q}\vec{v} & \gamma \\ \gamma & -\alpha + q^2 D - q^4 E + i\vec{q}\vec{v} \end{pmatrix} \qquad (24)$$

where $\alpha = \frac{\partial NL_+(\rho_+^0,\rho_-^0)}{\partial \rho_+^0} = \frac{\partial NL_-(\rho_+^0,\rho_-^0)}{\partial \rho_-^0}$ and $\beta = \frac{\partial NL_+(\rho_+^0,\rho_-^0)}{\partial \rho_-^0} = \frac{\partial NL_-(\rho_+^0,\rho_-^0)}{\partial \rho_+^0}$. Due to the symmetry of the problem with respect to positive and negative dislocations it is reasonable to assume that $\alpha = \beta$ and a pattern forming instability occurs since positive growth rates may be associated with the eigenvalues of this matrix. Effectiely, for perturbations of wavevector orthogonal to the glide velocity, the larger eigenvalue reads $\omega = q_y^2 D - q_y^4 E$ leading to the development of patterns having the usual preferred wavelength $\lambda_c = 2\pi\sqrt{2E/D}$. On the other hand, the eigenvalues associated with perturbations of wavevectors parallel to the glide direction are $\omega = -\alpha + q_x^2 D - q_x^4 E \pm \sqrt{\alpha^2 - (q_x v)^2}$ and we see that the plastic flow has a stabilizing effect on these perturbations. The resulting patterns will then correspond to layered structures having their layers parallel to the slip planes. These structures being nucleated preferentially at inhomogeneities, will then propagate in the crystal perpendicularly to the slip planes, in a way which is strongly reminiscent of the nucleation of Lüders or kink bands [28].

Hence we see that the plastic flow may have different orientational effects. When it participates in the destabilizing process, it triggers structures with their walls perpendicular to the slip direction. On the contrary, when the instability is induced by other processes, it tends to orient the layers parallel to the slip planes, similarly to the flow field effects discussed in section 2.

It is now clear that the competition between dislocation motion and interactions in stressed and strained solids is able to destabilize uniform distributions and, as a result, induce strain localisation. Instabilities leading to oscillations and bursts are also observed in dislocation populations and patterning phenomena occur in other defect populations such as interstitials and vacancies in irradiated materials and lead to the destabilization of solid solutions, to the nucleation of voids and of void lattices, ... Hence a coherent description of materials instabilities associated with the spatio-temporal organisation of defects will hopefully lead to a deeper understanding of these phenomena. Due to the strong nonequilibrium conditions under which they occur, classical mechanical or thermodynamical considerations are not sufficient and we need an important input from nonlinear dynamics and instability theory. Effectively, despite the huge complexity of the defect kinetics, the reduced dynamics near instability points leads to a possible description of the pattern selection and stability properties in the post-bifurcation regime. The selected structures strongly depend on the competition between the dominant nonlinearities and the underlying crystal structure, but also on the experimental procedures. For example, slow or fast increase of the constraints may induce different final structures even for systems where the dynamics is potential (in this case, the system may be locked in local minima of the potential for time scales which are much longer than any experimental ones). By the combination of the results of bifurcation analysis, amplitude equation formalism and numerical simulations we hence expect significant breakthroughs in the understanding and prediction of the effects of materials instabilities on the macroscopic behavior of driven or degrading solids.

Acknowledgments. It is a pleasure to thank E.C. Aifantis, P. Coullet, N. Ghoniem and L.P. Kubin for stimulating discussions. A NATO grant for international collaboration in research (082/84) is also gratefully acknowledged.

References.

1. M.Lowe and J.P.Gollub, *Phys.Rev.* **A31** (1985), p. 3893.
2. P.Coullet, *Phys.Rev.Lett* **56** (1986), p. 724.
3. T.C.Lubensky and K.Ingersent, in *"Patterns, Defects and Microstructures in Nonequilibrium Systems,"* D.Walgraef ed., Martinus Nijhoff, 1987, p. 48.
4. A.Pocheau, V.Croquette, P.Le Gal and C.Poitou, *Europhys.Lett.* **3** (1987).
5. S.C.Müller, T.Plesser and B.Hess, in *"Physicochemical Hydrodynamics: Interfacial Phenomena,"* M.Velarde and B.Nichols,eds., Plenum, 1987.
6. H.Brand, *Phys.Rev.* **A35** (1987), p. 4461.
7. D.Walgraef, in *"Instabilities and Nonequilibrium Structures,"* E.Tirapegui and D.Villaroel, eds., Reidel, Dordrecht, 1987, p. 197.
8. N.Thompson, N.Wadsworth and N.Louat, *Philos.Mag* **1** (1956), p. 113.
9. F.Ackerman, L.P.Kubin, J.Lepinoux and H.Mughrabi, *Acta Metal.* **32** (1984), p. 715.
10. A.T.Winter, *Philos.Mag.* **30** (1974), p. 719.
11. G.I.Taylor, *Proc.Roy.Soc.London* **A225** (1954), p. 473.
12. B.I.Shraiman, *Phys.Rev.* **A36** (1987), p. 261.
13. A.C.Newell, in *"Lectures in Applied Mathematics,"* Vol. 15, M.Kac, ed., American Mathematical Society, 1974, p. 157.
14. D.Walgraef, G.Dewel and P.Borckmans, *Adv. Chem. Phys.* **49** (1982), p. 311.
15. L.M.Pismen, in *"Dynamics of Nonlinear Systems,"* V.Hlavacek ed., Gordon and Breach, New York, 1985, p. 47.
16. D.Walgraef and C.Schiller, *Physica* **D27** (1987), p. 423.
17. A.Pocheau, *"Structures Spatiales et Turbulence de Phase en Convection de Rayleigh-Bénard,"* Thèse de Doctorat, Université de Paris 7, 1987.
18. P.Huerre, in *"Instabilities and Nonequilibrium Structures,"* E.Tirapegui and D.Villaroel, eds., Reidel, Dordrecht, 1987, p. 141.
19. G.Dewel, P.Borckmans, D.Walgraef and K.Katayama, *J.Stat.Phys.* **48** (1987), p. 1031.
20. D.Walgraef, *"Patterns, Defects and Microstructures in Nonequilibrium Systems,"* NATO ASI Series E121, Martinus Nijhoff, Dordrecht, 1987.
21. T.Tabata, H.Fujita, M.Hiraoka and K.Onishi, *Phil.Mag.* **A47** (1983), p. 841.
22. D.Walgraef and E.C.Aifantis, *Int.J.Engng.Sci.* **23** (1986), 1351, 1359,1364.
23. D.Walgraef and E.C.Aifantis, *Int.J.Engng.Sci.* **24** (1986), p. 1798.
24. D.L.Holt, *J.Appl.Phys.* **41** (1970), p. 3197.
25. C.Schiller and D.Walgraef, *Acta Metall.* (1988) (to appear).
26. N.Y.Jin and A.T.Winter, *Acta Metall.* **32** (1984), p. 1173.
27. P.Coullet, C.Elphick and D.Repaux, *Phys.Rev.Lett.* **58** (1987), p. 431.
28. H.Neuhauser, in *"Mechanical Behaviour of Solids: Plastic Instabilities,"* V. Balakrishnan and C.Bottani eds., World Scientific, Singapore, 1986, p. 209.

INSTABILITIES IN NON-AUTONOMOUS SYSTEMS

M. SAN MIGUEL
Departament de Fisica
Universitat de les Illes Balears
E-07071 Palma de Mallorca
Spain

ABSTRACT. Examples of dynamical effects in instabilities with time dependent control parameters are analyzed. The role of fluctuations in these situations is emphasized. Examples include the case of delayed bifurcation with swept control parameter in the context of the laser instability and the case of periodic modulation of the control parameter with application to transient pattern formation in the Freedericksz transition.

1. Introduction

Instabilities and pattern formation are most often studied under static conditions, that is, for fixed values of the control parameters which govern the instability. A richer phenomenology appears when the control parameter is considered to be time dependent. In fact, the experimental determination of an instability point already requires considering the effect of a continuous change of the control parameter. A first question which arises when studying systems with time dependent control parameters is the determination and meaning to be given to an instability point. It turns out that the dynamical instability point or point at which the instability is observed does often not coincide with the static instability point. Such situations are here analyzed through examples in which the control parameter is either linearly swept through the instability point or modulated with periodic crossing of the instability point. The analysis given here is restricted to systems described by a single relevant macroscopic variable and the periodic modulation considered is that of the control parameter of the instability and not other possible periodic forcings.

The concept of a delayed instability occurs when the control parameter is swept through the instability point. This situation has been recently addressed in the context of laser instabilities[1] but it has also been considered in fluid instabilities[2] and kinetics of first order transitions[3]. Likewise periodic modulation has been considered in fluid instabilities[4] and phase separation problems[5]. The two forms of time dependence are discussed below[6, 7] with emphasis on the role of fluctuations in the determination of the dynamical instability point. In addition, the effect of periodic modulation is also discussed in an instability involving a transient pattern formation. Such instability is the magnetic Freedericksz transition in nematic liquid crystals. The dynamics of transient pattern formation in this context has been recently analyzed[9]. Here the periodically modulated instability is considered[10].

E. Tirapegui and D. Villarroel (eds.), Instabilities and Nonequilibrium Structures II, 285–295.

A unifying theme in the discussion given here of several situations is the concept of the Mean First Passage Time which measures the lifetime of an unstable state. This quantity is determined by fluctuations and it is useful to define the time at which a dynamical instability is observed. Its relative value with respect to the sweeping rate or to the period of the modulation determines different regimes of dynamical behavior. The delayed bifurcation is discussed in Sect. 2 in the context of the laser threshold instability. The observation time of the instability is determined. This provides a good example of non-autonomous system in which a crucial role of fluctuations can be studied in a linear approximation. Periodic forcing in a system with no relevant spatial dependence is studied in Sect. 3.1. No shift of the instability point is found in a deterministic analysis since the model contains a single variable. However, a stochastic analysis identifies a shifted instability point as the result of the interplay of noise, nonlinearities and periodic modulation. The shifted instability is defined considering an observation time and it is interpreted in terms of lifetimes of states. These lifetimes are largely changed due to the periodic modulation. The examples of Sects. 2 and 3.1 make clear how the consideration of fluctuations in nonautonomous system change dramatically the dynamical behavior obtained in a purely deterministic description. The final example in Sect. 3.2 addresses questions of pattern formation. A shift of the instability point and the existence of a periodic pattern-forming state with large fluctuations are discussed. In addition, the possibility of delaying the onset time of pattern formation by a fast modulation is elucidated.

2. Delayed Bifurcation with Swept Control Parameter: Delayed Laser Instability.

The simplest example in which the concept of delayed bifurcation[1] can be illustrated is given by the model equation

$$\partial_t q = aq + N(q) \tag{2.1}$$

where q is the relevant variable, a the control parameter and $N(q)$ a nonlinear function of q. In a static sense an instability occurs for $a = 0$ where the stationary solution $q = 0$, which is stable for $a < 0$, becomes unstable. We now consider the situation in which he control parameter a is continuously swept in time through the instability point at a rate α:

$$a(t) = \begin{array}{ll} a_0 + \alpha t, & 0 \le t < t_1 \\ a, & t > t_1 \end{array} \tag{2.2}$$

The static instability occurs at a time $\bar{t} = (|a|/\alpha)$ such that $a(\bar{t}) = 0$. However, a linear stability analysis of (2.1) identifies an instability at the time $t^* = 2\bar{t}$ such that $\int_0^{t^*} a(t')\,dt' = 0$. This is interpreted as a delayed bifurcation or dynamical stabilization of the solution $q = 0$. The dynamical stabilization occurs because the solution $q(t)$ does not follow adiabatically the variation of the control parameter.

The natural question arises of the role played by fluctuations in such dynamical situations. The point of view adoped here is that stability in a dynamical sense is associated with the concept of lifetime of a state and lifetimes are determined by fluctuations. In a stochastic framework a lifetime is defined as a Mean Passage-Time. The passage-time is the time taken for $q(t)$ to reach some observable prescribed value. The identification of the Mean Passage-Time with the time at which the instability is observed gives an operational definition of the dynamical bifurcation point, and of the observed delay, which has unambiguous experimental meaning. The question of delayed bifurcation for the laser instability has been the subject of recent experimental studies[11-13]. A calculation of passage time statistics in this context is described below[6].

As a starting pont I consider the appropriate laser model equation for a single mode on resonance, near threshold and in the good-cavity limit:

$$\partial_t E = a E - |E|^2 E + \sqrt{\epsilon} \, \xi(t) \tag{2.3}$$

E is the electric field complex amplitude, $E = E_1 + i E_2$, $a = \Gamma - K$ where Γ and K are, respectively, the gain and loss parameters. The complex random term $\xi(t) = \xi_1(t) + i \xi_2(t)$ models spontaneous emission fluctuations of strengh ϵ. It is taken as Gaussian white noise of zero mean and correlation

$$< \xi_i(t) \, \xi_j(t') > = \delta_{ij} \, \delta(t-t'), \quad i = 1,2 \tag{2.4}$$

The decay of the unstable zero intensity state $I = |E|^2 = 0$ when the laser is switched-on involves different stages of evolution. A first one is dominated by noise and linear dynamics. In a second regime saturation nonlinear terms become important and fluctuations are not essential. It is during the first stage that the system leaves the vicinity of the unstable state, so that its lifetime can be calculated linearizing (2.3). For comparison purposes it is interesting to consider first the case of an instantaneous change of the control parameter a. This corresponds to the limit $\alpha \to \infty$, $t_1 \to 0$ in (2.2). The solution for the linearized version of equation (2.3) is

$$I(t) = |h|^2(t) \exp(2at) \tag{2.5}$$

The complex process $h = h_1 + i h_2$ is defined by

$$h_i(t) = \sqrt{\epsilon} \int_0^t dt' e^{-at'} \, \xi_i(t') + E_i(0), \quad i = 1,2 \tag{2.6}$$

It is Gaussian bivariate and the variance of the modulus is

$$< |h|^2(t) > = (2\epsilon/a)(1 - e^{-2at}) + <I^2(0)> \tag{2.7}$$

The statistics of the passage time to an observable value of the intensity can be calculated considering times at $\gg 1$. For such times the process h(t) becomes a time independent random variable, so that (2.5) can be inverted giving t as a random function of a fixed prescribed value I. The statistics of t are determined by the random variable $|h|^2(\infty)$. The generating function of the passage time distribution is easily calculated as

$$W(\lambda) = < e^{-\lambda t} > = \Gamma((\lambda/a) + 1)(I/<|h|^2(\infty)>)^{-\lambda/2a} \tag{2.8}$$

so that

$$< t > = (1/2a) \ln(aI/<|h|^2(\infty)>) - (\Psi(1)/2a) \tag{2.9}$$
$$(\Delta t)^2 = < t^2 > - < t >^2 = (1/4a^2)\Psi'(1) \tag{2.10}$$

where Ψ is the digamma function. An important feature of (2.9) - (2.10) is that $< t >$ diverges as $\ln \epsilon^{-1}$ when $\epsilon \to 0$, while $(\Delta t)^2$ is independent of the noise strength. The consistency of (2.9) with the assumption on the time scale of interest (at $\gg 1$) implies that the calculation is valid in the asymptotic limit (ϵ / a) $\ll$ 1. Nonlinearities do not modify the essential ϵ-dependance in (2.9) - (2.10).

In the case of a time dependent control parameter a the calculation of the passage time statistics follows the same steps, with (2.5) and (2.6) replaced respectively by

$$I (t) = | h |^2 (t) \exp 2 \int_o^t a (t') d t' \tag{2.11}$$

$$h_i (t) = \sqrt{\epsilon} \int_o^t d t' \exp (\int_o^{t'} d t'' a (t'')) + E_i (0) \tag{2.12}$$

The calculation is carried out in the limit $\epsilon \ll \sqrt{\alpha} \ll a$. The first inequality represents the small noise limit and the second one corresponds to slow sweeping. The times of interest are in this case always smaller than t_1. The results obtained for the first two moments of the passage time distribution are

$$< t > - \bar{t} = (\alpha^{-1} T)^{\frac{1}{2}} + 0 (T^{-\frac{1}{2}}) \tag{2.13}$$

$$(\Delta t)^2 = (1 / 4 \alpha) T^{-1} \Psi' (1) \tag{2.14}$$

were $T \equiv \ln (I / < | h^2 | (\infty) >) \sim \ln \epsilon^{-1}$. The right hand side of (2.13) gives the delay in the observation of the instability. It is clear in the stochastic framework considered here that such delay is determined by the noise intensity ϵ and it is not given by $t^* = 2 \bar{t}$.

The dependence on ϵ of $< t >$ and $(\Delta t)^2$ are clearly different than the corresponding ones given by (2.9) and (2.10) for the case of an instantaneous change of the parameter a. Here $< t > \sim (\ln \epsilon^{-1})^{\frac{1}{2}}$ and $(\Delta t)^2$ does depend on ϵ. The dynamical bifurcation point can now be defined by the mean value of a (t) to reach an observable value I. From (2.13) this gives

$$< a (t) > = (\alpha T)^{\frac{1}{2}} + 0 (T^{-\frac{1}{2}}) \tag{2.14}$$

the $\alpha^{\frac{1}{2}}$ dependence of (2.14) coincides with the one found in Ref. 11. A stringent test of (2.13) and (2.14) is given by a direct measurement of passage time statistics as done in Ref. 12. In particular the α^{-1} dependence of $(\Delta t)^2$ is well confirmed. In addition, (2.13) - (2.14) give an overall good description of earlier numerical studies[14] and analogic simulations[15].

The calculation of passage time statistics can also be carried out in the limit $\epsilon \ll a < \sqrt{\alpha}$ of small noise and fast sweeping. In this limit one finds the modifications, for a finite α, to the results (2.9) - (2.10) found in the instantaneous limit $\alpha \to \infty$. One obtains that $(\Delta t)^2$ remains independent of ϵ and unchanged while

$$< t > = T - (\Psi (1) / 2a) + (a + | a_o |)^2 / 2a\alpha \tag{2.15}$$

T has the same formal expression than before, but $< | h |^2 (\infty) >$ is now obviously different. The result (2.15) is confirmed by experimental results in CO_2-lasers[13].

The above technique for calculating passage-time statistics might be used in other similar situations. It gives a way of characterizing a dynamic instability as given by the mean value of the control parameter $< a (t) >$ for which the instability is observed.

3. Periodic Modulation of Control Parameter

3.1. FLUCTUATIONS AND STABILITY LIMITS

As a simple model to study the dynamics of fluctuations in a periodically driven system[7] I consider the usual stochastic Landau model for a relevant variable q

$$\partial_t q = - \partial_q U + \xi(t) \tag{3.1}$$

where the potential U is $U(q) = (-a/2) q^2 + (b/4) q^4$ and the stochastic force $\xi(t)$ is assumed to be a Gaussian white noise of zero mean and intensity ϵ. An important effect of fluctuations in (3.1) is to restore the symmetry which is broken in the deterministic limit for a > 0. The stationary probability density associated with the process q (t) is given by

$$P_{st}(q) = N \exp - (U(q)/\epsilon) \tag{3.2}$$

where N is a normalization constant. The distribution $P_{st}(q)$ is peaked around the stationary states $\pm (a/b)^{1/2}$. Due to the symmetry of $P_{st}(q)$ the mean value < q > vanishes for any a. Nevertheless, for small fluctuations this symmetry restoring is a very slow dynamical process: as $\epsilon \to 0$, $P_{st}(q)$ becomes very sharply peaked. As a consequence, a distribution initially centered around one of the symmetric deterministic stationary states for a > 0 takes a long time to decay to $P_{st}(q)$. Such distributions represent very long lived metastable states with a lifetime given by Kramers escape time through the barrier of U (q) at q = 0. Therefore, only for very small values of ϵ quasi-stationary states with < q > $\neq$ 0 are of relevance and the deterministic bifurcation diagram remains meaningful. In the following I describe how this symmetry restoring effect of fluctuations is dramatically enhanced when the system is periodically swept through the instability point a = 0. The main effect of the interplay of the fluctuations and the periodic driving is a macrospic effective shift of the instability. This is due to a change in the limit of metastability caused by the new mechanism of symmetry restoring.

Under a periodic driving, a in eq. (3.1) becomes a periodic function of time a (t). For simplicity only instantaneous changes of a at times t_j (j = 0,1,2,...) are considered: a (t) = a + f (t) where f(t) is a periodic function of period T= t_{2j} - t_{2j-2} and amplitude R

$$f(t) = \begin{cases} R & t_{2j-1} < t < t_{2j} \\ -R & t_{2j} < t < t_{2j+1} \end{cases} \tag{3.3}$$

The problem has three important dimensionless parameters: μ= RT, σ= - a/R, ν=2ϵ/R^2. For | σ | < 1 the system is periodically driven through the instability. The deterministic (ν=0) solution of the problem can be obtained considering the varible x(t) = q'(0) / q'(t). For t → ∞, x(t) diverges for a < 0 and goes to a periodic solution x$^\infty$ (t) for a > 0

$$x^\infty(t) = [2q^2(0)b/(1-\exp 2\mu\sigma)] \int_0^t ds \exp \{ (2\mu\sigma s/T) - 2 \int_t^{t-s} ds' f(s') \} \tag{3.4}$$

Two symmetric solutions $q_\pm^\infty$ are obtained from x$^\infty$ (t). Defining a time averaged quantity as $\bar{q}_\pm^\infty = (1/T) \int_t^{t+T} dt' q_\pm^\infty (t')$ it is found that $\bar{q}_\pm^\infty = \pm (a/b)^{1/2}$. These values of $\bar{q}_\pm^\infty$ and the change of behavior of x (t) at a = 0 indicate that the instability of eq. (3.1) is not esentially modified by the periodic driving in the absence of fluctuations. In order to understand the effect of fluctuations it is instructive to study (3.4) in particular limits. For $\mu \ll 1$, q_+^∞ (t) oscillates with a small amplitude around $(a/b)^{1/2}$. For $\mu \gg 1$, q_+^∞ (t) follows the periodic modulation going at the end of each semiperiod to a value close to the corresponding stationary value: For $\mu \gg 1$, $\sigma <$- 1, q_+^∞ (t) reaches a value close to to $[(a - R) / b]^{1/2}$ at the end of a first semiperiod and close to $[a+R)/b]^{1/2}$ at the end of the second semiperiod. For $\mu \ll 1$ and $-1 < \sigma < 0$ the behavior is the same in the first semiperiod but $q_\pm^\infty \simeq 0$ at the end of the second semiperiod in which a (t) < 0 :

$$[q_\pm^\infty (t_{2j-1})]^2 \approx [(a+R)/b] [1-(2/(\sigma+1)) \exp 2\mu\sigma]; \quad \mu \gg 1, \mid \sigma \mid < 1, \sigma < 0 \qquad (3.5)$$

$$[q_\pm^\infty(t_{2j})]^2 \approx (a/b) [(1 - \sigma^2)/2\sigma] \exp\text{-}\mu(\sigma+1); \quad \mu \gg 1, \mid \sigma \mid < 1, \sigma < 0 \qquad (3.6)$$

The effect of small fluctuations in the system is easy to understand from eq. (3.6). For $\mu \gg$ 1 and $\mid \sigma \mid < 1$ the system is periodically brought close to the unstable state q = 0. A small fluctuation present at this point can lead from a q_+^∞ (t) trajectory to a q_-^∞ (t) trajectory and viceversa. This is the new mechanism which restores the broken symmetry. The difference with Kramers mechanism is that it does not need a large fluctuation to overcome a barrier: the + and — solutions can be here regarded as symmetric metastable solutions which decay to a state of zero mean value. This decay is the result of a phase space mixing caused by fluctuations. The important question is to know when this mechanism is effective, that is, when it leads to a slow or a fast symmetry restoring. The effect of small fluctuations changing a + into a — trajectory cannot be described by a perturbation expansion around the deterministic solutions. The breakdown of perturbation theory gives an estimate of the point in parameter space beyond which the fluctuations are expected to lead to a fast symmetry restoring. This point is the one in which $[q_\pm^\infty (t_{2j})]^2$ is of the order of the strength of fluctuations. This criterion gives from eq. (3.6)

$$(1-\sigma^2/2) \exp\text{-} \mu(\sigma+1) = \nu \qquad (3.7)$$

Asymptotically for large μ and small ν

$$\sigma_c \simeq -1 - (\ln \nu/\mu) \qquad (3.8)$$

The above ideas which are confirmed by a numerical simulation of (3.1) lead to the interpretation of σ_c as a shifted instability point in the following sense: For a given finite observation time $\tau \gg$ T a stochastic trajectory shows small deviations around the deterministic trajectory for $\sigma < -1$. For - $1 < \sigma < 0$ fluctuations occassionally produce large deviations from the determinsitic trajectory as described above. This also happens as t $\rightarrow \infty$ for $\sigma < -1$ due to the symmetry of the problem. This is summarized defining an order parameter $\overline{m^\tau}$ = $\overline{m_\pm}$ (t = τ), where

$$\overline{m_\pm (t)} = (1/T) \int_t^{t+T} dt' <q_\pm(t')> \qquad (3.9)$$

$q_\pm$ (t) indicates respectively a stochastic trajectory with a finite positive or negative initial condition. The quantity $\overline{m^\tau}$ is the stochastic counterpart of the deterministic q^∞. It is expected to be largely independent of τ. The limit of metastability is defined by the value of $\sigma = \sigma_c$ for which $\overline{m_\pm^\tau}$ becomes zero. For a time τ smaller than Kramers time, and in the ab sence of modulation, the metastability limit differs from a $= 0$ in a quantity of order ϵ. Due to the presence of modulation a shift of this limit to a value σ_c given by (3.8) is observed. In summary, for $0 > \sigma > \sigma_c$ fluctuations mix positive and negative trajectories and long lived metastable states do not exist. For $\sigma < \sigma_c$ long lived metastable states which correspond to the ones present in the absence of modulation still exist. This is the contents of the effective macroscopic shift of the instability point.

An exact analytical solution of the problem can be given in the spherical limit[7] of (3.1). Defining y (t) $= (m (0) / m (t))^2$, where m (t) is the mean value of a component of a vector $\bar{q}(t)$, it is found that[7]

$$y\ (t_{2j}) = \alpha\ \lambda_+^j + \beta\ \lambda_-^{\ j} \tag{3.10}$$

where the two eigenvalues $\lambda+$ and λ_- can be calculated. The eingenvalue λ_- is always smaller than 1. For large times only λ_+ survives in eq. (3.10). It plays the role of an effective decay rate which allows a discussion of the limit of metastability of the system. In the deterministic limit $\nu = 0$, $\lambda_+ = 1$ for $\sigma < 0$ and $\lambda_+ = e^{2\sigma\mu}$ for $\sigma > 0$. This indicates the presence of an instability at $\sigma = 0$: For $\sigma < 0$ there is a very slow decay of m (t) to zero (metastable state), while for $\sigma > 0$ it decays very fast. A numerical analysis of the result for λ_+ shows that, for large μ and small ν , λ_+ changes quite drastically, in a narrow region of σ, from a value near 1 to a very large value. The value of σ for which λ_+ starts to grow abruptly is identified with the shifted instability point σ_c . For $\sigma < \sigma_c$ there exist very long lived metastable states with a decay rate $\lambda_+ \simeq 1$. For $\sigma > \sigma_c$ fluctuations restore the broken symmetry leading to m(t) $= 0$ in a very short time. A better analytical understanding of the problem is obtained considering the limit behavior of λ_+ for $\mu \gg 1$, $\nu/ (1+\sigma)^2 \ll 1$ with $\mu\nu\sigma/ | 1 - \sigma^2 |$ $\ll 1$. In this case

$$\lambda_+ = \frac{1 + (2\nu/ (1 - \sigma^2)^2)\ \exp \mu\ (\sigma+1)}{1 + \nu\mu\sigma/(1-\sigma^2)} \quad , \quad \begin{array}{l} -1 < \sigma < 0 \\[6pt] \sigma < -1 \end{array} \tag{3.11}$$

For $\sigma < -1$ there is only a small correction to the deterministic value, but for $\sigma > -1$ there is an exponential amplification of the effect of fluctuations. An estimate of the value of σ_c can be obtained from eq. (3.11) as the value of σ for which λ_+ becomes significantly different from one. In the asymptotic limit in which eq. (3.11) is valid, this estimate coincides with the one in eq. (3.8)

3.2. TRANSIENT NONEQUILIBRIUM PATTERN UNDER MODULATION

The molecules of an homogeneous nematic liquid sample reorientate when a magnetic field larger than a critical one Hc is applied perpendicular to the director field. This instability is known as the Freedericksz transition. The decay of the homogeneous state which becomes unstable is accompanied by the formation of a transient pattern which consists in parell stripes of a well defined wavelength. An interesting possibility to be examined is the stabilization of this transient pattern by periodically changing the magnetic field from values below to above Hc . In view of the results of sects. 2 and 3.1 other natural questions to be addressed are the existence of a shift in the instability point and the possible delay on the onset time of the spatial pattern.

Pattern formation in the Freedericksz transition is described by the equations of stochastic nematodynamics[9] in which the director field is coupled to the velocity flow. The dynamical coupling gives rise to an effective wave-number dependent viscosity coefficient which produces an instability at a finite wave number. Those equations are here considered when the magnetic field becomes time dependent in a form similar to (3.3): $H(t) = H_o + h(t)$, and $h(t)$ takes periodically values $\pm H_1$. The question of the intability shift can be addressed in a mean field approximation in which hydrodynamic effects do not appear. In this approximation a reorientation angle Θ obeys the equation[9]

$$\partial_s \Theta(s) = (h^2(s) - 1) \Theta(s) - \frac{1}{2} h^2(s) \Theta^3(s) + \xi(s) \tag{4.1}$$

where s is a dimensionless time, $h(s)$ is a reduced magnetic field $h(s) = H(s)/H_c$ and $\xi(t)$ is a Gaussian white noise of zero mean and intensity ϵ. The main difference between (4.1) and (3.1) is that the time-dependent control parameter $h(s)$ appears quadratically in (4.1). In terms of the parameters $\mu = r_1 T$, $r_1 = 2 h_0 h_1$ and $r_0 = 1 - (h_0^2 + h_1^2)$ it is seen that the static instability is periodically crossed when $|r_0| / r_1 < 1$, while the dynamic instability occurs at $r_0 = 0$. The meaning of the dynamic instability is that the deterministic solution of (4.1) goes for large times to a periodic solution for $r_0 < 0$ while $\Theta(s) \to 0$ (no reorientation) as $s \to \infty$ for $r_0 > 0$. The instability at $r_0 = 0$ corresponds to $H_{inst} < H_c$. In this sense a deterministic shifted instability is found as a consequence of the quadratic dependence on the control parameter. When including fluctuations an additional shift occurs as discussed in Sect 3.1. The latter is a stabilizing shift which acts in an opposite sense of the deterministic one found here. It occurs for $r_0 < 0$ and slow modulation $\mu \gg 1$. In the following a different effect due to fluctuations will be considered for $r_o < 0$ in the opposite limit of fast modulation $\mu \ll 1$.

Questions concerning pattern formation require the consideration of the spatial dependance of the orientational fluctuations. These are described by a time dependent structure factor $C(Q, s)$ for a wave-number Q. In a linearized approximation $C(Q, s)$ obeys the equation.

$$\partial_s C(Q, s) = \frac{2}{f(Q)} [h^2(s) - 1 - \frac{K_3}{K_2} Q^2] C(Q,S) + \frac{2\epsilon}{f(Q)} \tag{4.2}$$

where K_3 and K_2 are elastic constants and $f(Q)$ incorporates the effect of the hydrodynamic coupling as a wave number dependent viscosity coefficient

$$f(Q) = 1 - \alpha / (1 + \eta Q^{-2}) \tag{4.3}$$

with α and η given system parameters. The solution of (4.2) approaches a time-periodic state as $s \to \infty$ for $r_0 > 0$ while it diverges for a range of Q- muodes for $r_0 < 0$. The periodic solution corresponds to the case in which $\Theta^2(s = \infty) = 0$ and no global reorientation occurs. For $r_0 < 0$ the system finally reorientates and the linear approximation is only meaning ful during a transient state. In the periodic solution for $C(Q,s)$ nonlinearities play no essential role. Such solution describes large periodic fluctuations which are clearly displayed in the limit of large modulation $\mu \gg 1$. In this limit and for $s \to \infty$

$$C_i = C_{eq} (h^2 = h_o^2 + h_1^2) F_i \qquad i = 1, 2 \tag{4.4}$$

where $C_{1,2}$ is the structure factor at the and of the semiperiod in which

$$(h_o + h_1)^2 < 1, (h_o + h_1)^2 > 1$$

respectively, C_{eq} is the equilibrium structure factor associated with the mean value of h^2 and F_i

$$F_1 = \sigma/(\sigma + 1) \tag{4.5}$$

$$F_2 = \begin{cases} \sigma/(\sigma - 1) & , \sigma > 1 \\ [2\sigma/(1 - \sigma^2)] \exp [f^{-1}(Q)(1 - \sigma)\mu] & , \sigma < 1 \end{cases} \tag{4.6}$$

where

$$\sigma(Q) = [r_o + (K_3/K_2) Q^2]/ r_1 \tag{4.7}$$

The parameter σ plays the same role than σ in Sect. 3.1. For $\sigma(Q) > 1$ the mode Q is always stable, while for $\sigma < 1$, $r_0 > 0$ the system is globally stable, but the mode Q is unstable during the semiperiods in which $(h_0 + h_1)^2 > 1$. As a consequence, for $\sigma > 1$ equilibrium fluctuations are modulated by a finite periodic quantity; but for $\sigma < 1$ fluctuations are greatly enhanced with a modulating function which grows exponentially with μ as shown in (4.6). In this periodic regime a pattern is formed when the maximum of C (Q, s) occurs at Q ≠ 0. This happens under certain conditions[10] on the parameters of the system for large μ. The pattern appers and disappears periodically in time. Since this happens for Ho < Hc, it might be understood as a dynamical stabilization of the transient pattern occuring when the magnetic field is changed from an initial value Hi < Hc to a final fixed value Hf > Hc. Decreasing the value of the period of modulation the pattern may not be formed in the periodic solution described here. Physically it corresponds to situations in which the period T becomes comparable with the lifetime of the unstable state as measured by a Mean First Passage Time.

The divergent solution of (4.2) for $r_o < 0$ makes sense up to that lifetime of the unstable state. Therefore such solution is only of interest for $\mu \ll 1$, that is fast modulation. When h^2 (s) crosses periodically the instability $h^2 = 1$, there are Q - modes which are periodically unstable. The emergence of a spatial pattern is then associated with the existence of a mode Q ≠ 0 of maximum systematic growth. Fig. 1 shows the evolution of the maximum of the structure factor for different values of the period of modulation as computed from (4.2). It describes how the transient evolution proceeds via the formation of a spatial pattern. The characteristic wavelength of the pattern is determined by Q_{max}. The most noticeable feature of Fig. 1 is that the emergence of the pattern can be significantly delayed by reducing the period of modulation. The emergence of the pattern is identified with the time at which Q^2_{max} becomes different from zero. The fast modulation of the control parameter slows down significantly the process of pattern formation. The delay in the onset time for pattern formation is seen to occur when the semiperiod of modulation becomes comparable with this onset time in the absence of modulation, given by a Mean First Passage Time. In these circumstances the system has no time to decay from its unstable state during the semiperiod of instability.

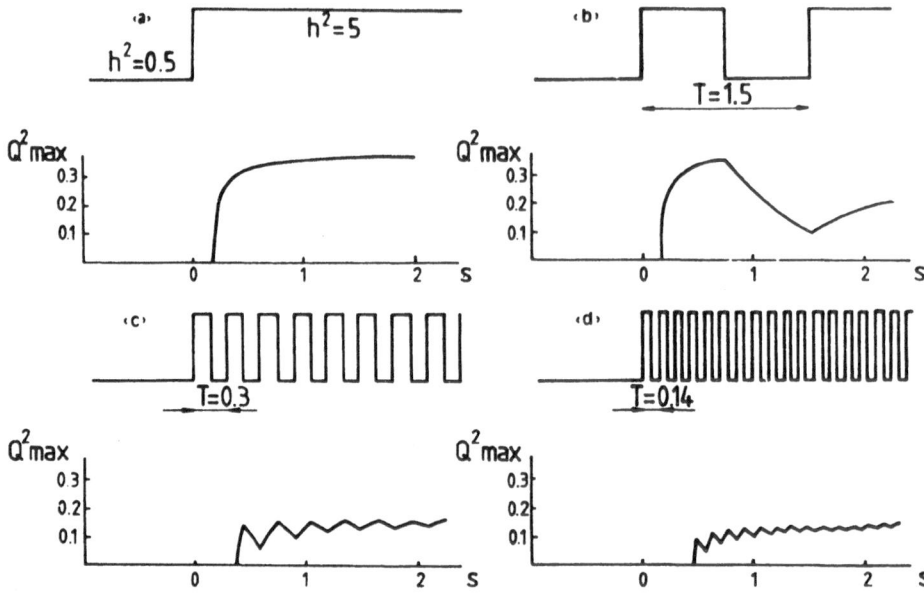

Fig. 1. Evolution of the maximum of the structure factor vs time
     for different values of the modulation period. For the para-
     meters used the Mean First Passage Time ≈ 1.7.

ACNOWLEDGMENT:

The original research work summarized in this lecture is the result of collaborations with F. Arias, F. De Pasquale, Z. Racz, F. Sagués, P. Tartaglia and M.C. Torrent. Financial support from Dirección General de Investigación Científica y Técnica Project PB-86-0534 is acknowledged.

REFERENCES

1. P. Mandel and T. Erneux, Phys. Rev. Lett **53**, 1818 (1984); Phys. Rev. A **30**, 1893, 1902, (1984).
2. G. Ahlers, M. C. Cross, P. C. Hohenberg, and S. Safran, J. Fluid Mech. **110**, 297 (1981).
3. H. O. Carmesin, D. W. Heermann, and K. Binder, Z. Phys. B **65**, 89 (1986).
4. G. Ahlers, P. C. Hohenberg, and M. Lucke, Phys. Rev. Lett. **53**, 48 (1984); Phys. Rev. A **32**, 3493, 3519 (1985)
5. A. Onuki, Phys. Rev. Lett. **48**, 753 (1982); M. Joshua, W. I. Goldburg, and A. Onuki, Phys. Rev. Lett. **54**, 1175 (1985).
6. M. C. Torrent and M. San Miguel, Phys. Rev. A **38**, 245 (1988).
7. F. De Pasquale, Z. Racz, M. San Miguel, and P. Tartaglia, Phys. Rev. B **30**, 5228 (1984).
8. E. Guyon, R. Meyer, and J. Salan, Mol. Cryst. Liq. Cryst. **54**, 261 (1979); F. Lonberg, S. Fraden, A. J. Hurd, and R. B. Meyer, Phys. Rev. Lett. **52**, 1903 (1984); Y. W. Hui, M. R. Kuzma, M. San Miguel and M. M. Labes, J. Chem. Phys. **83**, 288 (1985); M. R. Kuzma, Phys. Rev. Lett. **57**, 349 (1986).
9. M. San Miguel and F. Sagues, Phys. Rev. A **36**, 1883 (1987); F. Sagues, F. Arias, and M. San Miguel, Phys. Rev. A **37**, 3601 (1988).
10. M. C. Torrent, F. Sagues, F. Arias, and M. San Miguel, Phys. Rev. A **38** (1988).
11. W. Scharpf, M. Squicciarini, D. Bromley, C. Green, J. R. Tredicce, and L. M. Narducci, Opt. Comm. **63**, 344 (1987).
12. A. Mecozzi, S. Piazzolla, A. D'Ottavi, and P. Spano, Phys. Rev. A (1988).
13. F. T. Arecchi, W. Gadomski, R. Meucci, and J. A. Roversi, Phys. Rev. A (1988).
14. G. Broggi, A. Colombo, L. Lugiato, and P. Mandel, Phys. Rev. A **33**, 3635 (1986).
15. R. Mannella, F. Moss, and P. V. McClintock, Phys. Rev. A **35**, 2560 (1987).

POLYNOMIAL APPROXIMATIONS FOR NONEQUILIBRIUM POTENTIALS NEAR INSTABILITIES

O. Descalzi and E. Tirapegui
Departamento de Física,
Facultad de Ciencias Físicas y Matemáticas,
Universidad de Chile, Santiago.

ABSTRACT. Polynomial expansions for potentials of Markov processes describing macroscopic systems are discusssed together with conditions for their existence. We consider local expansions near instabilities and study two general situations where a complete analysis can be done.

Macroscopic systems can often be modelized by Markov processes. Let $x = (x_1, \ldots x_M)$ be the gross variables. In many cases the transition probability $\tilde{p}(x, t | x', t')$ obeys a master equation in canonical form [1] (we omit (x', t') unless confusion arises)

$$\eta \partial_t \tilde{p}(x, t) = \sum_{\alpha \geq 0} \eta^\alpha L_\alpha(\eta \partial, x) \tilde{p}(x, t). \tag{1}$$

where η is of the order of the inverse of the size of the system ($\eta \ll 1$) and measures the intensity of the noise. One has

$$L_\alpha(\eta \partial, x) = \sum_{k \geq 1} \sum_{\mu_j=1}^{M} (-1)^k \prod_{j=1}^{k} (\eta \partial_{\mu_j}) A^{(\alpha)}_{\mu_1 \ldots \mu_k}(x; \{\lambda\}). \tag{2}$$

with $\partial_\mu = \partial/\partial x_\mu$ and $\{\lambda\}$ is a set of external parameters. For $\eta \to 0$ eq. (1) reduces to a deterministic equation [2] $\dot{x}_\mu = A^{(0)}_\mu(x; \{\lambda\})$. We shall speak of a bifurcation in (1) when the vector field $A^{(0)}_\mu(x)$ is singular. For $\{\lambda\} = \{\lambda_c\}$ we are in a bifurcation point of the equilibrium $x = 0$ if $A^{(0)}_\mu(0; \{\lambda_c\}) = 0$ and if the operator $\mathrm{D}A^{(0)}_\mu(0; \{\lambda_c\})$ with matrix elements $L_{\mu\nu} = \partial_\nu A^{(0)}_\mu(0; \{\lambda_c\})$ has eigenvalues with real part equal to zero. Suppose we have n of these eigenvalues and let $\{\gamma_1, \ldots, \gamma_N\}$,

E. Tirapegui and D. Villarroel (eds.), Instabilities and Nonequilibrium Structures II, 297–306.
© 1989 by Kluwer Academic Publishers.

Re$\gamma_j \neq 0$, be the other $N = M - n$ eigenvalues. Then by a nonlinear change of variables $x \rightarrow C = (C_1, \ldots, C_M) \equiv (A_1, \ldots, A_n, B_1, \ldots, B_N)$ we can put the vector field $A_\mu^{(0)}(x; \{\lambda\})$ in normal form in the neighborhood of $(x = 0; \{\lambda_c\})$. If non resonance conditions are satisfied among the eigenvalues this normal form is [3] $H'_\mu(C) = H'_\mu(A)$, $1 \leq \mu \leq n$, and $H'_{n+\alpha}(C) = B_\alpha(\gamma_\alpha + F_\alpha(A))$, $1 \leq \alpha \leq N$. The master equation for $P(C, t)dC = \tilde{p}(x, t)dx$ will be

$$\eta \partial_t P(C, t) = [\sum_{\ell \geq 1} \sum_{\mu_j=1}^{M} \frac{(-1)^\ell}{\ell!} (\eta \partial_{\mu_1}) \ldots (\eta \partial_{\mu_\ell}) H'_{\mu_1 \ldots \mu_\ell}(C)$$
$$+ \sum_{\alpha \geq 1} \eta^\alpha L'_\alpha(\eta \partial, C)] P(C, t) \tag{3}$$

with L'_α of the form (2). We assume now that Re$\gamma_\alpha < 0$, then the manifold $B_\alpha = 0$ is the central manifold for the deterministic system $\dot{C}_\mu = H'_\mu(C)$ associated with (3). In the limit $\eta \rightarrow 0$ (weak noise limit) the Fokker-Planck approximation to (3) gives the dominant contribution, and then the stochastic process defined by (3) is equivalent to a system of stochastic differential equations. In this situation, and for times $t \gg |\text{Re}\gamma_\alpha|^{-1}$, one can show that [4], [5]

$$P(C, t) = p_s(B|A)p(A, t) \tag{4}$$

$$p_s(B|A) = \frac{\sqrt{\det\theta}}{(2\pi\eta)^{N/2}} \exp(-\frac{1}{2\eta} \sum_{\mu=1}^{N} B_\mu \theta_{\mu\nu} B_\nu). \tag{5}$$

In (5) the matrix θ satisfies $\theta L_s + L_s^T \theta + \theta Q_s \theta = 0$ where $(Q_s)_{\mu\nu} = H'_{n+\mu,n+\nu}(0)$ and L_s is the Jordan matrix of DA$_\mu^{(0)}(0; \{\lambda_c\})$ in the subspace of the eigenvalues $\{\lambda_\alpha\}$ (L_s^T is the transposed matrix). If one can diagonalize in this subspace $(L_s)_{\mu\nu} = \gamma_\mu \delta_{\mu\nu}$. In fact we made this assumption, which involves no loss of generality, to give the previous form of $H'_{n+\mu}(C)$, and in this case $(\theta^{-1})_{\alpha\beta} = -(\gamma_\alpha + \gamma_\beta)^{-1}(Q_s)_{\alpha\beta}$. The result (4,5) states that $P(C, t)$ is the product of a gaussian $p_s(B|A)$ in the stable variables B_α centered in the central manifold $B_\alpha = 0$ by a reduced probability $p(A, t)$ depending only on the critical variables A_ν. We can replace now (4) in (3) and integrate over dB_α to obtain an equation for $p(A, t)$

$$\eta \partial_t p(A,t) = [\sum_{\ell \geq 1} \frac{(-1)^\ell}{\ell!} \sum_{\mu_j=1}^n (\eta \partial_{\mu_1}) \ldots (\eta \partial_{\mu_\ell}) \tilde{H}_{\mu_1 \ldots \mu_\ell}(A) +$$
$$\sum_{\alpha \geq 1} \eta^\alpha L''_\alpha(\eta \partial, A)] p(A,t) \tag{6}$$

with L''_α of the form (2) and $\tilde{H}_{\mu_1 \ldots \mu_\ell}(A) = H'_{\mu_1 \ldots \mu_\ell}(A, B = 0)$. One can associate to (6) a Hamiltonian

$$\tilde{H}(A,p) = \sum_{\ell \geq 1} \sum_{\mu_j=1}^n \frac{1}{\ell!} p_{\mu_1} \ldots p_{\mu_\ell} \tilde{H}_{\mu_1 \ldots \mu_\ell}(A) \tag{7}$$

which determines the $\eta \to 0$ limit in the sense that the conditional probability $p(A, t | A', t')$, which is a solution of (6), has the expansion [6], [7]

$$p(A,t|A't') = P_{WKB}(A,t|A',t')(1 + \sum_{k \geq 1} \eta^k I_k) \tag{8}$$

with $P_{WKB} = \eta^{-n/2} \overline{N} \exp(-\frac{1}{\eta} S(A,t|A'.t'))$, where S is the classical action calculated with $\tilde{H}(A,p)$ for the indicated boundary conditions. The correction terms $\{\overline{N}, I_k\}$ are independent of η and one can give explicit expressions for them [8] in terms of the mechanical problem defined by $\tilde{H}(A,p)$. This shows that the stationary probability $p_{st}(A) = \lim P(A, t | A't')$, $t' \to -\infty$, has the expansion $p_{st}(A) = \exp(-\eta^{-1} \psi(A,\eta))$ with $\psi(A,\eta) = \phi + \eta \phi^{(1)} + \eta^2 \phi^{(2)} + \ldots$. Here $\phi(A) = \lim S(A, t/A', t')$, $t' \to -\infty$, is called the potential [9]. Since S satisfies the Hamilton-Jacobi equation $\partial_t S = -\tilde{H}(A, \partial \phi)$ one has that ϕ satisfies

$$H(A, \partial \phi) = \sum_{\ell \geq 1} \frac{1}{\ell!} \sum_{\mu_j=1}^m \tilde{H}_{\mu_1 \ldots \mu_\ell}(A) \partial_{\mu_1} \phi \ldots \partial_{\mu_\ell} \phi = 0 \tag{9}$$

which is the equation we must try to solve. Up to now the critical variables $A = (A_1, \ldots, A_n)$ are real. It is useful to introduce complex variables to treat (9) since this will diagonalize the linear part of $\tilde{H}_\mu(A)$ for Hopf bifurcations without Jordan blocks. For example if we have a simple Hopf bifurcation $n = 2$ and $\dot{A}_1 = \tilde{H}_1(A; \mu) = \mu A_1 - \Omega A_2 - (\alpha A_1 - \beta A_2)(A_1^2 + A_2^2) + 0(A^5)$, $\dot{A}_2 = \tilde{H}_2(A; \mu) = \mu A_2 + \Omega A_1 - (\alpha A_2 + \beta A_1)(A_1^2 + A_2^2) + 0(A^5)$, and we introduce $\{q_1 = z = A_1 + iA_2, q_2 = \overline{z} = A_1 - iA_2\}$ to write $\dot{q}_1 = \dot{z} = H_1(q; \mu) = (\mu + i\Omega)z - (\alpha + i\beta)z|z|^2 + o(z^5)$, $\dot{q}_2 = \overline{H}_1(q; \mu)$

(the bar means complex conjugate). Suppose we have done this change of variables $(A_1, \ldots, A_n) \to (q_1, \ldots, q_n)$ which transforms (9) to ($\{\mu\} = \{\lambda - \lambda_c\}$)

$$H(q, \partial\phi) = \sum_{\ell \geq 1} \frac{1}{\ell!} \sum_{\mu_j=1}^{n} H_{\nu_1 \ldots \nu_\ell}(q; \{\mu\}) \partial_{\nu_1}\phi \ldots \partial_{\nu_\ell}\phi = 0 \qquad (10)$$

We study now two general situations in which polynomial expansions for ϕ can be discussed rather extensively: a) The equilibrium $q = 0$ ($H_\alpha(0; \{0\}) = 0$) is persistent in a neighborhood of the critical point $\{\mu = 0\}$ and $H_\alpha(q; \{\mu\}) = \sum_\beta B_{\alpha\beta}q_\beta + O(q^2)$ has diagonalizable linear part, i.e. $B_{\mu\nu}$ is diagonal; b) the dynamical system $\dot{q}_\alpha = H_\alpha(q; \{\mu\})$ can be considered as a perturbation of a Hamiltonian system. We study case a) [10] where the general instability is $(m\varsigma)(\Omega_1 \ldots \Omega_p)$. This notation means that the matrix $B_{\alpha\beta}(\mu = 0)$ has eigenvalues $\{\sigma_\alpha, 1 \leq \alpha \leq n\}$ where $\sigma_\alpha = 0$, $1 \leq \alpha \leq m$; $\sigma_{m+2j-1} = i\Omega_j$, $\sigma_{m+2j} = -i\Omega_j$, $1 \leq j \leq p$, $m + 2p = n$. Here $\{q_1 \ldots q_m\}$ are real and correspond to the m zero eigenvalues while $\{q_{m+2j-1} = z_j, q_{m+2j} = \bar{z}_j, 1 \leq j \leq p\}$ are associated to the eigenvalues $\pm i\Omega_j$. One has $H_\alpha(q) = \mu_\alpha q_\alpha + F_\alpha(q_1, \ldots, q_m, |z_1|^2, \ldots, |z_p|^2)$, $1 \leq \alpha \leq m$, $H_{m+2\alpha-1}(q) = (\mu_{m+\alpha} + i\Omega_\alpha)z_\alpha + z_\alpha G_\alpha(q_1, \ldots, q_m, |z_1|^2, \ldots, |z_p|^2)$, $1 \leq \alpha \leq p$, $H_{m+2\alpha} = \bar{H}_{m+2\alpha-1}$. Here $\{\mu_\alpha \in \mathbb{R}, 1 \leq \alpha \leq m+p\}$ are the unfolding parameters (the instability is of codimension $m+p$) and we assume that the frequencies $\{\Omega_\alpha\}$ are non resonant. We try to solve (10) putting $\phi = \sum_{r \geq 2} \phi^{(r)}$, where $\phi^{(r)}(q)$ is a polynomial of degree r in q and the sum starts with $r = 2$ since $q = 0$ must be an extremum of the potential, i.e. $\partial_\alpha\phi(0) = 0$ [11]. We use the notation $(\ldots)^{[r]}$ for the terms in $(\ldots)$ of order r in q. Then from (10) $\{H(q, \partial\phi)^{[r]} = 0, r \geq 2\}$. For $r = 2$, putting $\phi^{[2]} = \frac{1}{2}\sum_{\mu,\nu=1}^{n} q_\mu \Lambda_{\mu\nu} q_\nu$, we obtain $\Lambda B + B^T \Lambda + \Lambda Q \Lambda = 0$, $Q_{\mu\nu} = H_{\mu\nu}(0)$. One has $B = A + \sum_{j=1}^{m+p} \mu_j B^{(j)}$ and it is easy to see that Λ vanishes when $\mu_j \to 0$ and has the expansion $\Lambda = \sum_{j=1}^{m+p} \mu_j \Lambda^{(j)} + O(\mu^2)$ if $Q_{ij} = Q_{ii}\delta_{ij}$, $0 \leq i,j \leq m$. This last condition is the first of a series of conditions which must be satisfied in order to have a polynomial expansion in q and in the unfolding parameters $\{\mu_j\}$ for the potentials. We assume the validity of this condition and determine the matrices $\Lambda^{(j)}$. Then $H(q, \partial_\alpha\phi)^{[r]} = 0$ for $r \geq 3$ gives equations of the form

$$\mathcal{L}\phi^{[r]} = I^{[r]}, \quad \mathcal{L} = \sum_{\mu,\nu=1}^{n} L_{\mu\nu}q_\nu \frac{\partial}{\partial q_\mu} \qquad (11)$$

where $L = B + Q\Lambda$ and $I^{[r]}$ depends only on $\{\phi^{[s]}, s < r\}$ which shows that we can proceed by recursion in r to determine ϕ. Since we want polynomial solutions in

$\{\mu_j\}$ we write (omitting terms $O(\mu^2)$), $\phi^{[r]} = \phi_0^{[r]} + \sum \mu_j \phi_j^{[r]}$, $I^{[r]} = I_0^{[r]} + \sum \mu_j I_j^{[r]}$, $L = L^{(0)} + \sum \mu_j L^{(j)}$, $L^{(0)} = A$, $L^{(j)} = B^{(j)} + QA^{(j)}$, and $\mathcal{L} = \mathcal{L}^{(0)} + \sum \mu_j \mathcal{L}^{(j)}$ with $\mathcal{L}^{(j)}$ defined as in (11) with L replaced by $L^{(j)}$. One has

$$D \equiv \mathcal{L}^{(0)} = \sum_{j=1}^{p} i\Omega_j \left(z_j \frac{\partial}{\partial z_j} - \bar{z}_j \frac{\partial}{\partial \bar{z}_j} \right) \tag{12}$$

and equations (11) give

$$D\phi_0^{(r)} = I_0^{(r)}, \tag{13a}$$

$$D\phi_j^{[r]} = -\mathcal{L}^{(j)}\phi_0^{[r]} + I_j^{[r]} \equiv J_j^{[r]}, \quad 1 \le j \le m + p \tag{13b}$$

which are our basic equations. Let $\mathfrak{H}$ be the space of formal power series in $(q_1, q_2, \ldots, q_n)$ and $\mathfrak{H}^{[r]}$ its subspace of polynomials of degree r. The linear operators $\mathcal{L}, D, \mathcal{L}^{(j)}$ act on $\mathfrak{H}$ and leave $\mathfrak{H}^{[r]}$ invariant. We define in $\mathfrak{H}$ a scalar product $< \cdot, \cdot >$ such that the monomials $\prod_{j=1}^{n} (m_j!)^{-1/2} q_j^{m_j}$, m_j entire numbers, are orthonormal [3], [10], then the adjoint of the operator q_j (multiplication by q_j) is $\frac{\partial}{\partial q_j}$, i. e. $q_j^+ = \frac{\partial}{\partial q_j}$, $D^+ = -D$, and $\ker D = \{f \in \mathfrak{H} : Df = 0\} = \ker D^+$ is the space of functions $F(q_1, \ldots, q_m, |z_1|^2, \ldots, |z_p|^2)$. In (13) we have equations in $\mathfrak{H}^{[r]}$ and $I_0^{[r]}$ is known (we are solving by recursion in r). The condition to have a solution $\phi_0^{[r]}$ in (13a) is $I_0^{[r]}$ orthogonal to $\ker D^+$, i. e. $< f, I_0^{[r]} > = 0$, $f \in \ker D^+$, and this imposes relations between the coefficients of the $\{\phi_0^{[s]}, s < r\}$ and those of the master equation. Once this solvability condition is satisfied we can solve (13a) as $\phi_0^{[r]} = \chi_0^{[r]} + \psi_0^{[r]}$, where $\chi_0^{[r]}$ is a particular solution of (13a) and $\psi_0^{[r]}(q) \in \ker D$ is a polynomial in $(q_1, \ldots, q_m, |z_1|^2, \ldots, |z_p|^2)$ with arbitrary coefficients. The next step is to replace $\phi_0^{[r]}$ in (13b) and impose again the Fredhölm alternative $< f, J_j^{[r]} > = 0$, $f \in \ker D^+$. This will give a set of relations which are linear in the arbitrary coefficients of $\psi_0^{[r]}$ which overdetermines them in the sense that one has in general more relations than coefficients and this imposes again relations between the coefficients of the master equation which must hold in order to have polynomial solutions for ϕ. For details and explicit examples see [10], [12], [13].

We turn now to case b). The general method is as follows. We assume that the origin is a persistent equilibrium and that the deterministic equation is of the form $(n = 2p)$

$$\dot{q}_\mu = H_\mu(q) = \sum_{\nu=1}^{2p} \sigma^{\mu\nu} \frac{\partial E(q)}{\partial q_\nu} + \varepsilon F_\mu(q) \qquad 1 \leq \mu \leq 2p, \qquad (14)$$

where $\sigma^{\mu\nu} = 0$, $\sigma^{\mu,p+\nu} = \delta_{\mu\nu}, \sigma^{p+\mu,\nu} = -\delta_{\mu\nu}, \sigma^{p+\mu,p+\nu} = 0$, $0 \leq \mu,\nu \leq p$, $E(q)$ starts with quadratic terms and $F_\mu(q)$ with linear terms. We see then that (14) is a Hamiltonian system with Hamiltonian $E(q)$ perturbed by $\varepsilon F_\mu(q)$. The potential $\phi(q)$ satisfies (10)which takes here the form (sum over repeated indices is to be understood from now on)

$$\sigma^{\mu\nu}\partial_\mu E\partial_\nu\phi + \varepsilon F_\mu(q)\partial_\mu\phi + \frac{1}{2}H_{\mu\nu}(q)\partial_\mu\phi\partial_\nu\phi + \cdots = 0 \qquad (15)$$

We look for solutions ϕ with polynomial dependence in (ε, q). Consequently we put $\phi(q) = \varepsilon\phi_1(q) + \varepsilon^2\phi_2(q) + \ldots$. From (15)we obtain at order ε

$$\mathcal{L}\phi_1 = 0, \quad \mathcal{L} \equiv \sigma^{\mu\nu}\partial_\nu E\partial_\mu \qquad (16)$$

Let $\{I^{(1)}(q), \ldots, I^{(s)}(q)\}$ be conserved polynomial quantities for the Hamiltonian system, $I^{(1)}(q) = E(q)$. Then the solution of (16) is

$$\phi_1 = \chi_1(I^{(j)}) = \lambda_j I^{(j)} + \lambda_{jk} I^{(j)} I^{(k)} + \ldots \qquad (17)$$

since $\sigma^{\mu\nu}\partial_\nu E\partial_\mu I^{(j)} = \{I^{(j)}, E\} = 0$ ($I^{(j)}$ is a constant of motion and $\{\cdot, \cdot\}$ stands for Poisson bracket). The $\{\lambda_j, \lambda_{jk}, \ldots\}$ are unknown constants. Eq. (15) at order ε^2 gives

$$\mathcal{L}\phi_2 = -F_\mu(q)\partial_\mu I^{(j)} \frac{\partial\chi_1}{\partial I^{(j)}} - \frac{1}{2}H_{\mu\nu}(q)\partial_\mu I^{(j)}\partial_\nu I^{(k)} \frac{\partial\chi_1}{\partial I^{(j)}} \frac{\partial\chi_1}{\partial I^{(k)}} \qquad (18)$$

Eq. (18) is of the form $\mathcal{L}\phi_2 = J_2(\lambda_j, \lambda_{jk}, \ldots; q)$ and if there is a solution it is of the form $\phi_2(q) = \chi_2(I^{(j)}) + \psi_2(q)$ where $\chi_2(I^{(j)})$ is of the form (17) and $\psi_2(q)$ is any particular solution of (18), i.e. $\mathcal{L}\psi_2 = J_2$. The existence of a solution ψ_2 to this last equation implies that the solvability condition $J_2 \in \text{Ran}\mathcal{L} = $ range of the linear operator $\mathcal{L}$, has to be satisfied. If this condition determines the coefficients $\{\lambda_j, \lambda_{jk}, \ldots\}$ one has a polynomial solution. The coefficients of $\chi_2(I^{(j)})$ will be determined when we consider the solvability condition of eq. (15) at order ε^3 which will give an equation of the form $\mathcal{L}\phi_3 = J_3[\chi_2(I^{(j)})]$, and so on. Let us look at an explicit algebraic method to impose $J_2 \in \text{Ran}\mathcal{L}$. We put $I^{(j)}(q) = I_1^{(j)} + I_2^{(j)}$, where

$I_1^{(j)}(q)$ is the quadratic part (for $j = 1$ we put $E = E_1 + E_2$, $I_1^{(1)} = E_1$). Then $\mathcal{L} = D + R$, $D = \sigma^{\mu\nu}\partial_\nu E_1\partial_\mu$, $R = \sigma^{\mu\nu}\partial_\nu E_2\partial_\mu$, and $\mathcal{L}\psi_2 = J_2$ can be written as

$$D\psi_2(q) = -R\psi_2 + J_2 \equiv K_2 \tag{19}$$

We put $\psi_2(q) = \sum_{r\geq 2} \psi_2^{[r]}$, where $\psi_2^{[r]}$ is of degree r in q. Eq. (19) gives at order r

$$D\psi_2^{[r]} = K_2^{[r]}[\psi_2^{[s]}, s < r] \tag{20}$$

since D preserves the polynomial order in q and R increases it. We can try then to solve (19) by recursion in r. Putting $I_1^{(j)} = \frac{1}{2}B_{\mu\nu}^{(j)}q_\mu q_\nu$ and $D = \sigma B^{(1)}$ one has $D = D_{\mu\nu}q_\nu\partial_\mu$. Let $(\mathfrak{H}, \mathfrak{H}^{(r)})$ be as before (see after eqs. (13)), then (20) is an equation in $\mathfrak{H}^{(r)}$. Defining the same scalar product the solvability condition for (20) is $K_2^{[r]}$ orthogonal to $\ker D^+$ with $D^+ = D_{\mu\nu}^+ q_\nu\partial_\mu = (\sigma\tilde{B}^{(1)})_{\mu\nu}q_\nu\partial_\mu$, where $\tilde{B}^{(j)} = \sigma B^{(j)}\sigma$. Let $\{V^{(1)},\ldots,V^{(u)}\}$ be a complete set of linearly independent quadratic polynomials in $\ker D$, i.e. $DV^{(j)} = 0$. The polynomials $I_1^{(j)}$, $1 \leq h \leq s$, satisfy $DI_1^{(j)} = 0$, but in general $u > s$ since $\{E, I^{(j)}\} = 0$ implies $\{E_1, I^{(j)}\} = 0$ but not conversely. Put $V^{(j)} = I_1^{(j)}$, $1 \leq j \leq s$; $V^{(j)} = \frac{1}{2}B_{\mu\nu}^{(j)}q_\mu q_\nu$, $1 \leq j \leq u$; then $\{\tilde{V}^{(j)} = \frac{1}{2}\tilde{B}_{\mu\nu}^{(j)}q_\mu q_\nu$, $\tilde{B}^{(j)} = \sigma B^{(j)}\sigma$, $1 \leq j \leq u\}$ satisfy $D^+\tilde{V}^{(j)} = 0$, and the elements in $\ker D^+$ are functions of $\{\tilde{V}^{(j)}\}$. From (20) one has $D\psi_2^{[2]} = J_2^{[2]}$ and $J_2^{[2]}(\lambda_j; q)$ depend only on q and $\{\lambda_1,\ldots,\lambda_s\}$. The solvability conditions give u relations $< \tilde{V}^{(j)}, J_2^{[2]} >= 0$ which in general overdetermine the $\{\lambda_1,\ldots,\lambda_s\}$ unless $u = s$. If one cannot determine the $\{\lambda_j\}$ there is no polynomial solution, if it is possible one proceeds in the same way with eqs. (20) for $r \geq 3$ to determine $\{\lambda_{jk}, \lambda_{jkl}, \ldots\}$. The preceeding method is largely inspired in works of Graham and Tél [13], [14]. Although in our presentation the method refers to the research of polynomial potentials it can in some cases be used for more general forms of the potential. We treat now as examples of this technique the nondiagonalizable ξ^2 instability (it contains a Jordan block in the linear part) with and without reflexion symmetry. In the first situation the normal form of the deterministic equations is [3] $\dot{x}_1 = x_2$, $\dot{x}_2 = \nu_1 + \nu_2 x_2 + x_1^2 + bx_1 x_2$, where (ν_1, ν_2) are the unfolding parameters of the codimension two bifurcation which arises at $\nu_1 = \nu_2 = 0$. The interesting case here is $\nu_1 < 0$.

Doing a translation and a scaling $x_1 = \varepsilon^2(q_1 + \mu_1/2)$, $x_2 = \varepsilon^3 q_2$, $\nu_1 = -\varepsilon^4\mu_1^2/4, \nu_2 = \varepsilon^2(\mu_2 - \mu_1 b/2)$, $t \to t/\varepsilon$, one obtains

$$\dot{q}_1 = q_2 \equiv H_1(q), \dot{q}_2 = \mu_1 q_1 + q_1^2 + \varepsilon q_2(\mu_2 + b q_1) \equiv H_2(q) \qquad (21)$$

Although the method can be used for an arbitrary master equation we shall take for simplicity $H_{\mu\nu}(q) = Q_{\mu\nu}$, $Q_{11} = Q_{12} = 0$, $Q_{22} = Q_2$, $H_{\mu_1...\mu_\ell}(q) = 0, \ell > 2$, which reduces the master equation to a Fokker-Planck equation. The Hamilton-Jacobi equation (10) is

$$q_2 \frac{\partial \phi}{\partial q_1} + (\mu_1 q_1 + q_1^2 + \varepsilon q_2(\mu_2 + b q_1)) \frac{\partial \phi}{\partial q_2} + \frac{Q_2}{2} (\frac{\partial \phi}{\partial q_2})^2 = 0 \qquad (22)$$

Here $E(q) = E_1 + E_2$, $E_1 = \frac{1}{2}(q_2^2 - \mu_1 q_1^2)$, $E_2 = -\frac{1}{3}q_1^3$. We put $\phi = \varepsilon \phi_1 + \varepsilon^2 \phi_2 + ...$, $\phi_1(q) = \chi_1(E) = \lambda_1 E + \lambda_{11} E^2 + \lambda_{111} E^3 + ...$.

One has $\mathcal{D} = q_2 \frac{\partial}{\partial q_1} + \mu_1 q_1 \frac{\partial}{\partial q_2}$, $\mathcal{D}^+ = q_1 \frac{\partial}{\partial q_2} + \mu_1 q_2 \frac{\partial}{\partial q_1}$. Ker$\mathcal{D}$ will be functions of $V^{(1)} = I_1^{(1)} = E_1$, i. e. $F(q_2^2 - \mu_1 q_1^2)$, ker $\mathcal{D}^+$ will be functions of $\tilde{V}^{(1)} = \frac{1}{2}(\mu_1 q_2^2 - q_1^2)$, i.e. $F(q_1^2 - \mu_1 q_2^2)$. At order 2 one has $\mathcal{D}\psi_2^{[2]} = -\frac{Q_2}{2}\lambda_1^2 q_2^2 - \mu_2 \lambda_1 q_2^2 \equiv J_2^{[2]}$, and the solvability condition is $< q_1^2 - \mu_1 q_2^2, J_2^{[2]} >= 0$ which determines $\lambda_1 = -\frac{2\mu_2}{Q_2}$.

Proceeding in the same way one has to go up to $\mathcal{D}\psi_2^{[6]} = K_2^{[6]}$ to determine $\lambda_{11} = -\frac{b}{2\mu_1^2 Q_2}$ and $\lambda_{111} = \frac{55b}{108\mu_1^5 Q_2}$. Up to this order the potential is then $\phi = \varepsilon \phi_1$ with

$$\phi_1(E) = -\frac{2\mu_2}{Q_2} E - \frac{b}{2\mu_1^2 Q_2} E^2 + \frac{55b}{108\mu_1^5 Q_2} E^3 \qquad (23)$$

From this expression one can discuss the behavior of the deterministic system (21). Since $E = \frac{1}{2}q_2^2 - \frac{\mu_1}{2}q_1^2 - \frac{1}{3}q_1^3$ limit cycles occur for $-\frac{\mu_1^3}{6} < E < 0$, $\phi_1'(E) \equiv \frac{\partial \phi_1}{\partial E} = 0$. The boundary values of E give: a) $\phi_1'(E = 0) = 0 \Rightarrow \mu_2 = 0 \Rightarrow \nu_2 = b\sqrt{|\nu_1|}$, b) $\phi_1'(E = -\frac{(\mu_1^3)}{6}) = 0 \Rightarrow \mu_2 = 0.105\mu_1 b \Rightarrow \nu_2 = \frac{55}{70} b\sqrt{|\nu_1|}$ which is very near to the value $\nu_2 = \frac{5}{7}b\sqrt{|\nu_1|}$ obtained by Graham and Tél [13]. In the case with reflexion symmetry $x \to -x$ the normal form is $\dot{x}_1 = x_2$, $\dot{x}_2 = \mu_1 x_1 + \mu_2 x_2 - a x_1^3 + b x_1^2 x_2$. After the scaling $\mu_1 = \varepsilon^2 \nu_1$, $\mu_2 = \varepsilon^2 \nu_2$, $x_1 = \varepsilon q_1$, $x_2 = \varepsilon^2 q_2$, $t \to t/\varepsilon$, one has

$$\dot{q}_1 = q_2 \equiv H_1(q), \dot{q}_2 = \nu_1 q_1 - a q_1^3 + \varepsilon q_2(\nu_2 + b q_1^2) \equiv H_2(q). \qquad (24)$$

Now $E = \frac{1}{2}q_2^2 - \frac{\nu_1}{2}q_1^2 + \frac{a}{4}q_1^4$, $\mathcal{D} = q_2\partial_1 + \nu_1 q_1 \partial_2$. Taking again $H_{\mu\nu}(q) = Q_{\mu\nu}$, $Q_{11} = Q_{12} = 0$, $Q_{22} = Q_2$, $H_{\mu_1...\mu_\ell}(q) = 0, \ell > 2$, one obtains now

$$\phi_1 = -\frac{2\nu_2}{Q_2} E + \frac{b}{2\nu_1 Q_2} E^2 + ... \qquad (25)$$

We recall finally that the potential ϕ is a Lyapounov functional for the dynamical system $\dot{q}_\mu = H_\mu(q)$ in the sense that if we put $H_\mu(q) = -\frac{1}{2}Q_{\mu\nu}\partial_\nu\phi + r_\mu(q)$, then $\frac{d}{dt}\phi(q(t)) = -\frac{1}{2}Q_{\mu\nu}\partial_\mu\phi\partial_\nu\phi + r_\mu\partial_\mu\phi \leq 0$ if $Q_{\mu\nu}$ is positive semi-definite and $r_\mu\partial_\mu\phi = 0$, but this last condition is equivalent to the Hamilton-Jacobi equation $H_\mu\partial_\mu\phi + \frac{1}{2}Q_{\mu\nu}\partial_\mu\phi\partial_\nu\phi = 0$ (for a complete discussion see [15]).

Acknowledgments

The authors thank FONDECYT (Chile) for financial support.

REFERENCES

(1) N.G. van Kampen, *Stochastic Processes in Physics and Chemistry*, North Holland, (1984)

(2) E. Tirapegui, *Stochastic processes in Macroscopic Physics, lectures at ELAF'87*, Proceedings edited by J.J. Giambiagi et al World Scientific, (1988)

(3) C. Elphick, E. Tirapegui, M.E. Brachet, P. Coullet, G. Iooss, *Physica*, 20D, 95 (1987).

(4) C. Elphick, E. Tirapegui, *Normal forms with noise, in Instabilities and Nonequilibrium Structures*. Eds. E. Tirapegui and D. Villarroel, Reidel (1987)

(5) F. Baras, P. Coullet, E. Tirapegui, *J. Stat. Phys.*, 45, 745 (1986) and C. Elphick, M. Jeanneret, E. Tirapegui, *J. Stat. Phys.*, 48, 925 (1987).

(6) R. Kubo, K. Matsuo, K. Kitahara, *J. Stat Phys.*, 9, 51 (1973).

(7) F. Langouche, D. Roekaerts, E. Tirapegui, *Functional Integration and Semiclassical Expansions (Chapter IX)*, Reidel (1982)

(8) F. Langouche, D. Roekaerts, E. Tirapegui, *Nuovo Cimento*, A64, 357 (1981)

(9) R. Graham, *Weak Noise and Nonequilibrium potentials of dissipative dynamical systems, in Instabilities and Nonequilibrium Structures*, eds. E. Tirapegui and D. Villarroel, Reidel (1987), and references quoted there.

(10) O. Descalzi, E. Tirapegui, *Nonequilibrium potentials near instabilities*, to appear in J. Stat. Phys.,

(11) R. Graham, T. Tél, *Phys. Rev.*, <u>A31</u>, 1109 (1985) and <u>A33</u> 1332 (1986), R. Graham, D. Roekaerts, T. Tél *Phys. Rev.* , <u>A31</u> 3364 (1985).

(12) H. Lemarchand, G. Nicolis, *J. Stat. Phys.*, <u>37</u>, 609 (1984); H. Lemarchand, *Bull. Acad. Royale Belg.*, <u>70</u>, 40 (1984); E. Sulpice, A. Lemarchand, H. Lemarchand, *Phys. Lett.*, <u>A121</u> 67 (1987)

(13) R. Graham, T. Tél, *Phys. Rev.*, <u>A35</u> 1328 (1987)

(14) T. Tél, *J. Stat. Phys.*, <u>50</u>, 897 (1988).

(15) H.R. Jauslin, *J. Stat. Phys.*, <u>40</u>, 147 (1985) and <u>42</u> 573 (1986).

Instabilities and Nonequilibrium Structures
(Volume I)

Enrique Tirapegui and
Danilo Villarroel (Eds.)

ISBN 90–277–2420–2, MAIA 33
1987, x + 337 pp.

Physical systems can be studied both near to and far from equilibrium where instabilities appear. The behaviour in these two regions is reviewed in this book, from both the theoretical and application points of view. The influence of noise in these situations is an essential feature which cannot be ignored. It is therefore discussed using phenomenological and theoretical approaches for the numerous problems which still remain in the field.

Audience

This volume should appeal to mathematicians and physicists interested in the areas of instability, bifurcation theory, dynamical systems, pattern formation, nonequilibrium structures and statistical mechanics. Chemists and biologists working in these areas should also find this work of interest.

For the Table of Contents of this book, p.t.o.

Instabilities and Nonequilibrium Structures (Volume I)

TABLE OF CONTENTS